The Rural Efficiency Guide
Volume 3 – Agriculture Book

by C.G. Williams, L.L. Rummell and Z.P. Metcalf

with an introduction by Roger Chambers

Self Reliance Books

Get more historic titles on animal and stock breeding, gardening and old fashioned skills by visiting us at:

http://selfreliancebooks.blogspot.com/

Introduction

I am pleased to present yet another title in the *U.S. Department of Agriculture's "Rural Efficiency"* book series. These incredible books, now a century old, give us an insight into life in rural America in a days gone by, a glimpse into an era that sometimes feels like another world.

These books are a great read for anybody interested in homesteading or farming of any sort, long-term survival solutions, or in rural American history in general.

The work is in the Public Domain and is re-printed here in accordance with Federal Laws.

As with all reprinted books of this age that are intended to perfectly reproduce the original edition, considerable pains and effort had to be undertaken to correct fading and sometimes outright damage to existing proofs of this title. At times, this task is quite monumental, requiring an almost total "rebuilding" of some pages from digital proofs of multiple copies. Despite this, imperfections still sometimes exist in the final proof and may detract from the visual appearance of the text.

I hope you enjoy reading this book as much as I enjoyed making it available to readers again.

~ Roger Chambers
PNW, 2017

C.G.Williams

AUTHORS' PREFACE

TWENTIETH CENTURY agriculture has made stringent and urgent demands upon the American farmer. He now stands facing the fact that he can no longer take up new lands to farm, and yet at the same time he has more mouths to feed than ever before. Our population has doubled itself about every 25 years, and today about two million persons are added annually. Moreover, the nations of Europe are looking to the American farmer at this present crucial moment for sustenance. Increased production of foodstuffs on our farms therefore become highly imperative to meet this situation.

This can be done in three ways: First, larger yields may be attained; second, costs may be reduced; and, finally, what is produced may be saved in large part, if not entirely, by preventing and overcoming insect and disease attacks. It is the primary purpose of this volume to show the farmer how these ends may be reached in a practical way, without needless expenditures, but with measures wholly within his reach. Experimental work carried on by experts in various sections of the United States and Canada, under varying conditions of these regions, has been the main foundation upon which our recommendations are based. This knowledge has been garnered in many cases from tests extending over a quarter of a century or more. Then personal association with the farm itself, as well as with farmers, has added a wealth of valued knowledge obtained through such experience, and has served to indicate what the actual needs of the farmer are.

The average yield of winter wheat in the United States in 1916 was only 13.8 bushels to the acre. Only ordinary farm practices, under proper management, are needed to double that yield in winter wheat sections. Crop rotations, fertilization, drainage and good seed make almost the entire difference. Ability in carrying out details widens the margin of profit. Often only a comparatively simple matter like the saving of manure would mean $650,000,000 more added to the pocketbooks of American farmers than if this valuable product were not protected; and, again, testing the germination of seed corn will pay $5 or more an hour for time so spent. Such practical hints, generally understood, often overlooked and always most profitable when carried out, are here presented in language sheared of all technical phrases.

By increasing the productive capacity of the land, moreover, crop yields are possible at lower cost. If the average yield of wheat is doubled, cost of production should be reduced a third. The farm operations that are most effective from a financial standpoint to increase the output of the farm and lower costs are set forth briefly and, we trust, in practical form in this book for ready reference. Soil preparation and tillage, the use of manure, fertilizers and lime, rotation and successful methods of culture of field crops, fruits and vegetables have been given considerable attention.

In addition, weed destruction and control of insects and plant diseases are treated from the standpoint of eliminating losses from these sources. Weeds every year cause losses to the American farmer that amount well into the millions of dollars. A billion dollars probably would not cover the amount lost annually on our farms because of insects and diseases. The loss even to the individual farmer is enormous, and, if prevented in large part, might be the difference between success and failure. To overcome such useless waste is the further object of this book.

Special credit is due Dr. E. R. Allen and W. J. Green, of the Ohio Agricultural Experiment Station, for services rendered in reading parts of the manuscript; to the Department of Botany of the same station for illustrations of tobacco, and to the South Carolina Agricultural Experiment Station for illustrations on cotton.

C. G. WILLIAMS,
L. L. RUMMELL.

Ohio Agricultural Experiment Station, Wooster, Ohio, June 15, 1917.

TABLE OF CONTENTS

PART ONE

MAINTAINING SOIL FERTILITY

PART TWO

FIELD CROP PRODUCTION

INSECT PESTS AND PLANT DISEASES

PART ONE
MAINTAINING SOIL FERTILITY

CHAPTER I.

THE SOIL AND ITS CONSTITUENTS.

Upon the few inches of the earth's crust, commonly spoken of as the soil, have the destinies of men and nations been determined. Animals and plants have always been dependent upon the soil constituents for their existence; and dying, return to the soil to complete their cycle. All the structures of human ingenuity—the growth of our vast population and of our mighty cities —have been made possible only through the soil or the products originating therein. Why, then, should it not justly be called the world's greatest asset?

Origin of Soil.

Original Rocks.—But before we can go ahead to understand the physical and chemical properties of soils, their differences and their proper treatment, we should obtain an idea of how these soils originated. With such knowledge the farmer may know better how to treat his land.

Geologists say the earth was once a molten mass, and upon cooling hardened into rocks. In these original rocks were contained elements of what we now know as plant food, such as oxygen, silicon, aluminum, iron, calcium, magnesium, sodium, potassium, carbon and hydrogen in various amounts and combinations. The compounds of these elements were locked up in insoluble forms in these rocks and were unavailable to plants.

Fortunately many agencies have been at work through countless ages to tear down this bed rock and gradually build up a soil in which plants thrive. Soil, therefore, is a product resulting from the decomposition and disintegration of the earth's crust, and the addition of plant and animal remains. Granite, diorites, diabase and basalts (all being igneous rocks) were the chief constituents of the parent, or original rock. Certain changes, principally heat and pressure, transformed these to metamorphic, while solution and suspension and redeposition gave rise to sedimentary rocks, such as sandstone, limestone, dolomite, shale and clay. Of these different forms sedimentary rocks are of greatest agricultural importance.

Different rocks on weathering give rise to different soils, and frequently the nature of the parent rock determines largely the nature of the soil. Thus a sandstone weathers ordinarily into a sandy soil, or, if it is very fine-grained, into a soil high in silt (intermediate between sand and clay); and a shale usually weathers into a clayey soil, often fine-grained, cold and hard to work. Again, the igneous rocks are mentioned above in the order of their content of silicon, the chief constituent of sand. A soil formed from granite is therefore more likely to be higher in sand than one formed, for instance, from diabase, which contains a higher content of softer minerals and which in turn tends to the production of clayey and silty soils. The weathering of limestones and dolomites is accompanied by a large amount of chemical solution; indeed, the soil which is formed is largely a residue composed of the impurities of the original rock. Limestones and dolomites give rise to silty and clayey soils, almost never to sandy soils. A much thicker layer of limestone must also be weathered than is the case with less soluble rocks, if the same depth of soil is formed.

While the parent rock is important in determining the nature and properties of the soil formed, the conditions under which the material weathers, especially with regard to rainfall and drainage, may give rise to even greater differences than those resulting from variations in the parent rock. Thus, with the variety of rocks on the earth's surface and with the number of processes which each may undergo, it is not surprising that there is such an almost endless array of soils.

Agencies that Formed Soils.—In this process of rock weathering several agencies have been and are today constantly at work. The wind, for example, transports dust particles, which may be driven against rock faces, grinding off other particles and thus forming additions to the soil mass. Water likewise carries a large amount of soil-forming material, filling depressions and removing elevations, reducing the soil particles to a finer state; and upon freezing in cracks in rocks it splits them apart and breaks off particles. Alternate heat and cold will chip off pieces from rock faces. Then the great glaciers which in prehistoric times crept southward from the polar regions over the northern states, acting like a huge rasp upon the earth's surface, rounded off the hills, filled depressions and otherwise ground up the rock and soil and carried the drift, in many cases, to long distances. With all these agencies plants and animals have united to break down the original rock and build soil from it.

Chemistry, moreover, has played a great part in forming soils. Water dissolves many rock minerals, and when combined with various gases and acids is a still greater force in changing the composition of the earth's surface. Limestone caverns are a familiar example of this action of water.

Humus in the Soil.—It must not be understood that soil is made up of only ground and decomposed rocks; for if organic matter, coming from the decay of plants and animals, were not added too, farm crops could not thrive in it. This organic matter, as roots and leaves of plants and bodies of animals, is decomposed, about 5 per cent of the amount added remaining after complete decay. It is this decaying material, commonly called "humus," that has considerable influence in determining the fertility of a soil; and the addition and proper conservation of this humus constitute important problems in any system of profitable, permanent agriculture.

The organic matter varies greatly in amount. Peat and muck soils are composed largely of it, while in open, sandy soils it is almost lacking. In general, the amount decreases with depth, until in the lower subsoil the amount is hardly appreciable. It is the organic matter, or humus, that gives a soil its dark color.

Physical Properties of Soils.

Kinds of Soil.—From what has already been set forth, the idea can be gained that soils do not always remain near the rock from which they were formed. Water and wind transport the fine particles, tending to move the soil always to lower levels. Even the soil itself tends gradually to work down hillsides; hence, the soil is thinnest on top of hills and most fertile in valleys. Lakes and ponds receive large annual deposits of soil carried by streams, and they may in many years hence become useful for cultivation.

A gravelly soil that will be replaced in this pond by muck soil high in organic matter as animal and plant remains are mixed with sediment washed in.

Soils remaining where formed are called "residual," as distinguished from "transported" soils, and vary according to the nature of the parent rock. Glaciers have not covered areas of such soils. In the United States this class of soils is found south of a line drawn roughly from New York to Cincinnati and to St. Louis; thence up the Missouri River to the Dakotas and west to upper Washington. These soils as a general rule are not as productive as the glaciated soils.

Still another kind of soil is that formed by the gradual accumulation of plants and animals mixed with soil in shallow bodies of water. Vegetation may grow year after year and decay but slowly under water, thus forming the black, fertile muck and peat beds in old lakes, ponds, arms of streams and similar places. These soils when drained are unusually fertile, especially suited for growing truck crops like celery and onions. Frequently they require an application of potassium to be most productive.

Texture and Structure.—Soil texture, which has to do with the size of particles, and structure, or the arrangement of the particles in the soil, are dependent in some measure upon the mode of soil formation. Thus, clay soils are fine-textured and great care must be exercised in working them while wet so that they do not become too compact or puddled. On the other hand, sandy soils are coarse, easily worked and open in texture. The ideal soil has the small particles intermingled with the larger ones forming a granular, or crumb structure. It is the purpose of tillage to affect the structure of the soil so that this granular condition may be obtained.

Then, too, variations in the water content, the growth of plant roots, the amount of organic matter and the action of earthworms and other forms of lower animal life affect the soil structure. Of these, the growth of plant roots and the content of organic matter are within the control of man, while the others are of minor significance.

Soil Moisture.—Besides its organic matter and inorganic constituents, the soil contains moisture, which is essential for the growth of all plants.

In the plant it is a direct food, being in combination with particular elements as well as occurring in the free state. In the soil it dissolves the plant food, and in this form the small rootlets can absorb their nourishment.

Soils vary in their moisture content, their capacity being determined largely by the kind of soil. The supply of moisture is determined by precipitation, drainage and irrigation. Of these three factors the last two deserve especial consideration because they are controlled by man's influence. Excess water is drained away, while irrigation practices and proper methods of tillage, or cultivation, are resorted to when the supply is deficient.

Temperature.—Still another physical property of soil which has a direct bearing upon its capacity for crop production is temperature. This is determined in a large measure by the method of soil formation. Then, too, the amount of organic matter and moisture in the soil has an important influence on temperature. Black or dark-colored soils warm up earlier in the spring than light-colored soils. Cold soils are heavy soils, due to their high water-holding capacity. The addition of crop residues, of manure or of other forms of organic matter, and also proper drainage, will make soils warmer in the spring and thus hasten plantings.

It has thus been shown that the productiveness of soils, resulting from the breaking down and decomposition of the original rocks on the earth's surface, depends primarily upon the chemical composition of the original rock and also upon the organic constituents added by plants and animals.

Different kinds of soils have resulted from transportation, through various agencies, from the place where the soil was formed. These factors have caused important differences in soil texture and structure, water-holding capacity and temperature, which are limiting factors in crop production and therefore deserve the farmer's attention.

Essential Elements of Plant Food.

Elements in Soils.—To understand the feeding of plants and therefore the reasons for manuring, fertilizing and cultivation, let us consider the elementary constituents of which these soils are composed. There are in the world about 80 so-called elements, the simplest forms of all matter. Oxygen and nitrogen are common examples. These 80 elements combine chemically with each other in various ways to form compounds, simple and complex. Of this number only about 15 are found in animal and plant bodies, and of these only 10 are ordinarily considered essential; namely, phosphorus, potassium, carbon, sulphur, nitrogen, calcium, oxygen, hydrogen, magnesium and iron, while silicon, sodium and chlorine are also incidentally found in plants.

Oxygen, hydrogen and carbon are combined to form fat, carbohydrates and crude fiber in plants, and with nitrogen in addition, form protein. When the plant is burned, all these elements are driven off as gas, while the others remain in the ash. All 10 elements are highly necessary in the growth of every plant and no one can be substituted for another.

Four Elements Most Important.—While plants need all ten elements for growth, only four are likely to be deficient in the soil. These are phosphorus, potassium, nitrogen and calcium; although sulphur may possibly be needed. The other elements are used in smaller amounts and are abundant in nearly every soil; hence, they receive little attention in soil-fertility maintenance.

Phosphorus, the Keystone to Soil Fertility.—Phosphorus is considered the most important of all plant food elements, because it is used in large amounts and is most frequently deficient in soils. A 75-bushel corn crop would remove nearly 20 pounds of phosphorus, and a ton of clover 5 pounds. While an average silt loam soil contains in the first 7 inches of soil about 2,000 pounds of nitrogen and 30,000 pounds of potassium, it has less than 1,000 pounds of phosphorus per acre.

In cereal crops about three-fourths of the phosphorus is found in the grain. It plumps the kernel, gives it weight and hastens maturity. The bones of animals and milk are also comparatively rich in this element. Therefore, it is evident that grain farming, live stock production and dairying all tend to reduce the supply of phosphorus because the products are carried off the farm.

Commercial fertilizers, as discussed later, are used to make good this loss. Barnyard manure and other organic matter when added to the soil decay and act chemically upon the store of phosphorus locked up there in a form unavailable for the use of plants.

Potassium.—The same means may be used to increase the soil's available supply of potassium, which is the second important essential element. It is found in large amounts in all soils: 20,000 pound per acre 7 inches in sand, 30,000 pounds in silt loam and 42,000 pounds in clay. Unlike phosphorus it is found in correspondingly large amounts in the subsoil, or the layer between the surface soil and the bed rock. These large stores, however, can be little used in this condition by plants, but in the presence of decaying organic matter this potassium becomes available. For this reason farmers should plow under green crops, crop residues and manure to make this potassium useful to plants.

In the plant about three-fourths of the potassium is found in the straw or grass. It gives the stem strength and rigidity. Since potassium is not often sold off the farm, and because of the great store in the soil, this element is bought in fertilizers in much smaller amounts than phosphorus. Muck soils and light yellow sands, however, are extremely low in potassium and to them it must be supplied in large quantities for the best crops. Moreover, attention must be directed to the potassium supply if hay and straw are sold, or if a crop like tobacco, which consumes a large amount of potassium, is the chief industry.

Nitrogen.—In many respects nitrogen may be considered the most important plant food, for it is used in largest amounts by plants, is most costly in fertilizers and is most easily lost from soils. A 70-bushel corn crop contains more than 100 pounds of nitrogen, while 50 pounds is found in a ton of alfalfa. The legumes (as clover, beans, peas and alfalfa) are especially high in nitrogen. Because of its high cost, nitrogen in any rational system of farming should be secured from the air, which contains nearly 80 per cent of this element. The growing of leguminous crops in proper rotations, along with proper conservation of manure, are therefore discussed thoroughly so that the farmer will know how to avoid paying out large amounts for nitrogen.

Most crops get their necessary nitrogen from the soil. Yet in the soil nitrogen is present in rather small amounts. A silt loam has about a ton of nitrogen per acre in the first 7 inches, and about half this amount in the second 7 inches. Since nitrogen occurs mainly in the organic remains of plants and animals, muck soils are exceptionally rich in this element. When the stubble of a clover crop is plowed under, considerable organic matter is added to the soil. Countless millions of soil bacteria work upon this decaying material, changing the nitrogen into the form of a nitrate, so that it can be taken up by the plant rootlets. These are called "nitrifying" bacteria, and the process is termed "nitrification." The addition of organic matter, like manure or crop residues with an occasional crop plowed under, besides the growing of legumes, will solve this great problem of keeping up the nitrogen supply in the soil.

In the plant, nitrogen is found mainly in the leaves, being essential in the nucleus of every cell. It makes rank growth and gives plants their dark green color.

Calcium.—In the case of many soils calcium is of fundamental importance. It is a direct food for plants, being particularly abundant in the legumes. A ton of clover hay contains 20 to 30 pounds of calcium, and a ton of alfalfa hay about 40 pounds. Black clay soils generally contain about 8 to 9 tons of calcium in 7 inches over an acre, while clay and silt loam soils have only 2 to 3 tons.

The great use of calcium, however, is to serve functions outside the plant. The physical condition of the soil may be improved by its addition in the form of lime or limestone. Thus compact clay soils may be made more granular. Then, again, lime neutralizes the acidity of the soil and promotes the growth of the millions of bacteria that aid in plant growth. Hence, for soils deficient in this element applications of lime are highly necessary for the most profitable crop production.

Sulphur.—Within recent years sulphur, which is found in small quantities and is used in small amounts by all plants, has commanded special study. At the Ohio Experiment Station it has been noted that land receiving sulphur in addition to other fertilizers produces greater crop yields than soil similarly fertilized except that no sulphur is supplied. However, it is an element of minor importance, and will not be added directly as a fertilizer for many years to come.

Soil Analysis Not a Guide to Fertilizing.—Contrary to the common opinion that chemical analysis of a soil reveals facts to lead to its immediate treatment with respect to fertilization, such an analysis is not an index to the fertility of the soil. To the farmer nothing definite is presented, when a chemist tells him how much total nitrogen, phosphorus and potassium a soil contains, that will aid in determining how the soil should be fertilized. There are many other factors that influence the productivity of the soil.

It must be remembered that the soil is like a great storage battery, in which plant food is locked up in unavailable form, and then is gradually and slowly made soluble as the plant needs it. In this way waste of plant food is prevented, for if all of it were available it would have been leached out of the soil ages ago.

Three examples will serve to illustrate this point. Feldspar, a constituent of granite, contains about 14 per cent of potassium, but the element is so firmly held in complex chemical combination that plants cannot utilize it. Likewise, the beds of phosphatic rocks in the southern states contain an immense store of phosphorus, but before it is available for plants it must be either treated with sulphuric acid or added to manure or other decaying organic matter, the acids of which make this otherwise worthless material a valuable fertilizer. Muck soils are unusually rich in nitrogen, but here again the fertility element is gripped so firmly that as such this material is beyond the reach of plants.

It is thus easy to understand that a chemical analysis which reports the total supply of plant food, and not the available amount, cannot direct one accurately in fertilizing a soil. Soils often respond generously to applications of a fertilizing element even though their chemical analysis shows a high total content of the same element.

Soils differ in their **degree of availability of fertility elements.** Furthermore, common plants vary widely, both in their needs and in their abilities to utilize plant food in the soil. A soil analysis cannot tell whether available forms of plant food can be used with profit. One would think that a soil having nearly 17 tons of potassium per acre in the top 6 inches and 18 tons in the second 6 inches would be well supplied. Nevertheless, 40 pounds per acre of muriate of potash applied to such a soil in Ohio once in 3 years, has, for 8 years, increased the yield of corn 4.41 bushels per acre and oats following, 2 bushels.

For the information which a knowledge of the total plant food elements in the soil fails to reveal, one must therefore resort to a carefully conducted field test.

A knowledge of the fertility elements in a given soil does have a value, although it does not show the immediate needs of the soil. Such information should aid one in buying a farm just as the resources of a bank should be known before buying stock in it. The buyer might notice the productivity of the farm, but such observations should extend over several years; a single year's crops might be abnormal for the farm. With the knowledge of the total nitrogen, phosphorus, potassium and calcium, there is little danger of mistaking an apparently rich soil for one really rich. Chemical analysis, then, shows total resources, future soil capabilities and future treatment necessary but is not a guide in fertilizing the immediate crop.

Sources of Supply of Plant Food.

Making Elements Available.—It is the problem of the farmer to make this great store of plant food found in a large unavailable condition in the soil into such a form that it can be absorbed by the plant rootlets. It has already been shown that water, especially when carrying gases and more or less complex acids, brings about chemical changes in the soil setting free insoluble compounds and thereby increasing the amount of available elements in the soil solution for the use of crops. Moreover, the addition of organic matter, in such forms as the residues of crops, green manures and barnyard manures, was mentioned in connection with the setting free of such plant food elements as phosphorus and potassium, which are costly when purchased in the fertilizer sack. The amount of plant food rendered available by the ordinary methods practiced on the average farm depends largely on the season and the condition of the soil, particularly with respect to its organic matter. Such material in the soil decomposes through the action of bacteria; and the chemical reactions between this decaying matter, or humus, and the soil elements break down complex compounds and set free such essential elements as phosphorus and potassium. The bacterial activities change the humus nitrogen to available nitrates, and the carbon dioxide evolved tends to liberate both phosphorus and potassium. No one can say exactly what percentage of these plant food elements will be rendered available annually, but Dr. C. G. Hopkins estimates it roughly at 2 per cent of the total nitrogen, 1 per cent of the phosphorus and 0.25 per cent of the potassium. While the percentage of the last element is lowest, it should be remembered that the content of potassium, as a rule, is many times greater than that of the other elements mentioned.

The functions of tillage with regard to hastening the decomposition of organic matter in the soil will be discussed fully in the next chapter. It must be remembered in this connection that we are concerned here with the original store of plant food in the soil; it is merely made useful for plant use and no fertility is added. The other sources of supply of plant food are concerned with its direct addition to the soil.

Crop Residues.—An important and constant source of supplying plant food to the soil is the roots and stubbles of the ordinary farm crops. This supply becomes of even greater importance if the crops are legumes, for they are able to take nitrogen from the air and store it in the soil for the use of succeeding crops. It is customary with some farmers to devote an occasional crop to soil improvement. The second clover crop, cowpeas or soybeans are particularly well adapted for this purpose; or again, such nonleguminous crops as rye or buckwheat may be plowed under. This material decomposing in the soil tends to liberate some of the insoluble mineral elements, and at the same time acts as a sponge to increase the water-holding capacity. Still, it serves a most important function in adding elements directly to the soil which may be utilized by the succeeding crops.

The practice of adding green manuring crops in order to increase the supply of fertility elements and of organic matter, however, is one that is not to be generally recommended if these crops may be fed practically to live stock. Tests at the Ohio Experiment Station show that the loss of organic matter from clover cut and allowed to remain on the surface of the soil for nearly 7 winter months is about the same as the loss by feeding. A large loss results even by plowing the clover under early in the fall for spring crops. Hence, it is more profitable, except for badly run-down farms, to harvest the crop, feed it during the winter and save the manure carefully, the same results in maintaining soil fertility being obtained and in addition the feeding value of the crop is secured.

Barnyard Manure.—This matter brings us up to the addition of barnyard manure as a source of supply of plant food in the soil. Manure in itself is a valuable fertilizer containing large amounts of easily available fertility elements, especially nitrogen and potassium, in both solid and liquid excrement. The value of this manure depends largely upon the food consumed by the animals in question, upon the age and work of the animals, and upon the handling of the manure.

Fresh manure contains, approximately, 10 pounds of nitrogen, 2 pounds of phosphorus and 8 pounds of potassium per ton. While these elements may not be as available as in ordinary, high-grade, soluble fertilizers, and consequently extend their usefulness over a longer period of time, the organic matter in manure must not be lost sight of. The total result from the use of manure after many years is more satisfactory. Since most farmers can obtain these necessary elements far more cheaply by producing manure than by buying commercial fertilizer, the proper handling of this material is one secret of success on a well-managed farm.

Commercial Fertilizers.—The last important means by which fertility elements are added directly to the soil is commercial fertilizer. In the fertilizer sack are purchased nitrogen, phosphorus and potassium, the first of which is most costly and used in smallest amounts. The second for nearly all soils is most important, as previously explained. It is the phosphorus which is carried off the farm in largest amounts in grain, in the bones of animals and in milk; it is the same element which is found in smallest amount in most soils; hence, its addition is of greatest importance in keeping up the fertility of the soil. The nitrogen and potassium can largely or entirely be maintained by other means than by commerical fertilizers, as by proper crop rotation and the use of manure.

Summary.—This discussion of the fundamentals of soils and soil fertility sets forth the tools with which the farmer works. Beside having an immense store of mineral elements derived from the original parent rock from which it was formed, soil contains the organic remains of plants and animals, additional fertility constituents which are equally essential for crop growth. Crop yields are materially affected by changes in the structure, moisture content and temperature of the soil.

Adding elements directly to the soil, besides making available what are already there, completes, in general, the management of the land for successful farming so far as suitable plant food for crops is concerned. Of all the many elements, only nitrogen, phosphorus and potassium are sold commercially as plant food, while calcium is also added to neutralize soil acidity. Crop residues, barnyard manure and commercial fertilizers are the main sources of supply of the fertility elements, while liming corrects acid conditions.

CHAPTER II.

THE MANAGEMENT OF THE SOIL.

Why Have Abandoned Farms?—In our present day of agriculture there is a familiar cry of depleted soils and run down and abandoned farms. Why does this condition exist, and why can we not raise the crops our forefathers harvested? While the contributory causes are legion, a study of the more important ones will explain this troublesome question and show how the soils, robbed of their virgin fertility, may be covered again with fields of golden grain and verdant meadows, and sleek cattle will again graze on fertile pastures.

The most serious loss has been in the humus content. Before man first tilled the soil it was stocked with decayed and decaying vegetation constantly accumulating, but later all this movement was reversed and plant food was taken from instead of added to the soil. Manure and straw were not utilized; in fact, they were often considered worthless rubbish. Then as the soils were drained and cultivated, air circulating through them hastened the decomposition of the organic matter. Bacterial action was at its height and summer fallowing, often practiced, tended in the same direction.

Cultivated plants did not find a congenial home in such soils. The mechanical condition of the seed bed was poor, and the mineral resources were not abundant because of the depletion of the organic matter. As it disappeared so accordingly did the bacterial activity diminish. Little available plant food, no organic matter and no life inevitably meant no production after such exhaustive systems of cropping.

Our problem of today, therefore, is to replenish these impoverished fields. Organic matter must be restored, and commercial forms of plant food must be added to the soil to compensate for that carried off the farm in field crops, in live stock and in the drainage and surface waters. And, finally, the loss of lime by leaching and cropping has had its share in diminishing the productiveness of our soils. Each of these subjects is so important as to deserve special treatment, but at the same time the handling of the soil and the control of the moisture content are essential in successful farming. Tillage, drainage, irrigation and farming in semiarid regions are manifestly problems that have direct bearing upon the country's production of farm crops.

Tillage.

Tillage is a broad term, involving all operations of stirring the soil by plowing, pulverizing, harrowing and cultivating. Its objects have to do with physical, chemical and bacterial activities, together with weed control.

Tillage Increases Feeding Surface for Plants.—Soils are made more inviting to plant roots as the large lumps are broken into smaller particles. In the first place, the rootlets can penetrate the soil more easily, and with the finer structure more area is exposed to the action of the rootlets. In other words, tillage increases the feeding area of plants.

Tillage Increases Moisture Supply.—As the soil particles are broken up and made finer, the air can penetrate more readily to the plant roots. This air is necessary for the growth of all farm crops, and it is also useful in the soil by hastening the liberation of plant food. Besides having greater feeding ground and more thorough aeration, the plants are given more water by proper tillage. The moisture which is available to the plant surrounds the soil particles in a thin film; free water in the soil is not absorbed by plant roots. It is the film, or "capillary" moisture that carries the plant food and enters the small rootlets. As the soil is made finer by tillage, the number and therefore the amount of surface of the particles are increased. Hence, an increase in the surface area of the particles means an increase in the available supply of soil moisture.

Tillage also conserves the water in the soil. In a closely packed soil the capillary water moves freely; and, as the surface layer is dried out through the action of wind, sun and plants, fresh supplies are lifted from the subsoil. If the top layer of soil is broken up 2 or 3 inches deep and left loose by a harrow or cultivator, there is no longer a connection between the surface and subsoil water; and water can be lifted only to the stirred soil, or mulch, which is separated from the firm soil by large spaces. The deeper the loose layer, the more effective it will be. Still, excessively deep or too frequent cultivation is unnecessary. At the New Hampshire Experiment Station one field was cultivated 11 times more than another, with an increase of only 2 bushels of corn. Corn given shallow cultivation yielded 6 bushels more per acre than corn cultivated deep.

To maintain an effective mulch in the field, frequent shallow stirring is necessary, especially after a rain packs the soil, care being taken not to cultivate so soon afterward that the soil will become puddled. The dust mulch is less expensive than straw, hay, sawdust and other similar materials, is just as effective and incidentally kills the weeds.

Tillage Makes Plant Food Available.—With respect to increasing chemical activity in the soil, tillage hastens the liberation of plant food elements. By this stirring of the soil different particles are brought into contact with each other, resulting in new chemical reactions. Air contains the element oxygen, that changes some insoluble mineral elements to a form plants can assimilate. Thus tillage serves in two important ways in increasing the food available to plants.

Tillage Encourages Bacterial Life.—Indirectly it performs a third function in the same direction. It makes conditions more favorable for the activities of bacteria, which break down the organic matter, setting free plant food, especially changing nitrogen to soluble nitrates. This process can take place only in the presence of oxygen, and hence cultivation increases this action. In this connection, too, it should be remembered that the organic matter is thus more rapidly depleted and provision should be made for its maintenance.

Oxygen is equally essential to the bacteria that live in close relation to the legumes and take nitrogen directly from the air. It is therefore necessary that air circulate freely about the roots of such crops.

Tillage Controls Weeds.—Some soil experts contend that cultivation is of greater importance in checking weed growth than in any manner mentioned above. Moisture, light and plant food taken by weeds mean an equivalent loss to the cultivated crop. Weeds left on the ground mean little loss of total fertility, but mineral elements taken by them are beyond all reach of the present crop.

In one test at the Illinois Experiment Station, killing weeds increased the corn yield 38.6 bushels per acre over no cultivation. Killing weeds without stirring the surface gave as large corn yields as shallow cultivation, proving that cultivation should be done primarily to kill weeds, which rob the crop of light and plant food more than of water.

Time of Plowing.—The extent of the activities previously outlined depends in some measure on the time of plowing, and this factor has further influence in several ways on crop yields. Plowing should be done sufficiently in advance of seeding that the soil may settle to produce a firm seed bed, and incidentally some plant food will thus be made available at the surface where the young plants may easily reach it. A surface mulch by frequent harrowing will conserve moisture and prevent weed growth.

The Kansas Station affords an illustration of the value of timely plowing. Plowing land 7 inches deep July 15 gave a net profit of $16.87 an acre with wheat, while plowing the same depth in September returned only $8.60 per acre. A gain of 6 bushels of wheat per acre was found at the Michigan Station and a gain of 5 bushels at the Oklahoma Station by early plowing.

Early plowing, however, should not be the sole aim if there is any danger from handling wet soils. When the moisture content is excessive, puddling will result and large, hard clods will be formed.

Plowing in late fall or early winter for the next spring's crops should as a rule be encouraged in nearly all the states of the Middle West, because it distributes labor more evenly throughout the year, favors earlier seeding, improves soil structure and kills many insects. In the semiarid and arid sections the practice is of even greater importance, as the moisture supply is ordinarily the limiting factor in crop production. On soils subject to washing and blowing by wind, on fine clay soils that would be puddled by winter rains, and on sandy soils, the practice is not recommended.

Usually there is a rush at spring planting time. Farm labor can be used to better advantage in the fall, and crops may be seeded earlier in the spring. At the Ohio Experiment Station a difference of a month in time of planting corn has made a difference of 24 bushels in yield per acre, as an average of 7 years. Moreover, the early planted corn had about 10 per cent less moisture in it. At the Pennsylvania Experiment Station a difference of 4.4 bushels of corn and 199 pounds of stover per acre was noted in favor of fall plowing over early spring plowing.

Freezing and thawing during winter make plowed soils more granular. More moisture has been found in the spring in soils fall plowed than in unplowed land. On a Utah dry-farm nearly half a year's precipitation was saved by fall plowing. Under conditions of dry farming it is extremely necessary to start to conserve moisture for the next crop as soon as one is removed. The furrows as left by the plow are well adapted to hold winter snows and also act as a mulch to prevent the loss of capillary water.

The roller can be used to advantage to firm the seed bed for planting.

On hillsides and clayey soils, as in southern Indiana; on light, sandy soils subject to wind erosion and soil washing, as in upper Wisconsin; on sandy soils where the winter weather is open, as in Delaware; and where little alternate freezing and thawing occur in winter, as in the southern states, a cover crop will accomplish more for the subsequent crop than will fall plowing.

Harrow Immediately After Plowing.—Except when freezing weather follows plowing there is no time when a harrow will do as much good in as short a time as a few hours after ground is plowed. In certain cases, as immediately before a heavy rainfall, nothing is gained, but with dry weather after plowing many a farmer has spent days trying to put neglected clods in good mechanical condition only to find that a good seed bed was impossible because of his neglect to use the harrow right after the plow. If the ground is too wet the harrowing should be delayed a few hours, allowing the soil to dry out enough so that it will crumble nicely.

Deep Plowing Impractical.—Plowing 15 inches deep, either with a Spalding deep-tilling plow or with a subsoil plow following ordinary plowing,

The double cutaway disk harrow. The first operation after plowing is to break up any clods and make a smooth seed bed.

as compared with the usual 7½-inch plowing, has been found unprofitable on test plots at the Ohio Experiment Station. In this work a rotation of corn, oats, wheat and clover has been followed.

The expensive operation of deep plowing has returned about a bushel more wheat per acre than 7½-inch plowing, as an average of 4 years. Subsoiling has produced an increase of less than half a bushel to the acre. Corn yields have been increased only 1 to 2 bushels per acre by deep plowing. Clover, however, has shown no benefit from special plowing, while oats have yielded most with ordinary plowing.

At the Clermont County (Ohio) Experiment Farm 5-inch plowing yielded the same amount of corn and 800 pounds more soybean hay than 13-inch plowing.

Results at other experiment stations substantiate these conclusions. At Pennsylvania corn with ordinary plowing yielded from 65 to 395 pounds more to the acre than plowing with the Spalding deep-tilling machine, Little differences were noticed in the yields of barley. Oats produced 4.6 bushels more per acre by deep tilling in the fall, and 2.2 bushels less when the land was plowed in the spring. Alfalfa yielded 43 pounds more at the first cutting and 44 pounds more at the third cutting in favor of ordinary plowing. While the average draft of the Spalding deep-tilling machine for the three sets of trials was 1,289 pounds, that of the ordinary moldboard plow was only 486 pounds.

On a gray silt loam on tight clay in southern Illinois, a decrease of 2.7 bushels of corn per acre annually resulted during 8 years from subsoiling over ordinary plowing. The results did not warrant the use of deep-tilling machines. Tests in Oklahoma also show that subsoiling and deep plowing are unprofitable in that section.

Dynamiting as a Means of Tillage.—Within recent years manufacturers have advertised widely the supposed merits of dynamite to loosen subsoils to greater depths than can be accomplished by plows and other implements. Such claims as those of greater aeration, drainage, correction of acidity, increase of available plant food, etc., have been made. Experimental evidence has proved such assertions faulty.

Tests made by the Kansas Station shows that no greater moisture content, no greater formation of nitrates and only a slightly greater bacterial content resulted from dynamiting. The physical condition of the soil was poorer than before the blasting was done. In no instance was improvement sufficient to pay the expense.

Tests on a stiff Ohio soil showed no beneficial results after the first year. For tree planting on a stiff clay soil in southern Ohio dynamiting proved cheaper than hand-digging; but in orcharding in that state, too, it has been noticed that the soils become as compact as ever after the first year. For clearing woodland of stumps the Ohio Station has found blasting cheap and effective.

Soil Drainage.

Nature's efforts in supplying moisture for plants must usually be supplemented with man's ingenuity, either to carry off excess water or to overcome a deficiency. Tile drains are today considered more effective, less expensive in maintenance and less bothersome than open ditches. Still, tile drains in all cases are not a profitable investment, and on any farm are but supplements to other means of increasing crop production.

What Soils Need Drainage.—River bottom land that drains slowly, flat, swampy areas or basins surrounded by higher land, upland clay soils that hold water so long that wheat and grass freeze out, and flat soils formed in old lake beds generally return big dividends from drainage. Rolling clay soils are often profitably drained, particularly if underlain at shallow depths by an impervious layer.

In general, the prairie lands of Illinois (except in the southern part), Iowa and surrounding states, and the level lands from New York to Illinois may be profitably drained. Draining the cheaper lands of the South is also proving profitable with a more intensive culture.

Drainage Has Many Advantages.—A field well drained has several more or less noticeable advantages. The primary object in draining is to carry off excess water, or that which can be seen in about the first 3 feet of soil, in wagon or animal tracks or in the plow furrow. Then incidentally the tiles do other things. The rain passes through the earth instead of running off

Digging the ditch and laying the tile
by hand.

the surface, washing gullies and carrying with it a load of fertility. As the water leaves through the tile, air is drawn into the soil which will hasten the decay of organic matter; in the meantime a cold soil will be warmed up slightly. Hence, crops may be planted earlier, and they grow more rapidly because the soil structure is more inviting and available plant food is in greater abundance. Their roots penetrate deeper in search of food and moisture, and they can withstand more easily any subsequent drouth.

Kind of Tile to Use.—Select good, hard-burned clay tile that give a clear ring when struck, are strong, free from impurities, smooth and straight, with square ends. Collars are unnecessary. Cement tiles are satisfactory but usually cost more and break easily.

The size of tile depends upon the amount of water to be drained away, but needlessly small tile should not be selected so that the water cannot be drained off in at least 48 hours after a heavy rain. It never pays to lay anything smaller than 3-inch tile, as this with a good fall will drain about 5 acres; a 4-inch tile, about twice as much.

Hints on Laying Tile.—It is best to lay tile at least 3 feet deep, although depth varies with the kind of soil. The deeper tile are laid, the farther apart

Laying tile after a ditching machine.

the drains can be, although extreme depth is unnecessary. The distance apart of drains also depends upon the amount of water and kind of soil. Clay soils need drains closer, especially if a hardpan makes it necessary to lay the tile shallow. Place the drains 50 to 100 feet apart for farm crops, so that the fields may be worked in due season. If one can tell by the crops just where each drain is, the field needs more tiling.

There should be a fall of at least 2 inches per 100 feet, and 3 or 4 inches would be better. This fall should be uniform and never to a less steep grade or dirt will accumulate. Going from a medium to a steeper grade is satisfactory. It is economical on rolling land to follow natural courses, ditches, swales, etc., but not to the extent of making a crooked drain. It is better to lay tile down the slope of a hill rather than around it. Laterals should join the mains at oblique rather than right angles to prevent sediment from lodging.

Ditching machines are now being widely and satisfactorily used. Whatever the method, care taken in having a well-planned system in obtaining sufficient, uniform fall, in laying tile and in filling in around them will be paid

Filling the drains.   Protecting outlet to tile drain.

for many times in many years of service, where poorly laid tile would be constantly annoying, ineffective and expensive.

The outlet should be guarded by stones or concrete and covered with grating to keep out animals. An iron pipe or a few sewer pipes at the outlet are better than the ordinary drain tile.

Tile Drainage Does Not Always Pay.—The only experimental work with drained vs. undrained land in Ohio is in Clermont County in the southwest. As a 4-year average, corn in a rotation of corn, soy beans, wheat and clover has returned from 1 to 13 bushels, and wheat, during its only year of test, from 1 to 11 bushels more per acre on fertilized and manured crops on drained than on undrained land.

On the other hand, a loss resulted from tile-draining a gray silt loam on tight clay in southern Illinois. For 14 years an average of only 45 cents per acre at Dubois was realized, and $1.20 an acre would be necessary to pay 6 per cent interest on the investment, cost of drainage assumed to be $20 an acre. At Fairfield on a similar soil tile drainage paid $1.08 an acre, as an 8-year average, although its value is increasing in recent years. The Illinois soils are tighter in subsoil than those in Ohio, and the hardpan makes tile drainage uncertain.

Black soils of the glaciated limestone region invariably need under-drainage.

Surface Ditches.—Where tile drains are ineffective in such cases as men-

A convenient way of filling ditches.

tioned above, the fields should be plowed in small lands with dead furrows in the direction of greatest fall. On flat areas having no natural drainage channels, open ditches become necessary before the land can be cropped. Tile drains in adjoining fields may then be led into these open ditches.

Irrigation.

There is in the United States from 40,000 to 50,000 acres of arid land that can be reclaimed, or about one-half the amount of reclaimable swamp land. Of this area about one-third is now under irrigation. Although these large tracts in the western states are well stocked with plant food, vegetation is very scant unless water is artificially supplied during the growing season. Since reservoir or well water adds little or no fertility, the principles of soil fertility maintenance must be applied to these lands.

Irrigation is also practiced in humid regions, but only for the most intensive farming, as market gardening. The practice is most profitable where rainfall is less than 15 inches annually.

How Water Is Applied.—Small ditches in cultivated crops conduct the water from canals leading from streams, lakes or wells. For crops not intertilled the entire field is flooded; for crops like corn the ditches run between the rows. By thus controlling the water supply and distributing it evenly during the growing season, many areas have been made a source of great income. Thus, in Oregon for 7 years under study yields were increased 65 per cent by irrigation, and two crops a season can be raised.

Fundamentals in Cropping on Irrigated Land.—Since the growing season is generally short, early varieties of crops adapted to such culture should be selected. The irrigation should be thorough, and with intertilled crops cultivation should follow immediately after the ground is dry enough to work. Too much, too frequent and too early irrigation and too little cultivation mean smaller crops or failure. In the case of cereals most water is needed when the grain begins to form.

The Idaho Experiment Station reports seven or eight irrigations necessary for alfalfa in a three-crop season, making about $2\frac{3}{4}$ acre feet of water, while in Oregon this crop needs only about 6 acre inches per acre. For winter wheat one or two irrigations of less than an acre foot, and for potatoes, spring wheat, oats and barley, four applications totaling $1\frac{1}{2}$ to $1\frac{3}{4}$ acre feet is considered sufficient under Idaho conditions.

Early spring plowing and manuring, as well as rotations including legumes are beneficial for irrigated areas. Pastures of bluegrass, orchard grass, timothy and meadow fescue in mixtures have paid in Idaho even on high-priced land.

Dry Farming.

"Dry farming" is a term applied to an agricultural system in regions where the rainfall is too small to support the growth of ordinary farm crops every year with the usual farm practices. Selection of drouth-resistant varieties and the conservation of the moisture supply therefore become the problems to study for successful crop production.

Areas in United States.—Dry farming was first extensively practiced in Utah a half-century ago. This state is in the midst of one of the dry-farming areas. The Great Plains region, including part of Texas, Oklahoma, Nebraska, Kansas and the Dakotas, constitutes another, while the extreme western states follow a somewhat similar plan. In the last-named region most of the rainfall comes in winter; in the intermountain area, in late winter and early spring; and in the Great Plains area, in late spring and early summer, which are followed by weather conditions that favor rapid evaporation of soil water.

Conserving Moisture.—It is the problem of the farmers in these sections to conserve all the moisture possible. Lighter types of soil, like sandy and silty loams, are ideal, but the fine-textured clay soils absorb water slowly and become compact after rains. Likewise, shallow soils are undesirable because they do not encourage deep rooting.

To conserve all possible moisture the soil during the rainy season must be kept open, so that the rainfall will be absorbed and held; and cultivation must prevent the loss of water by weeds and evaporation. As soon as one crop is harvested, the soil should be prepared for the next. Fall plowing can be practiced in the coast states, early spring plowing in the intermountain region, and early spring disking followed by early summer plowing on the Great Plains. Thorough intertillage of crops always pays; frequent stirring of the soil not more than 3 inches deep, especially after rains, saves the moisture and keeps down weeds. Deep plowing is advisable. Listing corn is also successfully practiced.

Summer Fallowing.—Over much of the dry-land region not sufficient rainfall comes during the year to grow a crop. Hence, one season must be devoted to fallowing, so that the water needed by the crop may be stored in 2 years' time. Under such systems one is always assuming risks, and must provide for the years of scarcity and failure by his successful crops. Land agents often unduly exploit dry-farming prospects.

Summer-fallowed land in Kansas gave a yield of 21 bushels of wheat per acre; early fall-plowed, 11 bushels; and late fall-plowed, only 6 bushels. The practice of summer fallowing once in 3 or 4 years is profitable in central and western Kansas, the fallow following sorghum or corn and preceding wheat. In western North Dakota summer-fallowed land yielded only 7.4 bushels of wheat, 16.6 bushels of oats and 7.7 bushels of barley more per acre than continuous cropping, which indicates that the practice is doubtful in that section. No more than one-third of the rain can be saved by fallowing, and this stored moisture cannot mature a crop unless aided by fairly seasonal rainfall.

In the arid Southwest this practice has not been successful. Supplemental irrigation has been cheaper, and crops can be brought 4 to 8 weeks ahead of the season. The growing season there is so short that crops must be started ahead of the rainy period.

Cropping Systems.—For the Great Plains wheat is best adapted, because it makes its greatest growth during the season of rainfall and uses water economically. Turkey wheats are grown chiefly. Sorghum is also economical of water, resists drouth and can be grown profitably in this region. North of Kansas oats, barley and spring wheat are suitable, and potatoes after a year of fallow are usually successful. In the Southwest milo maize, peas and beans are considered the best grain crops, and corn as well as sorghum is grown for forage and silage. All these crops should be grown in rows and cultivated thoroughly.

CHAPTER III.

BARNYARD MANURE.

Among the sources of supply of plant food to the soil, barnyard manure was mentioned as a by-product material rich in fertility elements and capable of making mineral constituents available in the soil for the utilization of farm crops because of its high content of organic matter and bacteria. Its value from this standpoint is little appreciated in many cases: first, in extremely fertile sections where little attention need be given to maintaining soil fertility, and, second, on farms where the owners, not understanding its worth, allow it to be largely wasted. Wherever soils are being cropped and live stock kept or where the material can be bought and applied at a sufficiently low price, barnyard manure should be spread on fields so used.

In the discussion from here on the term "manure" will be used for "barnyard manure." Whenever "fertilizer" is mentioned, commercial forms of plant food are meant. "Green manures" are crops plowed under for soil improvement.

Production of Manure.—According to the Yearbook of the U. S. Department of Agriculture for 1915, there are in the United States at present the following number of animals: horses and mules, 25,731,000; cattle, 61,441,000; sheep, 49,162,000; hogs, 68,047,000. Experimental evidence has shown that a fair estimate of the amount of manure ready for field application which can be obtained from these animals is as set forth in the following table:

Kind of Livestock.	Average Daily Production	Average Annual Production.	Total Annual Production.
	Pounds.	Tons.	Tons.
Horses and mules	50	9.1	234,152,100
Cattle	43	7.8	479,239,800
Sheep	4	0.8	39,329,600
Hogs	9	1.7	115,679,900
Total			868,401,400

It now becomes of interest to know how much this manure would be worth if carefully preserved and used in the production of field crops. A ton of average fresh manure contains about 10 pounds of nitrogen, 2 pounds of phosphorus and 8 pounds of potassium. If the nitrogen is valued at 19 cents a pound, phosphorus at 13 cents and potassium at 6 cents—prices prevailing before the European war—a ton of manure fresh from the stable would be worth $2.64. The total value of the manure in the United States on this basis of fertility elements actually contained would therefore be $2,290,379,696.

Much of the Value is in the Urine.—The fertility elements in manure are divided between the solid and liquid excrement. The following table will show how valuable the urine really is, and why it should be carefully preserved:

Animal.	Pounds per Animal Daily.		Percentage Composition.					
			Nitrogen.		Phosphorus.		Potassium.	
	Solid.	Liquid.	Solid.	Liquid.	Solid.	Liquid.	Solid.	Liquid.
Horse....................	35.5	8.0	0.50	1.20	0.13	Trace	0.20	1.24
Cattle....................	52.0	19.4	0.32	0.95	0.09	0.012	0.12	0.79
Sheep....................	2.3	1.5	0.65	1.68	0.20	0.013	0.19	1.76
Hog....................	6.0	3.3	0.60	0.30	0.20	0.055	0.37	0.83

The table shows that the urine is richer in nitrogen and potassium than the feces, and that it contains nearly half these elements voided by the animal. Too much emphasis, therefore, cannot be laid on saving it.

The Value of Manure—Determining Factors.

Kind of Animal.—The value of a ton of manure depends upon several factors, and of these the class of live stock is logically the first consideration. A study of the table above shows the difference in percentage composition of the manure from common farm animals. The commercial values of a ton of these various manures varies more than 100 per cent, and if used where the plant food is needed the agricultural value will vary in the same proportion.

Even with cattle alone a difference is noted in the composition of manure produced. Dairy cows at the Ohio Station produced milk containing 46 per cent as much protein as they ate, but it was estimated that until they were 2 years old the meat they produced contained only about 18 per cent of the protein in the feed consumed. Analyses from a great many experiment stations show that manure from steers contains about 2 pounds more nitrogen, ½ pound more phosphorus and 1 pound more potassium per ton than manure from dairy cows. A farmer can therefore afford to pay more for a ton of manure from a fattening steer than from a dairy cow. The great difference in sheep manure as compared with horse or cattle manure is due chiefly to the difference in water content.

Kind of Feed.—Differences in manure values can also be attributed in large part to the feed consumed. Highly nitrogenous feedstuffs, like oilmeal, cottonseed meal, distillers' grains, etc., will, in addition to good milk yields, secure a further return in greatly increased value of the manure products when fed to dairy cows. Cows fed cheaply upon roughage, with little grain, will yield a manure product of correspondingly low value. The same thing is true with the steer or any other kind of farm animal, although the mature animal yielding no product, like milk or eggs, will extract a smaller proportion of the fertilizing value from the food consumed.

Nutrition studies at the Ohio Experiment Station showed that only a part of the fertility value of the feed consumed is returned to the soil in manure, while a small portion is retained and utilized in the animal body. Growing animals utilize more of the food constituents to build up their bodies than is needed by mature animals. Milk production involves even heavier drafts upon the fertility value of the feed. Growing hogs returned in their manure 74 per cent of the nitrogen, 87 per cent of the phosphorus and 86 per cent of the potassium contained in their feed. The excreta from milk cows contained only 67 per cent of the nitrogen, 72 per cent of the phosphorus and 74 per cent of the potassium in the ration.

As an average of all classes of live stock, 80 per cent is generally taken as the percentage of fertility constituents which may be returned in manure from feed consumed. The loss of organic matter is most serious in feeding, about one-third of that fed the animal being recovered in the manure. This is largely burned up in order to provide the animal with heat and energy. Then about one-fourth of the organic matter in the manure as voided is further lost by decomposition through the action of bacteria in the stable or yard.

Value of Bedding.—Besides supplying bedding for live stock, straw has value for the actual elements of fertility which it carries and for the organic matter which it adds to the soil. A ton of wheat straw contains about 10 pounds of nitrogen, 2 pounds of phosphorus and 10½ pounds of potassium; oat straw, 11½ pounds of nitrogen, 2½ pounds of phosphorus and 29 pounds of potassium per ton. At current prices wheat straw would therefore be worth about $2.80 a ton and oat straw, $4.25.

It should be said in addition that straw has a value on most farms that cannot be measured by its value as plant food. One can buy nitrogen, phosphorus and potassium in fertilizers, but it is not always convenient to get organic matter, either in manure or in green manuring crops.

Scarcity of straw is forcing many live stock farmers to find substitutes for it. From the standpoint of fertility, straw is to be preferred. It contains more actual plant food than shavings and sawdust and is an active, easily decaying form of organic matter. Only in exceptional cases of high prices, or where it can be bought back along with the excreta of animals, should it be sold off the farm.

Treatment Given Manure.—The treatment that manure receives both in the stable and in open yards will determine largely its value for increasing crop yields. Often the losses thus sustained by improper treatment equal half the value of the manure.

Fermentation Causes Loss of Nitrogen.—Manure is a home of millions of bacteria that start to decompose as soon as voided. This decomposition is most rapid in sheep and horse manure, usually called "hot" manures, as distinguished from "cold" manures from hogs and cattle. Hog and cattle manures contain more water and are more compact than the other kinds and hence ferment less.

This fermentation, or decomposition, is accompanied by heat, and is often called "fire fanging". It is undesirable because the organic matter is burned out, and nitrogen is liberated as a gas. The smell of ammonia is characteristic

The shed to the right was built to keep manure under cover, protected from rains.

of stables with decomposing accumulations of manure. Tramping, wetting and the use of preservatives are discussed later to show how this loss may be largely overcome.

A Cement Floor Soon Pays For Itself.—Since the urine contains a large proportion of the fertility value of manure, and all the plant food in it is soluble or available for plant utilization, it is evident that tight floors in the stable are imperative for its conservation. The Ohio Station fed 28 steers on cement floors and 30 on earth floors, with a difference in value of manure produced per steer of $4.48 a year in favor of the cement, or enough to pay for the cost of concreting.

Leaching In Open Barnyards.—Still greater losses in the value of manure occur in open yards. A single heavy rain will replace most of the urine with water, leaving the weight about the same, or even greater. Chemical analyses made during 5 years at the Ohio Experiment Station shows that manure loses about one-third of its nitrogen, one-fourth its phosphorus and one-half its potassium when exposed to the weather for 3 winter months. At prices pre-

The loss by leaching of manure, as shown here, is so great that on all farms where live stock are kept, a shed to preserve this valuable by-product would pay for itself in a year or two. A coffee-colored stream running from every such yard tells a tale of lost fertility.

vailing before the European War, the fertility constituents in a ton of manure would be worth about $2.65. The loss by leaching 3 months would then amount to about 95 cents. During the period of the war this loss would be nearly $2 for potash alone.

Field experiments in Ohio also prove this claim. As an average of 255 comparisons during 19 years, with manure used alone and when reinforced with acid phosphate, raw phosphate rock, gypsum and kainit, an application of 8 tons of stall manure per acre on corn has produced 3.9 bushels of corn, 1 bushel of wheat and 476 pounds of hay more than the same amount of manure left in an open barnyard for 3 winter months. With corn at 70 cents a bushel, wheat at $1 and hay at $10 a ton, this increase would be worth 75 cents for keeping a ton of manure under shelter or for hauling it directly from stable to field.

Amount Used Per Acre.—The rate of application is another factor that determines the value of a ton of manure. It cannot be expected that increasing the amount per acre will result in corresponding increases in yield. The law of diminishing returns explains that as the size of the dose is increased the rate of increase in yield is somewhat lessened. Four tons of manure to the acre on both corn and wheat in a 5-year rotation of corn, oats, wheat, clover and timothy at the Ohio Station has produced an increase of 15.4 bushels of corn, as an average of 22 years. Eight tons of manure on the same crops in the same rotation has increased the corn yield only 23.7 bushels during this period. In other words, doubling the amount of manure has increased the corn yield by only 54 per cent. Moreover, doubling the amount of manure has increased the wheat yield only 48 per cent.

Applying raw rock phosphate as an absorbent and reinforcing material for barnyard manure.

Results at the Pennsylvania Station likewise show that greater returns per ton of manure come from spreading it thinly. Under the conditions of the test an application of 12 tons per acre returned $2.16 per ton; 16 tons, $1.60 and 20 tons, $1.44.

Trucking involves considerable cultivation and consequent oxidation of organic matter in the ground. Commercial fertilizers cannot take the place of manure to keep up the organic content in such cases, although they are valuable supplements to manure. For truck crops from 15 to 20 tons per acre are frequently found profitable.

Balancing Manure With Phosphorus.—It is seen from the tables above, giving analyses of manure, that this material is not a balanced food; that is, it does not supply the plant the essential elements in the right proportion. It is lacking in phosphorus, which is carried off the farm in grain, milk and the bones of animals. Hence, this element should be supplied in commercial form to meet the deficiency; such procedure is called reinforcement.

The value of a ton of manure, then, is determined in some degree by the fact whether it is used alone or combined with phosphorus. At the Ohio Station an annual return of $2.76 per ton for yard manure untreated used at the rate of 8 tons per acre in a rotation of corn, wheat and clover has been

Manure is a great food for corn. Plot on left received 8 tons of stall manure; that on right, nothing.

realized during a period of 18 years, as compared with a return of $3.43 for yard manure reinforced with acid phosphate.

Kind of Crop to Which Manure Is Applied.—At the Ohio Station 8 tons of manure per acre per rotation in a 5-year rotation of corn, oats, wheat, clover and timothy has given a return of $3.06 per ton, as a 20-year average value of the total increase for the rotation. With the same application in a 3-year rotation of potatoes, wheat and clover, the return per ton has been $2.70.

Perhaps a still more striking example of the variation in return per ton of manure from its use upon crops is to be found in the continuous cropping plots of the same station. Five tons of manure per acre used on corn has returned $3.14 per ton; on oats, $1.19; on wheat, $1.95, if corn is valued at 70 cents a bushel, oats at 40 cents and wheat at $1.

Place in the Rotation.—The three-year rotation of potatoes, wheat and clover at the Ohio Station affords some interesting data regarding the variation in the return from a ton of manure when applied to different crops in the rotation. One plot receiving 8 tons of manure per rotation on wheat has returned $2.70 a ton, while another receiving the same application on potatoes has returned $3.39, all three crops being considered in each case for twenty years.

Fertility of Soil.—The return per ton of manure varies widely with the fertility of the land upon which it is used. The richer the land the less the increase in crop yield will be from the use of manure. In some cases in Ohio, manure returns 50 cents a ton and in others more than $6. Its net value per ton in 5-year rotation at Wooster has been $3.26 for 22 years, and only

Phosphorus and manure add bushels to the corn crop. Plot on left received phosphated manure, that on right, nothing.

$1.77 for 20 years in the same rotation and with the same treatment at Strongsville, about 100 miles north.

At the Wooster farm unfertilized corn in a 3-year rotation of corn, oats and clover has averaged 51.5 bushels per acre for 9 years; in a 3-year rotation of corn, wheat and clover, 32.9 bushels during the same years. The type of soil is the same. The two series are within 40 rods of each other. This wide variation in productiveness is due to previous treatment of the land.

The poorer soil receives 320 pounds of floats per rotation and the other series 1,000 pounds, in addition to 8 tons of manure. On the poor soil a return of $3.67 per ton of manure has been found, and on the rich land only $1.18 a ton. This comparison is slightly unfair because of the different amounts of floats, but still the returns from manure would seem to be magnified by the conditions. It shows unmistakably the tendency of manure to return greatest dividends when applied to the poorest soil on the farm.

How Far to Haul.—Other factors that will influence the value of a ton of manure are distance of hauling, prices received, distance to market, etc., but the last two named are not within the farmer's control. Often he must haul manure for such a distance that this cost alone is $2 or more. It is often doubtful whether he is justified in hauling manure such distances. It is possible that a judicious use of commercial fertilizers, coupled with green manuring, would furnish him plant food and organic matter cheaper.

Handling Manure.

Prevent Losses in Stable and Yard.—From what has been said heretofore it is evident that watertight cement floors are essential to prevent seepage of urine, the most valuable part of manure. Plenty of bedding is also necessary to take up this liquid portion. Spreading directly from the stable to the field, of course, will reduce this loss to a minimum.

When it is impossible to go on the land to spread manure, care should be exercised in storing it. To prevent leaching by rains, a covered yard should be provided. Never before in agricultural history would such a shed pay for itself so quickly as today. A single winter may cost a farmer more in manure losses than the expense of a cheap shed would be to keep rains off this product.

In the covered yard manure from horses and sheep should be mixed with that from cattle, and it should all be tramped well, so that fire fanging will be reduced as much as possible. Again, wetting horse and sheep manure will be profitable, but it increases the amount to handle, and too much water may cause a loss in fertility elements in seepage. Hence, it is better to mix hot and cold manures and keep them well tramped; in this way oxidation with its consequent serious loss of organic matter and nitrogen will be prevented. It can often be arranged advantageously for live stock to spend a few hours each day in covered manure sheds, where they will keep the manure more or less compact. This would hardly be possible except where large amounts of bedding are used. The manure should not be disturbed until hauled. Every time it is forked over the destruction of organic matter is increased.

Manure Cistern Impractical.—Many farmers have tried to solve the problem of saving liquid manure by constructing large cisterns into which the seepage drains, but many such plants have been abandoned after a short period of service. One must handle an extremely large amount of pretty thin liquid during the year. It is a better policy to protect the manure pile from leaching until it goes to the field. This can be done by immediate application, or by storing in a covered yard where it is kept moist and compact. Such investments will pay better than expensive manure cisterns; and it must be remembered that expense does not stop with building the cistern.

Reinforcing Manure.—Since phosphorus is carried off the farm in relatively large amounts in grain crops, in the bones of animals and in milk, some commercial form of this element must be added to manure to make it a balanced plant food if the fertility of the soil is to be maintained and full

Phosphorus is the key to fertility. Plot on right, received nitrate of soda and acid phosphate; that on left, only nitrate of soda.

value realized from the manure applied. Acid phosphate and raw phosphate rock are commonly used to supply this deficiency. Acid phosphate in Ohio has returned more than three times its cost in crop yields, even when used in connection with liberal applications of manure. For most of the states of the Middle West phosphorus is the plant-food factor that limits crop production. By this is meant that phosphorus is the first element of plant food needed, and when supplied in proper proportion will prove cheaper than any other fertility element, with the possible exception of calcium, which is used primarily to neutralize acidity. Of course, there are exceptions to this general conclusion, as many soils may be strikingly deficient in some other one element. For instance, particular Ohio soils have returned greater profits from applications of potassium than from those of phosphorus. In the old lake bed in northwestern Ohio, phosphorus has not paid for itself. However, such instances are exceeding rare and individual isolated cases can be solved only by particular methods.

Since phosphorus, then, is in most cases the first element to be lacking in the soil, it is wise economy to add it to manure. Especially is this true when one considers that a ton of average manure contains about 10 pounds of nitrogen, 2 pounds of phosphorus and 8 pounds of potassium—much less phosphorus than either of the other elements. Manure may give a result of value, but still its full effect can never be realized until phosphorus is used

Acid phosphate encourages the growth of clover as a mulch in old apple orchards. This cut shows the division line where its use was discontinued.

before or with it. The value of a ton of manure in increase of crops, on soil of average fertility, may be increased $3 a ton by adding acid phosphate or floats costing only half this amount.

Acid phosphate in field tests at the Ohio Station conducted for nearly 20 years has proved to be not only a more effective but also a cheaper carrier of phosphorus than raw phosphate rock under conditions which render the freight charges a relatively large part of the cost of the fertilizer. These tests have included experiments with these two fertilizing materials by themselves, and in connection with other fertilizers, with lime and with manure.

Acid phosphate is made by treating raw phosphate rock or floats with an equal weight of sulphuric acid. The percentage of phosphorus in the acidulated material is about half that of the raw rock, but the solubility of the fertilizer is increased by the process. The acid phosphate therefore costs more and the freight on it is also higher. Still, the tests show that while raw phosphate rock may be used with profit on land that is deficient in available phosphorus, acid phosphate has produced increases in crop yields more than sufficient to meet the larger cost.

Acid phosphate is ordinarily used at the rate of 40 pounds per ton of manure. It may be spread over the manure in the shed or yard or mixed at the time of hauling. It may also be sprinkled over the stable floor or gutter behind the animals, about one pound being used daily for each 1,000 pounds live weight of animal. In the latter case, it also acts as an absorbent of the liquid manure.

For open barnyards raw phosphate rock seems preferable to acid phosphate, because the loss of phosphorus through leaching by rains is less with it than when the latter material is used. It has been found that 17 per cent of the acid phosphate applied to manure at the rate of 40 pounds per ton was lost after 3 months' exposure in an open barnyard. When raw phosphate rock was used under similar conditions only 4.5 per cent was lost by leaching.

Other preservatives, such as land plaster (gypsum) and kainit have been tested as materials to reduce manure losses and improve its quality. None of these, however, has been as effective as phosphatic materials, although in Ohio each has been profitable. Kainit has proved superior to gypsum.

Application of Manure.

Where Should It Be Supplied?—Manure, as a general rule, is applied in one of three places in the ordinary rotation of corn, oats, wheat, clover and timothy. In this rotation, it is most profitably applied to corn, a fact that also applies to the money crop in any other rotation. There are also arguments for putting a light top-dressing on wheat during the winter, and on new seedings in the fall or early winter.

Application of manure to meadows tends to cause coarse straw being raked up in the hay as well as to promote weed growth when the stand of grass and clover is thin. Thin stands would be promoted by this practice for the reason that the new seeding would be farther removed from the application of manure than if it were applied a year later, directly preceding the corn crop. Strawy manure on meadows may be raked up with the hay, thus injuring its quality and lowering its feeding value. When so used it should be spread before March, although well-rotted manure can be spread during April and perhaps a little later in the northern states, but it would be of little value to the immediate hay crop.

More manure is probably applied to corn than to any other crop in the rotation and this appears a satisfactory method of procedure. Corn is a gross feeder and will make good use of these coarse materials. For many farmers it is the most profitable crop in the rotation and the returns from the manure are soon bankable. Under such a system, poor crops of hay, when the manure is applied to corn in a 3 to 5-year rotation, are seldom seen if other factors influencing fertility are attended to.

New seedings of clover frequently respond liberally to applications of manure. While this manuring helps, it may not be altogether what the new seeding needs. A close diagnosis of the case may show a lack of lime. An application of 1 to 2 tons of limestone, preferably to the corn crop, may then result in clover and grass growing in luxuriance. In Ohio clover and timothy crops, all that can be kept from spoiling, are harvested from land treated with phosphated manure and lime on corn, no fertilizer on oats, and a fertilizer rich in phosphorus with little nitrogen and potassium on wheat. Manure applied to such seedings would be more likely to damage than to improve them.

There is the reasonable objection that applying manure to sod for corn frequently delays plowing for that crop. Generally there is little excuse for any man to allow the year's production of manure to accumulate until the land is dry enough in the spring to haul it upon corn ground. The first of

The manure spreader is a prime essential on every farm.

March should find little stable manure about the premises if it is at all avoidable. That which accumulates after this date may be held over and applied in August. If hauled out during the winter as soon as made, it offers no serious objection to plowing.

Where manure is produced in sufficient abundance on the farm, it may be profitable to spread it lightly on wheat fields during the winter. In this way snow will be held on the land; plant food will be held ready for the use of the wheat plants in early spring; and less spring fertilizer need be bought. For such applications, a manure spreader is more practicable than hand spreading as no lumps will be left that might smother the wheat.

How Much to Apply.—It has already been pointed out that spreading manure lightly is more profitable than applying the same amount thickly over a smaller area. The returns per acre, of course, will be greater from heavy applications.

The manure spreader is a valuable implement having a place on nearly every farm for this reason: not only can the manure be spread thinner on the ground than is possible by hand spreading, but it can be applied uniformly and generally without lumps.

Little Fertility Lost When Manure is Directly Applied.—It will be recalled that leaching of the soluble compounds, particularly nitrogen and potassium, and decomposition resulting in setting free gases are the main leaks in fertility value of manure in the stable and barnyard. It is apparent, then, that such losses would be insignificant when manure is spread upon the land. Any leaching would be taken care of at once, while the drying out of manure will speedily put an end to any decomposition. Experiments conducted at the Canadian Experimental Farms, in which different sorts of manure were spread out in thin layers and allowed to dry, showed losses of only 2 to 5 per cent of nitrogen. It is doubtful whether manure can be handled in any other way that would show less loss.

Spread or Piled Manure.—It is much better in applying manure at any time to spread it at once. When it is piled up, even in small heaps, there is always fermentation and consequent loss of nitrogen. Then the soluble part which leaches out is unevenly distributed, causing serious loss. If spread out in a thin layer, there will be no heating; the soluble constituents will soak evenly into the soil and there be held. While the water in the manure will, of course, evaporate, no valuable ingredients will thus be lost. There would also seem to be some saving in labor in spreading the manure at once.

The Use of Hen Manure.—The value of poultry dropping is ordinarily little appreciated. A hen will produce about 13 pounds of dry manure (30 pounds of fresh manure) on the roost at night during the course of a year, and this dry material contains about 2 per cent of nitrogen, 1 per cent of phosphorus and 1 per cent of potassium. A ton therefore has a commercial value above that of any other class of live stock.

Coal ashes may be used along with the hen manure, although they add no fertility. Acid phosphate would add much value and would prevent the escape of ammonia, although it is not as effective a dryer as raw phosphate rock or steamed bone, both of which are also effective in balancing the manure. A common practice that is condemned is mixing wood ashes and hen manure; this always occasions a serious loss of nitrogen in the droppings.

Besides balancing the manure, steamed bonemeal improves its mechanical condition. By mixing equal parts of hen manure and a 2.5-25 bonemeal, along with 1 pound of muriate of potash to 9 of the mixture, a 2.5-13-5 fertilizer may be made. The mechanical condition, however, will likely be poor for drilling.

Wood Ashes and Lime Should Not Be Mixed With Manure.—Since wood ashes ordinarily contain about 30 per cent of lime, they should not be mixed with any kind of manure, for the lime would liberate nitrogen in the form of ammonia. Either should be applied separately from manure or commercial fertilizer. The manure may be spread on the sod and plowed under; the ashes or lime may then be applied on the plowed ground and thoroughly incorporated in the soil.

Live Stock vs. Grain Farming.

The benefits resulting from the making and saving of manure and its importance in the maintenance of fertility are well known. The practice of feeding as many of the raw products of the farm as can profitably be fed, thereby saving their manurial value to the farm, is to be encouraged. Especially should every particle of refuse organic matter be converted into humus. The selling of the cheaper sorts of roughage, as straw or stover, for less than their manurial value, or a mere pittance above, is indefensible.

However, some products of the farm must be sold in some form or other, and the problem confronting the business farmer today is to adjust these sales so as to conserve fertility and expand his bank account. This is a complex problem and no iron-clad rules can be laid down for it. One can hardly say that it is never permissible to sell any one particular crop. This is largely a local matter, and each farmer should settle it for himself.

In the eastern states it is the general practice to harvest all the crops, sell most of the grain and feed the rest to live stock. The crop roots and stubble and the manure are thus returned to the land, and are supplemented with commercial fertilizer. In some cases a corn crop may be hogged down or only the ears harvested, but this practice is not general. Western farmers find such a means of harvesting corn more practical than feeding the stover. Labor supply and prices and the cost of utilizing the feed value of the stover force these farmers to allow the stalks to remain. In this way they conserve fertility and reap financial returns. Harvesting only the grain of the wheat crop with a header and allowing the straw to stand is a general practice throughout the West.

Manure Produced Per Acre.—No direct experimental evidence is at hand today to prove that fertility can be maintained much more effectually by live stock farming than by a system of grain farming. The Ohio Experiment Station has such a test well under way, but it has not been of sufficient duration to warrant definite conclusions. Some general facts are nevertheless evident.

The amount of manure that can be produced from an acre will vary with the fertility of the soil. Land yielding 70 bushels of corn will produce feed that will make a greater amount of manure than will land producing only 35 bushels. In the Ohio test soil of medium fertility is being used, cropped in a 4-year rotation of corn, soy beans, wheat and clover. All the crops are either fed to live stock or used for bedding, except the wheat grain. No other feed is used in this test except that produced on the measured area in question. When this feed is used, the test stops for that year. The feeding is done on a cement floor, so that the manure is saved to the best advantage. The total amount of manure made from a given acreage is thus known.

The crop production for the last 7 years on this area has been as follows: corn, 62.3 bushels; soy beans, 22.6 bushels; wheat, 35.6 bushels; and hay, 2.6 tons. The feeding of all these grains except wheat, and the use of all the

The livestock industry has been important in keeping up the fertility of the soil.

hay, straw and stover for either feed or bedding has resulted in the production of 3.28 tons of manure per acre, thus furnishing 13.11 tons for each acre of land once in 4 years. As all four crops are grown each year it has been possible to apply this amount to each acre planted to corn. To compensate for the phosphorus carried off in the grain and in the live stock 400 pounds of floats per acre is used with the manure applied to the corn ground, and 300 pounds of acid phosphate on wheat. As the land increases in fertility the amount of manure produced per acre will undoubtedly increase.

Livestock Farming and Fertility.—It is evident, then, that one practical way of returning to the soil the fertility elements removed by crops is live stock farming, by which all the manure is saved and properly used. However, the selection of live stock rather than grain farming should be determined largely by personal preference rather than with the idea of fertility maintenance. It must be remembered that by such a plan only four-fifths of the fertility elements of the feed can be recovered in manure even with the very best handling. The loss of organic matter by feeding is even more serious than that of fertility elements, as only about one-third of that in the ration is recovered in manure. Still, only about half the roughage of the farm is fed, the rest being used as bedding; and the organic content of the soil may be kept up by plowing under all crop residues and by growing clover in the rotation.

A typical rotation would be corn, wheat (or oats) and clover; or corn, oats, wheat and clover. In the system wheat grain is sold; other grain is fed, and straw used as bedding; and the second crop of clover may be left on the ground. By such a system of farming there would be a gain in nitrogen, and slight loss in both phosphorus and potassium, with no loss in organic matter if all the manure is properly applied. Purchased feeds or fertilizer would be bought to keep up the supply of phosphorus and potassium.

In sections where highly nitrogenous feeds are fed on dairy farms and where live stock is most easily marketed, such a system of farming will continue to be followed extensively. In the hands of the average farmer fertility may be maintained better by live stock than by grain farming.

Grain Farming and Fertility.—Nevertheless, by careful management fertility may be maintained by grain farming. The system may be explained by the plan followed at the Ohio Station in comparing live stock and grain farming. In the 4-year rotation of corn, soy beans, wheat and clover, the cornstalks, soy bean and wheat straw, and clover hay are returned to the land after the grain or seed is threshed. The first crop of hay is cut and left on the ground. An application of 400 pounds of floats per acre to the corn ground and 300 pounds of acid phosphate per acre on wheat serves to maintain the phosphorus supply, an element carried off in large amounts in grain. By this plan there is a greater drain on the fertility of the soil with respect to all three essential elements. Under a poorly managed system of grain farming, the organic matter would be rapidly depleted, as a farmer would sell the clover, hay and straw rather than return them to the land. It is for this reason that grain farming is commonly associated with soil robbing.

The growing of green manuring crops may keep up the supply of organic matter, but it is doubtful economy to spend a year's time raising a crop to plow under. Catch crops can often be grown with little extra cost except the seed. With all such crops, however, there is the objection that the poorest land grows the poorest crop and thus receives the least organic matter. The richest land unfortunately is benefited the most.

The organic matter may also be kept up by plowing under the clover crop, a practice of farmers following a 3-year rotation of corn, wheat (or oats) and clover. Whether this practice pays depends on labor, use made of the clover as hay and the benefits of the green crop as compared with manure for corn. Generally it would be better to feed the clover. Ohio tests show that clover left on the surface for 7 winter months loses as much organic matter as by feeding; and even 28.5 to 58.5 per cent when plowed under in the fall, depending on field conditions and rate of application.

Crop yields at Rothamsted Experiment Station in England have been maintained for more than 70 years by the use of commercial fertilizers. The same thing has been done for more than 30 years in Pennsylvania and for more than 20 years in Ohio. Yields can be kept up without any kind of livestock, but such a system requires wise management.

The yields in the Ohio comparison have been for the first seven years as shown below:

	Livestock.	Grain.	Gain for Livestock.
Corn, 7 years	62.33	55.95	6.38
Soy beans, 6 years	22.61	20.05	2.56
Wheat, 5 years	31.58	27.54	4.04
Hay, 7 years	2.64	....	....

Manure vs. Fertilizers.—At this point it is well to study the comparison of systems of farming under which fertility is maintained in one case by manure and in the other by fertilizers. It is recalled that manure is valuable for the plant food elements (nitrogen, phosphorus and potassium) which it carries. In this respect it is similar to fertilizers, but it has the further advantage of supplying bacteria and humus-making material to the soil, as previously explained in this chapter.

Farmers ordinarily can produce the fertility elements cheaper in manure than they can buy them in fertilizers. Results at the Ohio Experiment Station covering more than 20 years' work show that less than $2 invested in fertilizers at usual prices will purchase a fertilizer carrying a little less nitrogen and potassium, but more phosphorus than is contained in a ton of manure, which will be as effective in immediate crop production as the manure. This means that if manure should cost more than $2 a ton it is a doubtful purchase.

However, there is one important point to remember in connection with the use of manure, and that is its cumulative, or lasting effect. Fertilizer applied to both corn and wheat in a 5-year rotation of corn, oats, wheat, clover and timothy at the Ohio Station has returned, during 23 years, 65 percent of the total weight of increase in the two fertilized crops, and 35 percent of the increase has been realized in the other three crops of the rotation which receive no fertilizer except that left by the corn and the wheat. The same two crops treated with manure have returned only 52 percent of the total increased yield, while 48 percent has been realized in the oat and two hay crops which received no manure directly. This shows that full returns from manure cannot be expected immediately after treatment, and is an argument for tenants to pay only part of the cost of fertilizing and manuring.

One plot in this rotation gets 8 tons of manure per acre during 5 years, worth $16.50 at the price of plant food in fertilizers; while another plot is given $16.05 worth of fertilizer. During the first rotation the difference in value of the increase of crops on the manured and the fertilized plots was $8.35; during the second $11.63; during the third, $1.97—all in favor of fertilizers, but showing a slower rate of increase of crops from fertilizers after 10 years. During the fourth 5 years a difference of $1.77 was realized in favor of manure. This cumulative effect of manure is remarkable. Fertilizers are more quickly available, but if not used are lost from the soil faster than manure.

At Rothamsted, England, eight annual applications of manure are quite noticeable for more than 40 years after the use of manure was discontinued! On one field 15.7 tons of manure was applied annually for 20 years and then the treatment was discontinued. Even after 35 years the yields of barley on this land are twice those on land unmanured for the 55 years. Surely manure is a great enricher of the soil!

The Problem in a Nutshell.—The value of manure depends largely on the feed the animal eats. Highly nitrogenous grains, like oilmeal and cottonseed meal, make more valuable manure than coarse roughage, like silage or stover.

Generally less than four-fifths of the fertility value of a feed is recovered in manure even with the best methods of handling.

Straw has a value of $2.50 to $4.25 a ton and should be used liberally for bedding.

To prevent losses of a valuable plant food haul manure fresh from the stable or else keep it in a covered yard. A cement floor should pay for itself within a year. Half the value of manure is in the urine, and tight floors and plenty of bedding are necessary to save it. Manure loses half its value within three months if left to leach in an open yard. Keep the manure wet and compact, and mix horse and cow manure together.

Add phosphorus, as acid phosphate or raw phosphate rock, to make manure a balanced food.

Apply manure to the cultivated crop in the rotation, as corn or potatoes. If some is still left, put it on the meadows or winter wheat before March. Spread it lightly rather than thickly over a smaller area. A manure spreader will soon pay for its cost. Manure pays its greatest dividends on the poorest soil on the farm.

Manure should not be hauled out in piles. There is little loss of anything but water when it is spread thinly over the ground. Ashes or lime added directly to manure cause a loss of nitrogen as ammonia.

The value of manure cannot be fully realized except over a long period of time. Its effect is more lasting than that of fertilizers.

Fertility can be maintained by grain farming, although in the hands of the average farmer a system of livestock farming generally involves less soil robbing.

CHAPTER IV.

COMMERCIAL FERTILIZERS.

The use of commercial fertilizers as we know them today is a modern practice. Their use first attracted the attention of our census bureau in 1880, for the reason that little of these materials was used prior to 1870. From 1900 to 1910 the expenditure for fertilizers in the United States increased 113 percent. Under normal conditions of 1913 approximately $125,280,000 worth of such materials was used in this country. That this great investment is yielding a profitable return for many of the users can hardly be questioned. All the users cannot be fooled all the time, but much of the money so spent is used unwisely and wastefully.

The Necessity for Fertilizers.

Plant Food Removed.—It has previously been shown that the mineral elements of the soil came originally from the bed rock. An inventory of the resources of the soil shows that the plant food in it is by no means inexhaustible. The production and removal of farm crops, without any return of plant food to the soil, will soon reduce the stock of fertility to a point where crop production ceases to be profitable. This point, in fact, has already been reached in many of the older portions of our own country.

Even when many of the products of the farm are fed to livestock, and the manure is carefully saved and used in rebuilding the soil, plant food is lost from the land. Carcasses of these animals carry away elements of fertility, which must be brought back in some other form, or soil impoverishment will ultimately result. Even in the case of dairy farming there is a constant drain upon the fertility of the soil when milk is sold. Butter carries off little plant food, but feeding skimmilk to calves or hogs only postpones the loss of the fertility elements.

Then, added to such losses are those occasioned by erosion and leaching, resulting in many cases in the most valuable portion of the plant food being carried away in streams. Manure itself suffers loss by leaching, and often crop residues are burned, which causes further loss in nitrogen and organic matter. As a result of this crop production and sales in one form or another, along with natural losses and wasteful practices in handling the by-products, resort must be taken to some external sources of supply if crop production is to be maintained, much less increased.

For some farmers, these external sources may be foodstuffs—the products of other farms. Thus dairymen often grow only the roughage to feed their large herds, and purchase all the concentrated feeding stuffs. Still other farmers may be able to keep up fertility by the addition of manure, but even they must buy a commercial plant food (phosphorus) to supplement the manure. For nearly all farmers, therefore, the only available recourse lies in some form of commercial fertilizer.

A note of warning should be sounded that fertilizers are not to be considered means of correcting all ills of the soil. In other words, they cannot take the place of proper drainage, organic matter, rotations, liming and tillage. They should be but supplements to the other factors of crop production.

Only Three Elements Need Consideration.—As explained in the first chapter, ten elements are necessary in the growth of every plant, but of this number only a few give the farmer any concern. From the air we can get carbon, hydrogen and oxygen in inexhaustible quantities and without effort. Nitrogen, which constitutes the largest portion of the air, is an elusive gas that can be made available to plants by the growth of different species of bacteria on the roots of leguminous crops. Securing nitrogen in this way is not universally practiced, and hence it is often supplied in commercial forms.

Calcium, iron, magnesium and sulphur are required by plants in small quantities and are not likely to be deficient in soils as plant food elements, although calcium is often needed to correct acidity. This leaves only two mineral elements, phosphorus and potassium, which along with nitrogen are deficient in soils. These three elements are the ones that concern the farmer in his efforts to maintain or increase the fertility of his soil. He accordingly buys fertilizers that he may add one or more of these elements.

Fertilizer Terms Defined.—These three elements are more commonly referred to commercially as they are united with other elements. Nitrogen is the principal component of ammonia, which contains 14 parts of nitrogen and 3 parts of hydrogen. Fertilizer manufacturers express the nitrogen content of a fertilizer in terms of the compound ammonia, rather than the element nitrogen.

Phosphoric acid is a combination of phosphorus with oxygen, and in fertilizers means that the material in the sack carries sufficient phosphorus to make the quantity of so-called phosphoric acid indicated by the percentage given, if combined in proportion of 62 parts of phosphorus to 80 of oxygen. Potassium is likewise combined with oxygen, uniting in the proportion of 78.3 parts of potassium to 16 parts of oxygen to form what is called potash in fertilizers. To find the equivalent of nitrogen, phosphorus and potassium in fertilizers expressed as ammonia, phosphoric acid and potash, respectively, the following conversion factors may be used: to find nitrogen multiply the ammonia by 0.82; to find phosphorus, multiply the phosphoric acid by 0.44; to find potassium, multiply the potash by 0.83.

As an actual fact, however, fertilizers do not contain ammonia, phosphoric acid and potash as such, but instead they carry the elements nitrogen, phosphorus and potassium locked up in various complex forms. While numerous materials are used in fertilizers, only a few that are well known and easily obtained need be used for the ordinary crops and soils. These will be discussed with their sources of supply.

There are several other terms applied to the phosphorus content of fertilizers that need explanation before their value can be understood. Ordinarily the "total phosphoric acid" is expressed on the fertilizer sack and includes the soluble portion plus the insoluble. The latter is much lower in value than the soluble portion. When phosphatic rock is treated with sulphuric acid, phosphorus is made "available," and includes the phosphoric acid soluble in water and in soil solution. "Unavailable phosphoric acid" is of no immediate value to the plant, but by chemical actions in the soil, it may later be utilized, at least in part.

Sources of Supply.

There are available to the manufacturers of fertilizers certain natural products and certain waste or by-products of manufacturing industries for use as carriers of the three needed elements of plant food. Let us consider these sources briefly along with the relative values of these materials.

Carriers of Nitrogen.

Nitrate of Soda.—The most common source of nitrogen in fertilizers is nitrate of soda, a crystalline substance much like rock salt in appearance. It is obtained from large deposits in Chili, South America, which have probably resulted from marine deposits along with the excrement of birds. This product carries 15 to 16 percent of nitrogen, and is nearly 100 percent available. It will not be retained long by the soil on this account and should be applied only when plants can use it soon. Under these conditions, it is a cheap and satisfactory source of nitrogen. It takes up moisture in damp atmosphere and from dirt and cement floors. It often becomes too moist to drill satisfactorily and later gets to be lumpy. These lumps may be easily pulverized with a shovel or pestle and sifted through an ordinary sand sieve. Its mechanical condition can be improved by mixing fine dry material, like steamed bonemeal, with it.

Dried Blood.—Another important source of nitrogen is dried blood, a waste product of slaughterhouses. The blood, dried by artificial heat, is sold as a dry powder. It is not as soluble as nitrate of soda and can well be used where crop needs are not so immediate. The best grades carry about 14 percent of nitrogen.

Tankage.—Tankage is another slaughterhouse by-product dried and ground for fertilizer. It usually contains the contents of the stomach and intestines, some bone, flesh and hair waste of animals. It contains from 5 to 10 percent of nitrogen and from 2 to 8 percent of phosphorus. It is usually sold to contain ammonia and "bone phosphate," which is a combination of about 46 percent of phosphoric acid and 54 percent of lime found in bone.

A 7-30 tankage is a common grade—equivalent to 5.75 percent of nitrogen and about 14 percent of phosphoric acid.

Sulphate of Ammonia.—This is a by-product in the manufacture of gas and coke. It contains a higher percentage of nitrogen than any other material used in fertilizers, 20 percent or more. It is about 90 percent available, and it is not washed out of the soil as readily as nitrate of soda. Its use on other than strong limestone soils is not recommended unless the lime needs of the soil are well cared for, as it will soon put the soil in an acid condition. It is the only fertilizer in common use the tendency of which in this direction is unmistakably apparent.

Cottonseed Meal and Linseed Oilmeal.—While effective carriers of nitrogen, both these materials are highly valuable feeds for livestock. Their high price for feeding purposes is putting them out of the fertilizer markets.

Hoof and Horn Meal.—These waste products of the glue, comb and button factories carry 10 to 12 percent of nitrogen. Hoof meal is fairly available; horn meal, quite unavailable. Leather meal and wool wastes are still other animal by-products that carry small amounts of nitrogen. Their use should be avoided.

Tobacco Stems.—Tobacco stems, usually carrying about 2 percent of ammonia and 5 percent of potash, are regularly used in ready-mixed fertilizers. They are worth no more than phosphated stall manure. The same amount of elements in nitrate of soda and muriate of potash bring greater crop returns and can be bought at much lower cost either in these compounds or in manure.

Dried Peat.—Used by commercial fertilizer concerns as a source of nitrogen, this material makes a better showing in the laboratory than in the field. It carries 2 to 4 percent of nitrogen in a very unavailable form.

Carriers of Phosphorus.

Raw Phosphate Rock.—The largest and most important source of phosphorus is raw phosphate rock, commonly called floats. Deposits of such rock are found in various parts of the world, our supplies coming mainly from South Carolina, Florida and Tennessee. More recently deposits have been discovered in Arkansas, Wyoming, Idaho and Utah. These deposits are supposed to be largely fossil remains of prehistoric animals. The grades of rock phosphate vary in composition. High grades contain 30 to 36 percent of phosphoric acid. The more common grades on the market carry 28 to 32 percent, in a very unavailable form.

Raw rock phosphate has not been used to any considerable extent as a fertilizer until the last few years. When mixed with manure, or other decaying organic matter, it is about as valuable as acid phosphate because the organic acids in the decaying material render the phosphorus available. Where fertilizers are drilled in with or just before seeding the crop, this untreated phosphate rock is of but little value.

Acid Phosphate.—For use in fertilizers, such raw rock is ground to a fine powder and mixed with an equal part of sulphuric acid. This process of acidulation converts the insoluble compound made up of phosphorus and calcium, into a form soluble in soil water, in which form it can be taken up by plants, and is then called acid phosphate. The process reduces the percentage of phosphoric acid approximately one-half. The common grades of acid phosphate on the market carry 14 to 16 percent of phosphoric acid. The higher grades furnish the cheapest plant food.

As explained under the discussion of reinforcement of manure, acid phosphate, when used in comparison with the raw phosphate rock on the basis of equal amounts of carriers, has proved to be more effective and cheaper under conditions obtaining in Ohio and the East. There are, however, conditions of soil, coupled with varying prices due largely to freight haul which make it advisable to use raw phosphate rock.

The objection is frequently raised to the use of acid phosphate that it increases soil acidity. This is not fully sustained. It is true that unlimed land to which complete fertilizers have been applied with phosphorus in bonemeal and basic slag has produced more clover than land similarly treated except that acid phosphate was the carrier of phosphorus. However, applications of lime on land treated with bonemeal have returned greater increases in crop yields than those on land to which acid phosphate was applied, while clover yields have been greater on soils treated with acid phosphate than on unfertilized land.

Again, in orchard fertilization, acid phosphate, either alone or combined with muriate of potash, has encouraged the growth of clover on old sods. Nitrate of soda at the same rate of application brought a thick sod of timothy, redtop and bluegrass, no seed being sown in either case.

Bonemeal.—Raw bones from slaughterhouses are ground to a more or less fine meal. Raw bonemeal usually analyzes about 4 percent of nitrogen and 20 to 22 percent of phosphoric acid. It is somewhat slowly available because of the fat it carries, which protects the bone from decomposition and makes it impossible to grind the bone fine.

On this account the bones are frequently cooked, and the resultant steamed bonemeal, containing 1 to 2 percent of nitrogen and 26 to 30 percent of phosphoric acid, is thus made more effective as a carrier of phosphorus, although some nitrogen is lost by the steaming or boiling. The extraction of fat increases the percentage of the other constituents. The other slaughterhouse product, namely, tankage, is also a valuable carrier of phosphorus.

Basic Slag Phosphate.—In steel manufacture a by-product is formed by phosphorus uniting with the furnace slag. This material contains 15 to 18 percent of phosphoric acid in addition to about 30 percent of lime, and when finely ground is an acceptable carrier of both phosphorus and lime if sold at a reasonable figure. It is also called Thomas slag and is not to be confused with blast-furnace slag, which contains only a trace of phosphorus.

Carriers of Potassium.

Muriate of Potash.—The Stassfurt mines of northern Germany furnish nearly all the potash used in this country. Of these crude salts muriate of potash is most commonly used. It is a combination of potassium and chlorine and as imported carries about 50 percent of potash. Some experiments show that it affects injuriously the quality of potatoes, beets and tobacco. Pound for pound of potash, muriate is the cheapest carrier and the most economical for ordinary crops.

Sulphate of Potash.—Another German salt of equal concentration is sulphate of potash. It is free from chlorine and therefore not objectionable to use on any crop. Because of its higher cost, it is not used so extensively as the preceding salt.

Kainit.—Kainit is a crude salt containing about 12 to 14 percent of potash along with many impurities. On the basis of potash content, it is as good as the other carriers, but is used mainly in Germany and near the sources of its production.

Wood Ashes.—Until the discovery of the Stassfurt mines in 1857, wood ashes were used as a main source of potash. Unleached ashes contain about 6 percent of potash and 30 percent of lime. The amount of potash varies somewhat, being higher in hard woods. The potash is readily soluble in water and hence will be largely leached out if the ashes are not protected from rains. A fertilizer consisting of two parts of wood ashes and one part of either acid phosphate or steamed bonemeal and used at the rate of 400 to 500 pounds per acre may be used to advantage on spring crops, and the lime will be beneficial to acid soils. Wood ashes should not be used alone, because potassium cannot produce its full effect except in connection with phosphorus.

Indirect Fertilizers.

Gypsum, or Land Plaster.—Of the substances often applied to soils which affect crop yields favorably for a time although they furnish no actual plant food, gypsum is sometimes recommended. Its beneficial effects have undoubtedly been due to the liberation of unavailable phosphorus and potassium, or possibly from the stimulation from the sulphur which it carries. Its use can hardly be recommended except in stables as a preservative; still, acid phosphate will serve the same purpose and in addition supply phosphorus.

Common Salt.—Another soil stimulant is common salt. It may liberate potassium in soils rich in this element. In large quantities it is poisonous to plants and is often applied to kill weeds.

Lime.—Lime liberates plant food in soils rich in organic matter, phosphorus and potassium. It also is a direct food to plants, being used in largest quantities by legumes. For this purpose most soils contain sufficient calcium, and its chief function is to neutralize acidity. This subject is discussed fully in the following chapter.

The Selection of a Fertilizer.

In selecting a fertilizer, the farmer should take into consideration two important factors. The first of these is the needs of the particular soil, and the second, the needs of the crop.

Needs of the Soil.

Underlying Rock and Glacial Action.—Soils vary in composition depending upon the composition of the original rock from which they were formed, the action of glaciers in transporting and mixing the soil, and later treatment of the land. In some regions the glaciers spread over the surface a sheet of soil materials quite different from those beneath the drift. In such cases, we must look northward to find the strata of rock which has contributed to the surface soil. In other regions the glaciers ground up rocks similar to those underneath, and the bed rock and surface soil are much the same in composition. In the non-glaciated area the surface rocks have largely furnished the mineral constituents of the surface soil. In all cases, the character of the contributing rock will tell much regarding present soil conditions—whether lime and phosphorus supplies are short or fairly abundant; whether potassium compounds are present in moderation or in superabundance.

Physical Character of Soil.—Some light on soil needs may be shed by physical properties of soil. For instance, clay soils may be expected to have less need for potash than sandy soils. While sandy soils usually have less lime than clays, they can get along with considerably lower percentages and yet produce successfully the vegetation found on calcareous soils.

The color of the soil is indicative of organic matter and nitrogen. Black soils especially, and dark red soils to a less extent, indicate all-round fertility. Light gray and yellow soils are associated with low fertility.

Native Vegetation.—The capabilities and needs of a soil are indicated in a general way by the types of plants growing on it. Chestnut and hemlock trees are associated with a different class of soil from red cedar; white pine does not thrive on the same soil as blue ash and elm; horse sorrel and clovers are not seen thriving in abundance in the same fields.

Past Treatment.—One of the most important considerations in the selection of a fertilizer is the past treatment that the soil has received. If the farm has been under cultivation for a number of years with little consideration for keeping up the fertility it is likely that all elements will be needed. Likewise, if hay and grain have been sold off the farm, a different fertilizer will be needed than if they have been fed on the farm. Legumes grown regularly on the farm will solve the nitrogen problem, and this element would not need to be bought as in a system where legumes were grown infrequently or not at all.

Present Rotation.—With regard to the present rotation, one must consider the proportion of time the land is kept in intertilled crops and in sod;

whether legumes are grown, say once in 3 or once in 5 years. A corn or potato crop following a legume obviously would not need a fertilizer analyzing as high in nitrogen as when it followed another intertilled crop. One must also consider whether the soil is occupied all the time or is bare half of the year or more.

Plant Food Present.—As a final factor regarding the needs of the soil one must look to the chemical analysis. This is knowledge one should have before buying a farm; for while it tells little regarding the available plant food (the immediate soil needs), it does show the future possibilities of the soil and what should be the aim in fertilizing, viz., to supplement shortages of plant food and make available the large supplies of insoluble compounds. Even though the total fertility elements may be abundant, a little available plant food may be very profitable. A muck soil might need a fertilizer carrying phosphorus and potassium, or possibly only potassium; while with a soil also rich in humus steamed bonemeal or acid phosphate would be more profitable for such crops as wheat and corn. The fertilizer selected, therefore, depends largely upon the needs of the soil as indicated by its stock of fertility and the treatment it has received.

Needs of the Crop.

Character of the Crop.—In so far as the needs of the crop are concerned in the selection of a fertilizer, much will depend upon the particular crop in question. If grown for its stems and leaves, nitrogen is especially important since it is concerned with these portions of the plant. For this reason the production of hay calls for different fertilizers than that of cereal crops. Further, it is desirable that the nitrogen be in soluble form if the immediate crop is to be greatly benefited. A spring application of 100 pounds of nitrate of soda, or this amount plus 100 pounds of acid phosphate, on thin meadows will liven up the grasses.

An exception is afforded by the legumes, for while they are grown mainly for their leafy portion they call for large amounts of phosphorus and potassium. If thoroughly inoculated an alfalfa or clover crop will get its nitrogen from the air. Alfalfa calls for large amounts of potassium, and hence on sandy soils and where adequate provision has not been made for its liberation this element may be needed in addition to applications of 200 to 300 pounds of acid phosphate or steamed bonemeal.

Phosphorus is the important element in the growing of cereals, since it hastens maturity and promotes the development of the seed. An excess of nitrogen would result in weakness and softness of stem and consequent lodging of the small grains. Corn ground is usually treated with phosphated manure; but if this is impossible steamed bonemeal and acid phosphate in equal parts are ordinarily used on soils fairly well stocked with organic matter. Thin soils may need a little potassium in addition, but this element can largely be saved or made available by humus in the soil. Oats on fairly rich land may be profitably treated with 16 percent acid phosphate, and seldom does it pay to use anything else on poorer soils. At the Ohio Experiment Station nitrate of soda has seldom been profitable on corn or oats, but acid phosphate used at the rate of 80 pounds on corn and oats and 160 pounds on wheat in a 5-year rotation, in increased crop yields, has returned more than $6 per 100 pounds of fertilizer.

Fertilizers boost the corn crop. Plot on right received complete fertilizer; that on left. nothing.

Potassium is needed for the small grains, as it increases the stiffness of straw and resistance to disease, and in dry seasons favors the production of grain by prolonging the growing season. This element is also used largely by tobacco and by root crops, like sugarbeets and potatoes and other crops that store starch and sugar. Phosphorus is not called for in large amounts by such crops; but, since it is usually lacking in soils, it is ordinarily supplied in fairly large amounts. Except on muck soils potassium should always be used in connection with phosphorus if full effect is to be obtained. From 500 to 1,000 pounds per acre will often be advisable for these crops. Tobacco uses all three elements in abundance; 1,100 pounds of a 4-9-8 fertilizer made up of nitrate of soda, acid phosphate and muriate of potash has been most profitable in southern Ohio.

The root habits of any crop must also be taken into consideration in selecting a fertilizer. Shallow-rooted or surface-feeding plants need different fertilization from that of deep-rooted plants.

Place in Rotation.—A crop immediately following a clover or alfalfa crop will not require much if any nitrogen, certainly not as much as if farther removed from the nitrogen-gathering crop. Likewise, a crop following another heavily manured will need different treatment from that of one differently placed in the rotation.

Length of Growing Period.—Crops like wheat, which occupy the ground during many months, will require different fertilization from that of a crop like millet, which is made in 2 to 3 months. The period of most rapid growth

Effect of fertilizers on wheat. Plot on right received complete fertilizer; that on left, nothing.

is to be considered. A crop making its greatest growth in the northern states in April or May, when nitrification is slow, will not require the same treatment as a crop making its greatest growth in midsummer.

Make a Field Test.

The most accurate means of determining the fertilizer needs of the soil is to conduct a field test of long duration. Put phosphorus by itself on one strip of ground, phosphorus and nitrogen on a second, phosphorus and potassium on a third, and all three elements on a fourth. This will give a suggestion to future fertilization for that crop and soil, but one year's test, of course, is not conclusive because of differences in weather conditions.

Hints on Using Fertilizers.

Fertilizers Supplement Other Soil Treatments.—Other sources of plant food—namely, farm manures, green manuring crops, catch crops, sods, composts, etc.—must be considered before fertilizers are bought. They all influence fertility.

Neither fertilizers nor manure will give the full results of which they are capable until the soil is well supplied with lime. Acidity must first be corrected.

Rock fragments alone do not make soil, but organic matter is essential. . An abundance of lime and humus means a live, rich soil, the home of hordes of micro-organisms. Here fertilizers will produce their full effect.

Cooperative Buying Lowers Prices.—Financial profits alone make the use of fertilizers justifiable. They are influenced by buying and selling. Some farmers pay 20 to 40 percent more than others give for the same materials. By buying large quantities through cooperative organizations, like granges, improvement associations and farmers' clubs, lower prices and freight rates may be obtained.

Buy Nitrogen Only Temporarily.—Nitrogen should be bought only for a short time in ordinary rotation farming. With the inexhaustible supply of it in the air which can be obtained with slight effort, the general farmer should buy this element only as a temporary expedient, or in the growing of high-priced crops. If the ground will not grow legumes successfully, he should put it in such condition that it will; and then he can get nitrogen in this way, along with the conservation of animal manure. By such a system nitrogen in the cheapest element of fertility, but when bought in fertilizers is the most costly.

Phosphorus Is the Most Important Element.—As a general farm practice phosphorus must be added to the almost universal scanty stores in the soils kept long under cultivation. Only in extreme cases should it not be added to the fertilizer mixture. Steamed bonemeal or acid phosphate should be used for direct application depending upon relative prices. A good fertilizer scheme would be to use acid phosphate or raw phosphate rock, depending on relative prices, in connection with manure or clover for the intertilled crop of the rotation, like corn or potatoes, and then to use fertilizers on wheat, the other crops being left unfertilized or given only light applications.

Potassium May Become Available by Organic Matter.—Peaty or muck soils and sandy soils are usually deficient in potassium. Special crops, such as potatoes and sugar beets, call for especially large potassium supplies. In such cases the element must be supplied in commercial forms. On other soils, especially clays and clay loams, decaying organic matter, as manure and crop residues, should liberate the inert potassium therein contained. Moreover, manure should be carefully conserved and little potassium will thus be lost unless straw, stover and hay are sold.

Low-Grade Fertilizers Prove More Expensive.—Fertilizers carrying the highest percentages of fertility elements are generally the cheapest. There is no place in true farm economy for a 1-8-1 fertilizer, or a similar formula, in a ready-mixed form. If any nitrogen or potassium is needed higher percentages should be used; if not needed acid phosphate or steamed bonemeal would be better. Elements bought in low-grade, ready-mixed fertilizers usually cost twice what they do in farm manure. If one buys ready-mixed fertilizers he should get the highest grade obtainable.

Fertilizers Can Best Be Mixed at Home.—Home-mixed fertilizers are a success and put dollars in the bank account when many ready-mixed compounds take money out for wasteful use.

Evidence from experiment stations everywhere points to the fact that home-mixed fertilizers can contain the same amount of plant food as those mixed at the factories, at a material saving in cost, often $2 to $8 a ton. Home mixtures can easily be as thoroughly mixed, as fine, and in as good mechanical condition for drilling as the best brands of fertilizers on the market. The arguments for home mixing are: (1) One can know the source of his plant food; that is, whether he is getting muck or wool waste (very unavailable to plants) or nitrate of soda and dried blood (the best forms plants can utilize). (2) One can more readily adapt his mixtures to different soil and crop conditions. (3) There are fewer chances for adulteration. (4) The prices of most ready-mixed goods are higher than those of home mixtures. (5) Finally, the farmer gains a valuable education regarding soils and crops in planning his home mixtures.

How to Mix at Home.—To mix fertilizers on the farm, first break up any lumps in the materials with the back of a shovel or wooden pestle after spreading out on a hard, dry, tight floor or wagon box, passing them through a sand sieve with quarter-inch meshes. Shovel the ingredients over three or four times, or until the mixture has a uniform color.

Suppose we wish to make a 3-8-4 fertilizer; that is, 3 percent ammonia, 8 percent phosphoric acid and 4 percent potash; and we have on hand nitrate of soda, 16 percent acid phosphate and muriate of potash. Three percent of ammonia means 60 pounds in a ton; hence, to find the required number of pounds of nitrate of soda (which contains about 19 percent of ammonia) we divide 60 by 19 percent, which gives 316 pounds. The 8 percent of phosphoric acid means 160 pounds in a ton, which divided by 16 percent gives 1,000 pounds of acid phosphate needed. As muriate of potash is 50 percent potash, 160 pounds will be needed to supply 80 pounds (that is, 4 percent) of potash in the fertilizer. Thus we have 316 pounds of nitrate of soda, 1,000 pounds of acid phosphate and 80 pounds of muriate of potash, a total of 1,396 pounds.

"Filler" in Fertilizers Useless.—Our mixture is short 604 pounds of a ton. The manufacturers use some cheap filler, or make-weight, to complete the ton if they should use such high-grade materials as these. But this home mixture is ready for drilling, and needs no dirt or other filler; in fact, there is less to handle than would be the case with ready-mixed compounds. If 200 pounds of a factory-mixed fertilizer were used per acre, 140 pounds of this home mixture above would supply the same amount of plant food.

"Wet mixing" and "dry mixing" and the advantages of "fillers" are often talking points of agents. They are intended to confuse the farmer and mean nothing of value.

"Crop Specials" Are Misnomers.—"Wheat specials" and "corn specials" and all other like trade names of ready-mixed fertilizers claimed as specifics for different crops are misleading. While it is true some crops need more of a certain element, some other element should often be added in largest amount because the soil is deficient in that particular element. Wheat uses six times as much nitrogen and three times as much potassium as phosphorus; yet applications of phosphorus are usually the most profitable. The soil, not merely the crop, should be fertilized. Then, too, the mixtures recommended by various firms usually vary as much for the same crop as for different crops.

How to Apply Fertilizers.—Fertilizers may be broadcasted by hand or with an implement, then harrowed in, with satisfactory results. Most persons using fertilizers on cereals find it more economical to use a fertilizer grain drill, putting in both fertilizer and grain at one operation. On corn fertilizer should be applied over all the field, as the corn roots extend throughout the soil and can therefore use it better; and because succeeding crops, like wheat or oats, will make better use than if it were distributed in hills or drills.

Lime should not be mixed with fertilizers carrying nitrogen, as this element will be liberated by the chemical action with lime. If lime is mixed with acid phosphate it will unite with the soluble phosphorus causing it to change back to tricalcium phosphate, or the form in which it is mined. Such a reversion occurs in all soils, but before this happens the added phosphorus becomes well diffused through the soil. Lime should precede the acid phosphate a few days for this reason.

How Much to Apply.—Most fertilizers are applied in too scanty doses, and yet it is possible to waste them, especially the nitrogen. Soluble phosphorus and potassium will soon be taken care of in the soil and little of either will be lost. Soluble nitrates are easily washed out of the soil when there are no growing crops to utilize them. Smaller applications return more per ton (not per acre) than large doses, but just where any wasteful fertilization begins cannot be told for all conditions. Heavy applications of nitrogenous fertilizers are never advisable except for heavy-feeding crops during a rapidly growing season.

Can Fertilizers Be Held Over?—Mixed fertilizers if allowed to stand a long time are likely to undergo a chemical and physical change which renders them less valuable. Ammonia is often lost as a result, much as in the case of decomposing manure. Phosphoric acid may change from the soluble to the reverted form, in which case it is still available but not as diffusible in the soil. Old goods are likely to become lumpy, making it difficult to apply them.

If any acid phosphate is to be held over winter, it should be mixed at regular intervals and in the proper proportions with manure. From 40 to 60 pounds per ton of manure should be used and the two materials together will give better results than when used separately.

CHAPTER V.

LIMING THE LAND.

The importance of lime to agriculture has long been realized, its use antedating that of commercial fertilizers. Roman writers of 18 centuries ago recognized its value in increasing the "fruitfulness" of the land. The early records of English agriculture dating back 300 to 400 years show that liming was a regular routine among these farmers. The earliest liming undoubtedly consisted in the use of such soft materials as chalk, marl and sea sand (largely pulverized limestone). Later, in localities where these soft materials were not available, yet where lime rock was abundant, the burning of the stone was resorted to in order to make possible its distribution over the land. This was before limestone was reduced to a proper degree of fineness. In comparatively recent years only has limestone been ground sufficiently fine for agricultural uses and offered for sale in any large way.

Forms of Lime.

Chemically pure lime contains 71.5 percent calcium and 28.5 percent oxygen. The term "lime" commercially is used to designate a great variety of lime products, and these are offered on the market under different names.

Burned Lime.—Burned lime, chemically known as calcium oxide, is also caustic, quick and stone lime. It may be defined as the residues of any form of carbonate of lime (limestone, oyster shells and marl) after all volatile matters have been driven off by heat. Such lime must be used soon after burning and grinding, or else in slaking it bursts the sacks in which it is shipped. Lime increases in bulk two to three times in slaking. While used extensively in former years, it is being replaced largely by hydrated lime and limestone today, because it burns organic matter faster, is irritating to the skin of both man and animals on account of its caustic character, often costs more even in proportion, and has no decided advantages over the other forms.

Hydrated Lime.—When water is added to quicklime, hydrated, or water-slaked lime is formed, chemically called calcium hydroxide. It has no advantage over quicklime except that it may be stored without bursting sacks and can be more easily and conveniently handled in the field.

Limestone.—Raw limestone rock, oyster shells and marl are forms known chemically as calcium carbonate or carbonate of lime. This is the source of both the other forms mentioned. Thus, when 100 pounds of this rock is burned, 44 pounds of carbon dioxide gas is driven off, leaving 56 pounds of quicklime. Then when water is added to this amount of quicklime 74 pounds of hydrated lime will be formed.

The solubility of carbonate of lime in pure water is about one part in

73,000; in carbonated water (soil water), one part in 1,000. One part of hydrated lime is soluble in 600 to 800 of pure water and is less soluble in carbonated water.

Money Value of Lime.—A ton of the various forms of lime has the following equivalents:

		Equals in pounds of	
One ton of	Quicklime	Hydrated Lime	Limestone
Quicklime	2000	2640	3571
Hydrated lime	1513	2000	2703
Limestone	1120	1480	2000

From these figures we can work out easily the comparative values of the different forms of lime, thus:

If a ton of hydrated lime costs	Quicklime is worth	And limestone is worth
$6.00	$7.93	$4.44
5.00	6.60	3.70
4.00	5.22	2.96
3.00	3.97	2.22
2.00	2.64	1.48

Because of the impurities in the limestone a ton of burned lime is ordinarily considered equivalent to 1 1-3 tons of hydrated lime or 2 tons of limestone.

Air-Slaked Lime.—Air-slaked lime is a mixture of all three forms, for in slaking slowly in the air it takes up both water and carbon dioxide. After sufficient period of time it will all become carbonate of lime.

All these four forms are frequently called "agricultural" lime. Some firms advertise "fertilizer lime" or "lime fertilizer," terms which misrepresent the product because lime is not used primarily to supply plant food, and is not a fertilizer.

Dolomitic, or Magnesian Limestone.—The comparative value of these three forms of lime should be based primarily upon their content of calcium and magnesium. Limes may be classified as follows: (1) High-calcium, or calcareous lime, containing 80 to 95 percent of carbonate of lime; (2) dolomitic or magnesian limestone, containing some carbonate of lime and at least 30 percent of magnesium carbonate. So far as effectiveness in correcting soil acidity is concerned, the dolomitic is as satisfactory as the calcareous limestone. However, some experiments indicate that the proportion of calcium to magnesium in the soil should be such that the calcium is always in excess of the magnesium, and that excessive quantities of magnesian limestone will have an injurious effect on crop production. Often when lime is needed, a chemical analysis of the soil will show that magnesium is in excess of calcium present; and when lime is abundant, the reverse is true. Good limestone should contain at least 90 percent of calcium and magnesium carbonates.

Kind to Apply.—So far as the effectiveness in neutralizing acidity of the three forms (burned lime, hydrated lime and limestone) is concerned, there is little if any difference, provided equivalent quantities are used and the limestone is finely pulverized. Proof of this statement is afforded by many experimental tests, of which one from the Ohio Station may be cited.

In a 3-year rotation of corn, oats and clover, 1,000 pounds of burned lime and an equivalent amount of hydrated lime and limestone are used in connection with manure at the rate of 8 tons per acre on corn. The yields of corn and oats have not varied more than two bushels per acre in 10 years, and hay 165 pounds, in favor of limestone, differences no larger than can be expected with plots receiving the same treatment. The choice, therefore, should depend largely upon the cost, the one that can be laid down at the farm cheapest being the kind to select. Other things being equal a preference may well be given the use of ground limestone on account of its convenience and safety.

Limestone Should Be Fine.—The effectiveness of ground limestone depends upon its fineness as well as upon its purity. Limestone ground to the usual degree of fineness is satisfactory for correcting soil acidity; that is, 100 percent should pass through a sieve having 10 meshes to the linear inch, 50 to 60 percent should pass 50 mesh and 35 to 40 percent should pass 100 mesh. Material coarser than one-twentieth inch has little immediate effect, because it cannot be as thoroughly distributed throughout the soil and therefore cannot act as quickly in correcting soil acidity as more finely ground material. The Pennsylvania Station recommends all limestone to be finer than one-sixtieth inch.

Limestone siftings are often available, but are usually coarse. Yet when they can be bought at such a price that 4 or 5 tons per acre can be applied, their use may be practicable.

Gypsum.—Gypsum, or land plaster, chemically called calcium sulphate, is a lime by-product offered for sale. It has some agricultural value as an absorbent in the stable, but will not take the place of any of the other calcium compounds mentioned above for correcting soil acidity. Even in the stable it is being replaced by a phosphatic fertilizer or kainit as a preservative.

Wood Ashes.—As previously noted, wood ashes contain about 6 percent potash and 30 percent of lime when unleached. They should be saved on the farm, but usually command too high a price on the market to justify their use.

Blast-Furnace Slag.—Furnace slag has not been found as effective as lime and limestone for correcting soil acidity. Much larger amounts are necessary and the price is usually prohibitive. In one Ohio test eight tons of furnace slag was not as effective as four tons of limestone.

Marl.—Deposits of marl contain calcium and magnesium carbonates mixed with clay and other impurities, the composition varying from 10 to 90 percent. When used on the basis of the actual carbonates present, marl is as effective as limestone; that is, a 90-percent marl is worth the same as a 90-percent limestone.

Waste Lime Product from Factories.—In the manufacture of sodium carbonate limestone is used. After the process this limestone is thrown out, mixed with hydrate of lime, calcium chloride and common salt. This waste product is probably equal in value for land treatment to raw limestone, but is worth no more than limestone having the same content of carbonate of lime.

Likewise, by-product lime results from the manufacture of acetone, soap and beet sugar. All these forms have the same value as an equivalent amount of calcium carbonate in limestone.

Determination of Soil Acidity.

What Soils Need Lime.—Not all soils need lime, but those that have been cropped for a long time soon respond to lime treatment. The lime content is gradually depleted by cropping and leaching, and often soils in limestone sections do not grow good clover crops.

. Wherever lime is deficient no treatment can take its place. This shows division line on limed and unlimed plots.

Litmus Paper Test.—To determine the need of the soil for lime, the litmus paper test is the simplest, and when properly made is probably as good as any chemical test. This consists in placing blue litmus paper, which may be bought at drug stores, in contact with moist soil for half an hour. Place the paper in an opening made with a trowel or clean knife in the field, or put some soil in a cup and place the blue litmus paper in contact. Tests confirmed in the field and laboratory at the Ohio Station show that soils which turn the blue paper red in this time will be benefited by liming. Land that does not need lime will have but slight tendency to change the color of the paper. Neutral litmus paper has not proved satisfactory.

Hydrochloric Acid Test.—Another test consists in adding a few drops of hydrochloric acid to the soil. If any bubbling occurs, it is evident that carbon dioxide is given off and carbonate of lime is present, thus giving a negative answer to the question of acidity.

Clover an Indication of Lime Supply.—The behavior of red clover is generally the best indication of the lime supply. If after one gets a good stand in the spring it gradually becomes patchy, disappearing here and there, and gets worse as time goes on, with sour grass or sorrel coming in, it is a good indication that lime is needed.

The character of the plant growth on acid soils should be noted. Chestnut and sourwood trees and common sorrel prefer acid soils; while red cedar,

The effect of lime on sweet clover.

blue ash, chinquapin oak and most farm crops, especially legumes, prefer soils well supplied with lime.

Make a Field Test for Lime.—When in doubt about the advisability of liming, apply a strip or two across the field, noting the effects upon all crops of the rotation but particularly upon clover. Manure part of the corn ground; plow it under, and apply lime to a strip at right angles to the manured area.

The results will give positive evidence of whether lime or manure is needed, if either. Then if lime deficiency is plain, the sooner lime is added the better, for money and effort are wasted on an acid soil.

Benefits from Liming.

Lime as a Plant Food.—Calcium is a plant food, but is not added as such because most soils have sufficient amount to meet this purpose. Clover and alfalfa require especially large amounts of this element and under extreme conditions added lime may serve as a plant food for them.

Plant Food Liberated.—Pennsylvania experiments have shown that lime will increase the availability of potash from 6 to 55 percent upon different kinds of soil. Lime also promotes the decomposition of the organic matter

The effect of lime. Clover from unlimed and limed plots treated
with fertilizer.

and the liberation of nitrogen. Such use may be abused easily, with the supply of organic matter being gradually depleted with no attempt made toward its maintenance. For this reason, sod, green manuring crops or manure must be used somewhere in the rotation to keep up the humus.

Physical Properties of Soils Improved.—In addition, lime exercises an important influence in changing the physical properties of soils. Heavy, clays are lightened and made less sticky and more crumbly. Water and air have greater movement. Light soils are also benefited by an increase in water-holding capacity.

Acidity Neutralized.—By far the greatest effect of liming is realized outside the plant. Processes of fermentation and decay of organic matter are continually going on in the soil, resulting in the formation of acids of various kinds. Soils rich in lime do not need additional quantities to render these acids harmless to plant growth. Most soils, though, that have been long under cultivation respond favorably to liming.

Leguminous crops, such as clover, alfalfa, soy beans and cowpeas, do not thrive in an acid soil. The nitrogen-gathering bacteria, so essential to the growing of these legumes, will not develop normally in an acid environment. By supplying lime to neutralize acidity, the activities of these bacteria are increased; the nitrogen content of the soil is thereby enhanced and clovers grow in profusion; while succeeding crops also show marked effects of the lime treatment.

Lime Increases Crop Yields.—As a 12-year average, an expenditure of

The effect of lime. Clover from unlimed and limed plots treated with manure.

$5 for lime once in the 5-year rotation at the Ohio Station has resulted in an average gain of $16.47 an acre for all five crops. At first, 1 ton of caustic lime per acre was used but later from 1 to 2 tons of ground limestone has been applied to corn after plowing. Liming has returned 302 percent on the investment where nitrogen in complete fertilizer is carried in nitrate of soda. It has returned 261 percent in connection with manure and 234 percent in connection with acid phosphate alone and has even paid when used with basic slag. It has returned its greatest profit when used with complete fertilizer carrying nitrogen in sulphate of ammonia, $27.41 being returned from the expenditure of $5 for lime.

It is of particular interest to note that liming increases the yields of all crops, but particularly the clovers.

Application of Lime.

How Much to Apply.—It was shown in the discussion of the forms of lime that the kind selected depends mainly upon relative prices; that is, no one form is superior to another in correcting acidity provided an equivalent amount of calcium or magnesium is supplied. The form does determine the amount of application. A ton of quicklime or 2 tons of ground limestone is the usual application in a 4- or 5-year rotation after the land is once fairly well stocked with lime. For soils unusually deficient in lime, double this application will not be out of the way. Alfalfa may require a similar amount under certain conditions. It is more sensitive to a lack of lime than other crops.

Loss of Lime.—The continued application of lime is rendered necessary on account of its removal in crops and its loss by leaching. The passing

Applying lime with a special drill.

of rain water through the soil dissolves out various substances and carries them away in the drainage water. Chief among these is carbonate of lime. During a period of 40 years careful analyses in large numbers have been made by the Rothamsted Experiment Station, England, and Director Russell reports these losses as amounting to "no less than 800 to 1,000 pounds per acre each year at Rothamsted."

When to Apply.—It has been found most satisfactory under conditions in the northern states to apply lime to corn ground after it is plowed. The fitting of the ground preparatory to planting and the continued cultivation of the crop during the succeeding months secure almost ideal conditions. The objection to this plan is the interference of rushing spring work. Many farmers find it more convenient to lime for wheat, as there is more time for it at this season of the year in rotations which permit or call for midsummer preparation of seed beds. Liming under these conditions is permissible. It would be well to apply the lime some weeks in advance of seeding, if possible, and harrow the ground repeatedly after liming. This is essential in the use of burned lime, as it is likely to injure the germinating seed if applied at seed time. Carbonate of lime may be applied at any time without fear of injury to seed or crops.

Lime is often applied during winter. While it is better to use lime on acid soils then than not to apply any, the returns will not be as great as when this material is used after plowing for some cultivated spring crop. The lime can then perform its full function in promoting the growth of bacteria. These countless bacteria require both water and air for their existence, as well as a sweet or alkaline soil. Lime spread on the surface and not stirred into the soil can help them but little. They would die there for lack of moisture, and the lime would be dissolved and carried into the soil only after a long time.

Lime plowed under is also out of reach of the bacteria, which live chiefly in the upper 3 or 4 inches of soil where air is plentiful. When lime is plowed under, it remains in the bottom of the furrow, all the while tending to leach downward. It would be of practically no value to a clover crop until again plowed up, and then its value would be lessened.

It is not advisable to apply lime to oats after they are up. Limestone would not injure the plants, but it could not be mixed thoroughly in the surface soil. Moreover, the tramping of the horses and the running of the spreader would be injurious on the loose ground. It would be better to wait and apply the lime after plowing for the next crop.

How to Apply.—There are a number of satisfactory lime spreaders now on the market. A good woodworker and a blacksmith can easily make one, although prices may sometimes be as great as with factory-made spreaders. The manure spreader can be used advantageously, but a special lime spreader will be found better. The ordinary fertilizer drill is too slow and proves unsatisfactory. Lime should not be mixed with manure or fertilizers, and caustic lime should not come in contact with the seed.

Don'ts in Liming.

Don't pay more than twice as much per ton for burned lime or a third more for hydrated lime than for finely ground limestone if you can get the same amount of calcium delivered in one form as in another.

Don't buy limestone coarser than what will pass through a 10-mesh screen, with 50 percent of it through a 50-mesh and 35 percent through a 100-mesh.

Don't allow burned lime to remain long before using or it will burst the sacks.

Don't be misled by claims of agents or firms about "agricultural lime," "fertilizer lime" or "lime fertilizer."

Don't pay more for calcerous limestone than for a dolomitic limestone containing as much total calcium and magnesium carbonates.

Don't use gypsum for correcting soil acidity.

Don't waste wood ashes on the farm, and don't buy them at high prices when you can get as much lime in other forms cheaper.

Don't pay more for blast-furnace slag or any other by-product lime than is asked for an equivalent amount of carbonate of lime in other forms, freight charges considered.

Don't use lime until you have tested your soil by litmus paper, or noted that red clover does not thrive, or have made a field test.

Don't expect manure or fertilizers to take the place of lime, if lime is needed by the soil. Don't expect lime to take the place of manure, fertilizers or proper tillage.

Don't waste time and money cropping an acid soil.

Don't expect the same results from liming that another man on a different type of soil has obtained.

Don't apply more lime than necessary, but don't stop with less than a ton of burned lime or 2 tons of limestone once every 4 or 5 years if your soil lacks lime.

Don't put lime on ground and plow it under.

Don't put lime on oats, wheat, meadow or similar crops if you can apply it after plowing for an intertilled crop like corn. Don't put burned lime in contact with seed or young plants.

Don't apply lime just preceding a crop of cotton.

Don't add lime to manure or fertilizers.

Don't bother with a grain drill to apply lime in this day of cheap and efficient spreaders.

CHAPTER VI.

LEGUMES AND SOIL IMPROVEMENT.

Within comparatively recent years bacteria have been considered among the most important factors affecting crop production. Cultivation had long been considered a means of making plant food available in the soil for growing crops, of killing weeds and of fitting the soil to be a congenial place for the growth of plant roots. But later a secret of all these processes was revealed when it was found that tillage brought water, food and air to countless millions of microscopic plants within the soil, and that other methods of soil management, particularly liming, tended to favor the development and activities of these organisms we today know as bacteria.

What Bacteria Are and Do.

Nitrifying Bacteria.—A spoonful of soil may contain twenty million bacteria. Naturally they can be seen only with a high-power microscope, their average size being about 1-25,000 of an inch. Their functions in decomposing organic matter in the soil, changing ammonia to nitrites by one type and to soluble nitrates by another, has already been mentioned in this work. This is a continual and rapid process when conditions are favorable.

These favorable conditions have to do with food, air, water and removal of waste. Thus, these bacteria resemble in several respects many other plants. Their food is organic matter, or humus. For this reason manure and green manuring crops are beneficial for more reasons than merely the addition of plant food they contain. Black soils are normally more productive than light-colored soils because of their stock of plant food largely concerned with bacterial activity and humus.

Air is essential to those bacteria to do their work, and tillage is the chief means of meeting this need. Thus it is seen why cultivation is said to deplete the organic content of the soil. Well-drained soils are well ventilated, and are warmer than water-logged areas. Warmth is another requisite of bacterial activity, as is seen in the more rapid process of decay in summer than in winter.

Then, the excreta of these minute plants must be destroyed or the products thus formed will be disastrous to them. Carbonate of lime is most satisfactory to neutralize these acids or make them harmless. The use of barnyard manure, green manures, crop residues and systems of crop rotations, then, are dependent upon bacterial activities to change insoluble plant remains to a form available to growing crops. Drainage, tillage and liming are other practices that are concerned in large part with bacterial life in the soil.

Bacteria on Roots of Legumes.—Another kind of bacteria lives in the swellings, or tubercles, on the roots of leguminous plants, such as alfalfa, clover, peas, beans and vetch. Millions of these bacteria inhabit a single nodule, the size of it varying from a pinhead to larger than a pea.

A German named Hellriegel in 1886 brought to light a peculiar relationship existing between certain bacteria and the legumes. He proved that these minute organisms have the power to take up free nitrogen and cause it to unite with other elements forming compounds that plants can use. The legume itself cannot obtain nitrogen from the air; the bacteria in the nodules on its roots have this power, and the legume then draws upon this nitrogen stored in the nodules. These organisms are called "nitrogen-fixing" or "nitrogen-gathering bacteria." When one stops to consider that about seventy million pounds of nitrogen is present as a gas over every acre, he can readily understand the importance of legumes to any system of agriculture, as they furnish the home for bacteria that can fix this free nitrogen in the soil.

Legumes are large users of nitrogen, and they could not obtain sufficient of this element from ordinary soil to make such growth as they do if no bacteria existed to help them. As it is, a leguminous crop adds fertility to a soil of ordinary or little fertility.

This relationship between the green plants and bacteria is called "symbiosis." There is a mutual helpfulness, not a state of parasitism of the bacteria on the clover roots. The bacteria find a home there and thrive, grasping free nitrogen from the inexhaustible supply in the air about them and causing it to unite with other elements that form compounds suitable for plants, and store it slowly but constantly in the nodules on the roots of the legume. Then, the legume in turn lives on this nourishment, as well as on some taken directly from the soil.

Another peculiarity of these bacteria is that different kinds inhabit the nodules on roots of different species of legumes. For instance, the species found on alfalfa roots will not grow on red clover or soy beans. There is but one common exception definitely proved, and this is that the same species of bacteria live on alfalfa and sweet clover roots. It is thought that garden peas may have the same bacteria as vetch.

Nonsymbiotic Bacteria.—There is another kind of bacteria, found less frequently, that can take free nitrogen from the air, but they live independently of legumes, and are called nonsymbiotic, nitrogen-fixing bacteria. The nitrogen they gather may be stored in the soil to be used by crops later on. Because of their lesser activities and numbers, they are not of such importance in agriculture as the symbiotic type.

At the Rothamsted Experiment Station in England a gain of 25 pounds of nitrogen per acre annually for 20 years was found on wild grass lands where no nitrogen was applied by man or added by legumes. Of this 20 pounds, probably 5 might be brought down in rainfall, but the rest must have come largely if not entirely by nonsymbiotic bacteria working without any host plant.

Common Legumes.

Legumes, or pulses, are distinguished botanically by their alternate leaves and method of seed formation. Then, the distinguishing characteristic of the bacteria existing on their roots further differentiates this group of plants.

Nearly all the legumes prefer a soil well supplied with lime. Still, trefoil and white clover are indifferent to lime, and alsike is not as exacting in lime requirement as some other species. The lupines seem to be injured by the presence of lime.

The legumes as a class thrive only when the nodules are abundant on their roots. Even when well fertilized the crops do not grow well without these bacteria. In the latter case more nitrogen is taken from the soil mass itself for the support of the crop, and the plants contain a lower percentage of nitrogen. The nodules vary in size, shape and abundance, being characteristic of the species of legumes.

Alfalfa.—The greatest soil enricher from the standpoint of individual capabilities is alfalfa. It is not as commonly grown as clover, because it is unusually exacting in soil requirement (fertility, drainage, humus, lime, inoculation and freedom from weeds) and good stands are more frequently difficult to obtain.

Because of the large yields of hay and the large store of nitrogen in the roots and tops, it is alike the most valuable roughage and soil improvement crop for farms on which it can be grown. Its deep rooting system gives it an advantage over common varieties of clover. Because of its limited distribution the soil should be inoculated before seeding.

It can withstand severe drouth and maintain itself for several years in many regions without reseeding. Often when ordinary crops die from drouth, alfalfa is green because it can get its food and water at lower levels. The roots make the soil more porous and crumbly, and upon dying add greatly to the humus content of the soil. More than one-third of the nitrogen in the alfalfa plant is in the roots and stubble.

Clover.—There are nearly 250 species of clovers, the more important being red, mammoth, alsike, white, crimson, Japan and sweet. Sweet clover is more nearly like alfalfa in characteristics than red clover. The same species of bacteria thrive on the roots of both. It has a large rooting system, thus making an open, friable soil and adding to the organic content. It makes a luxuriant growth under favorable conditions, and is well adapted to be plowed under for soil improvement. Fields of young red clover may die out in seasons of little rainfall, while sweet clover in an adjoining field will remain green until late fall. Often its roots penetrate 2 feet or deeper into the soil. Unlike alfalfa, sweet clover grows on thin land, being commonly found along roadsides and on eroded hillsides.

Red clover is probably the most important legume in most sections, because it yields a hay about equal in feeding value to alfalfa when given the same care in harvesting and is not as exacting in soil and cultural requirements. It has the smallest nodules on its roots of all legumes. On young clover roots the nodules are globular, later becoming somewhat branched and lobed.

Mammoth clover is also a great enricher of the soil. It has a coarser stem than red clover, makes larger growth, and hence is suited for plowing under.

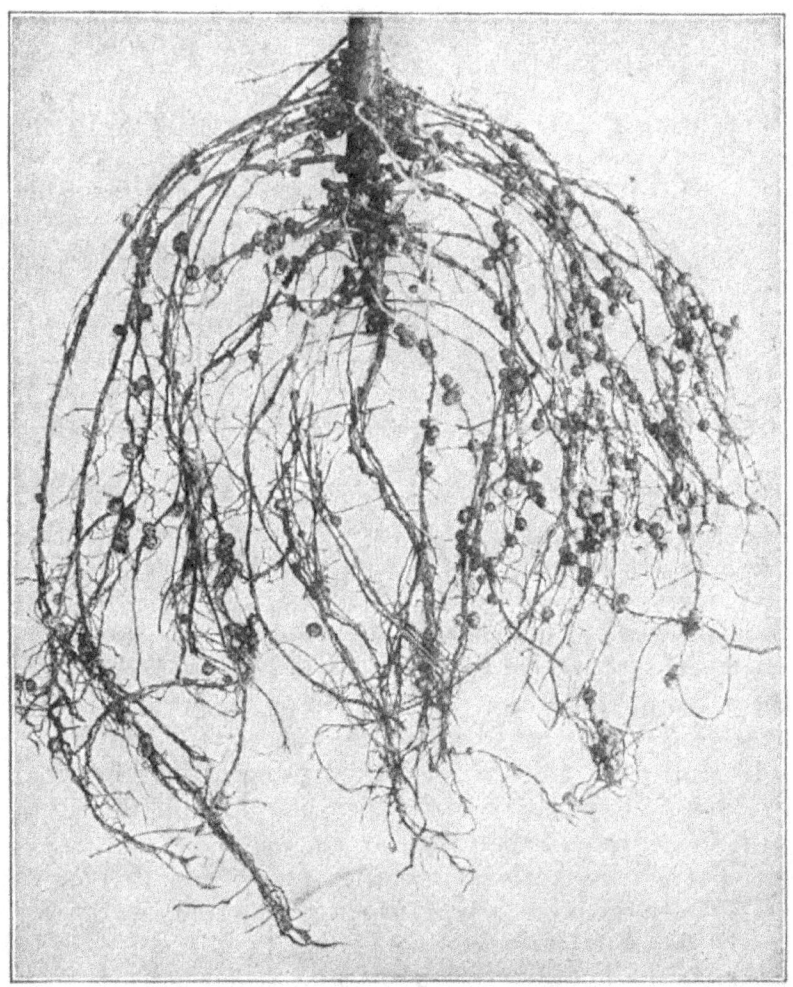

Soy bean roots showing nodules. The soil is greatly enriched by the nitrogen gathered from the air by the bacteria in these little nodules. This is the reason why clover, soy beans and other legumes enrich the soil.

Crimson clover is grown extensively as a green manuring crop in most parts of the South, and is well adapted for this purpose, although it makes no large growth. It is particularly difficult to cure as hay because it matures in the spring, but when plowed under it adds a large amount of organic nitrogen and humus-making material that will aid the corn crop materially. An especially large increase in the yield of tobacco after crimson clover was noted in tests in Virginia. It cannot well be seeded in corn or cotton at the last cultivation; hairy vetch is more suited for such purposes in the South.

Soy Beans.—The relative value of clover and soy beans depends upon the use made of the crops. If they are removed, no nitrogen is retained by the soil except that contained in the roots and stubble. Dr. C. G. Hopkins, of Illinois, states that a ton of soy bean hay leaves about 6 pounds of nitrogen in the roots and stubbles, or less than one-third the amount left by a ton of clover hay. But if the crops are plowed under the amount of nitrogen added to the soil is nearly equal. Soy beans have the advantage of quickness of maturity over clover, and may be seeded in late spring or in the last cultivation of corn for use as a green manuring crop.

Cowpeas.—The last-named advantage also applies to cowpeas, as compared with clover and alfalfa. While usually inferior as a farm crop to soy beans, cowpeas are better suited for green manuring because they can stand a more poorly drained and more acid soil. As a crop harvested in the rotation they cannot compare favorably with clover for soil enrichment. The nodules on the roots of cowpeas and soy beans are the largest of all legumes, often being as large as a pea.

Vetches, peanuts and locust trees also belong to the legumes and have the same characteristic of adding to the nitrogen store of the soil. Vetch has the largest and most branched clusters of nodules. It often makes an excellent crop to seed at the last cultivation of corn, to prevent washing and to plow under in the spring.

Legumes as Soil Builders.

Nitrogen Store Increased.—Farmers noted centuries ago that crops grew better after clover occupied the land, but they knew not why. Scientists tried to prove that clover does not enrich the soil, and it took many years of study before these certain microorganisms, termed bacteria, were discovered. Today the clover plant, together with other members of the great family of legumes, is accepted as our principal source of nitrogen. The legumes will be utilized more and more as we come to appreciate their value.

The amount of nitrogen left by a clover crop to be worked over by ordinary soil bacteria and made available for succeeding crops cannot be determined accurately. Some say that if the hay, or first crop of clover, is removed from the land, no more nitrogen will remain behind in the roots and crop residues than was taken directly from the soil by the clover plant. On a normally productive soil probably one-third of the nitrogen in the clover plant is taken from the soil itself, and not more than two-thirds secured from the air through the action of bacteria. Moreover, a great number of analyses show that about one-third of the nitrogen in the clover plant is in the roots and stubble. One would then infer that when the clover was removed there would be neither a gain nor a loss in nitrogen. This would not be the case with all soils. Fertile land would likely be left poorer in nitrogen, while poor soils would certainly be enriched by the legumes.

The Nebraska Station in its twenty-fifth annual report records a test with red clover which yielded a good hay crop the first summer after seeding and in late October was found to contain 41 pounds of nitrogen per acre in the roots and 136 pounds in the tops. Such an amount left on the land would furnish all the nitrogen needed for the grain of a 100-bushel crop of corn and a 50-bushel wheat crop.

Another illustration of the great benefit of clover upon succeeding crops is afforded by the work of the Rothamsted Station in England. In a rotation of clover, wheat, turnips (fertilized with phosphorus and potassium) and barley, during the clover year the field was divided, one portion growing clover and the other lying in bare fallow. Although a crop of 5,970 pounds of clover per acre was removed from one portion, the wheat crop following the clover was 22.7 percent greater than that after the fallow; the succeeding crop of turnips, 36.5 percent better; and the barley, 89.8 percent, as reported by Hall in "Fertilizers and Manures" (p. 33).

At the same station land cropped for more than 50 years yielded an average of 35 bushels of wheat per acre, 4,770 pounds of clover, 9.3 tons of turnips and 34.5 bushels of barley with no nitrogen ever added to the soil except by the clover crop. Moreover, analyses have shown that the nitrogen content of the soil has not diminished. It would seem, therefore, that clover must be given credit for leaving behind considerably more nitrogen in roots and residues than it has drawn from the soil.

Clover Aids Associate Crops.—Clover has an immediate effect on plants grown along with it by increasing the protein content. Analyses made at Cornell Experiment Station, in New York, showed that timothy contained more protein when it grew mixed in a field with alfalfa and with red clover than when grown alone. This effect would presumably be greater on soils where nitrogen is the limiting plant-food factor.

Soil Texture Improved by Legumes.—The legumes exert a beneficial effect on the texture of soils by making them more friable and easily worked. By nature of their deep rooting systems such crops as alfalfa and sweet clover break up compact soils, and make them more open. Then as the large roots decay openings may be left in the soil, permitting greater movement of air and water. Further, the stock of organic matter would thus be materially increased.

Legumes Alone Cannot Keep Up Fertility.—As with other factors previously discussed, the growing of legumes alone cannot make a soil fertile without supplementary farm practices. Legumes are but one link of the great fertility chain, joining lime, mineral plant food, humus, drainage and tillage with increased crops and profits. Taking a crop of clover alone from land year after year without returning in manure and fertilizers any of the fertility elements and organic matter removed will inevitably result in soil impoverishment. The legumes give their greatest profits when returned directly to the soil by being plowed under or indirectly by first being fed to live stock. From the standpoint of economy feeding is better, and when carefully saved nearly 80 percent of the fertility value of the feed will be recovered in the manure. Farmers selling legume hay off the farm are soil robbers just as much as the older southern planters following continuous cotton culture. But by careful management of clover grown in rotation and fed on the farm, with the manure returned to the land, the nitrogen store of the soil may be gradually increased and this element need not be bought in fertilizers. Then, by growing these leguminous crops, returning all organic matter and supplying lime, the potassium content of the soil may be taken care of on the average soil. Phosphorus and lime will then be the only materials to buy in commercial form.

Management of Legumes.

Place in Rotation.—Legumes are most frequently given a place in the rotation as a usual farm practice, as explained in the succeeding chapter. By such a system phosphated manure and lime are applied to the intertilled crop, like corn, potatoes or cotton, and fertilizers are used on the crops preceding the legume. This crop then is satisfied with the crumbs it can take from the table of the others. As the nitrogen content of the soil is increased through leguminous crops, this element need not be added to the cultivated crop following.

Legumes are also used for cover crops, catch crops and green manures. Clover and vetch are commonly used in the rotation and in orchard practice as cover crops to prevent washing and leaching during winter. Cowpeas and soy beans are frequently sown in early summer as substitutes for hay crops that have failed. Except on very poor soils green manuring is not recommended, for the use of the land for that year is lost and the feeding value of the crop might well be realized.

Liming.—Fertilizers, manure and lime are seldom applied directly to legumes. Other crops in the rotation respond better to fertilization and the lime can perform its full function when applied a few years ahead of the legume. Lime is necessary for two reasons: In the first place, the legume is a heavy feeder of calcium, as will be seen from the following table:

Crop	Calcium content (pounds per ton)
Corn (grain)	0.4
Oats (grain)	1.4
Wheat (grain)	0.8
Corn stover	7.0
Oat straw	6.0
Wheat straw	3.8
Red clover hay	29.2
Alsike clover hay	19.6
Alfalfa hay	36.4

With red clover and alfalfa calling for several times the amount of calcium that the cereals do, it is apparent why some soils may be slightly deficient in this element as a plant food. Yet many soils contain sufficient calcium for this purpose, but in such combinations as to afford no corrective influence upon soil acidity. Carbonate of lime is needed for this purpose; bacteria, so essential to the successful growing of legumes, will not thrive in an acid soil. Whenever the legumes fail to grow, attention should be directed toward the lime content of the soil.

Inoculation.—Even on a soil well stocked with lime a legume will fail if the proper species of bacteria is absent. Red and alsike clover, cowpeas and vetch are pretty generally distributed all over the United States, and for these legumes the introduction of proper bacteria into the soil is seldom necessary. Alfalfa and soy beans are less generally grown, and for them inoculation is necessary.

Inoculation is a means of introducing bacteria into the soil either with the seed or by adding soil from a field where the same legume previously grew. Unless this means is used, the alfalfa or soy beans will come slowly and scattered, if at all. The crop does better by inoculation particularly if the nitrogen content of the soil is deficient; and even if nitrogen is not a limiting factor, inoculation is profitable in order to put the plant in position to utilize the nitrogen in the air, thus reducing the drain upon the soil.

Some bacteria may be carried on the seed, but comparatively few of them are, and at best their growth is extremely slow in the soil. Several years may be necessary before there are sufficient bacteria present in the soil to become a factor in nitrogen fixation if they are introduced only by the seed. Of course, if one had the time he could scatter a few alfalfa seeds here and there in a field and in about 5 years there might be plenty of bacteria present to insure an alfalfa crop. Such a process usually is too slow and other methods are used.

Natural inoculation is probably the best means of introducing bacteria into the soil where they are needed for successful growing of a certain legume. This method consists in scattering soil from a field where the given legume thrived and had an abundance of nodules on its roots. For soybeans the soil should be taken from the surface 6 inches shortly before it is needed for use, dried in the shade if it needs drying in order to drill properly, passed through a sand sieve to remove small stones or any similar foreign matter that might interfere with drilling, then distributed through the fertilizer attachment of the grain drill when the soy beans are drilled.

From 200 to 500 pounds may be spread broadcast and harrowed or disked immediately afterward. Such operations should be followed on cloudy days, as sunlight is fatal to the bacteria. Broadcasting or drilling may be used with alfalfa, but with soy beans one operation may serve a double purpose. A little infected soil dropped into the furrow with the beans will give quicker and somewhat better results than several times the amount applied to the surface and harrowed in. An even distribution of 50 to 75 pounds per acre would be sufficient. Soil should be selected that is free from weed seeds and disease organisms. Soil from sweet clover fields may be used to inoculate soil to be seeded to alfalfa, or vice versa.

Artificial cultures are now sold commercially in liquid and in agar jelly. These are often cheap and convenient, but less certain and less successful than natural inoculation.

The introduction of bacteria is not all that is necessary to grow legumes. If soil conditions are unfavorable bacteria cannot develop rapidly. Soils well supplied with plant food, organic matter, lime and moisture (and yet well drained and well aerated) furnish the most favorable home for bacteria.

CHAPTER VII.

ROTATION OF CROPS.

Many important agricultural practices have come about as a result of more or less accidental discoveries, while some few owe their origin to scientific investigation. The rotation of crops is of the accidental sort. It would be difficult to determine when or who first discovered that soil which had become tired of one crop, would respond with renewed vigor when asked to grow another. For the last two to four centuries orderly successions of crops have been followed by the best farmers, particularly in the older sections of the country, where the necessity for so doing became most apparent.

Reasons for Rotation.

Rotation Maintains Supply of Humus.—Of the various reasons for growing crops in orderly succession, one of the most important is maintenance of the organic content in the soil. The continuous growing of a cultivated crop, like corn, potatoes or cotton, is extremely destructive of the organic matter of the soil. The supply can be kept up in many instances only by bringing crop residues from other fields in the form of manure or in the direct utilization of the bulkier portion of the crop itself; as for instance, putting back corn stover or straw of the smaller cereals. The matter of maintaining the humus content is greatly simplified by the introduction of such crops as clover and grass, which yield heavy sods that may be incorporated with the soil.

Alternating Crops Conserve Fertility.—Rotation aids in the conservation of fertility in that deep and shallow-rooted plants may be alternated, thus feeding from different levels. Deep-rooted plants not only have the power of appropriating fertility elements beyond the reach of shallow-rooted, but also leave portions of the food they gather within the reach of the latter upon the decaying of their roots and stubble. Temporary benefits of drainage and aeration also accrue.

Plants Use Elements in Different Proportions.—Rotation permits the succession of crops requiring plant food elements in different proportions. While rotation will cause a greater drain upon the soil for all fertility elements simply because much larger crops are thus grown, yet the mineral elements are being gradually liberated; and, as they are called for in different proportions, all-round larger production will result. Thus, in the corn-belt an average crop of clover calls for four times as much potassium as an average crop of wheat requires, although they use nearly identical amounts of phosphorus.

Rotation Keeps Soil in Good Tilth.—Any one who has plowed land that has been continuously in oats or wheat for 15 to 20 years, for instance, has discovered a sadness in the soil—a lack of life. Soils cropped in proper rotation are full of bacterial life.

Rotation Avoids Diseases.—Many diseases multiply under continuous culture that are held in check under a wise rotation of crops. Potato scab is an example. The micro-organisms of this disease live readily in the soil from one season to another; and when a soil once becomes infected, seed treatment is of no avail, unless the crop is shifted to another soil, or to soil that has not grown scabby potatoes for a few years. Club root in cabbage, and smut in onions and other crops, are additional examples. The susceptibility to disease is lessened for nearly every crop by proper rotation.

Rotation Avoids Insects.—In the same way insects that live over winter to attack the next crop are avoided by growing a crop that cannot serve as a host for them. Many insects find it difficult to migrate from one field to another, and perish when the plant upon which their existence depends is replaced by one they do not like.

Rotation Keeps Down Weeds.—Furthermore, weeds are kept down by an orderly succession of crops. In many of the northern states the Canada thistle and ox-eye daisy are great pests that frequently get full control in meadows and pastures, while in cornfields with clean cultivation they get a setback that holds them in check. Narrow plantain, common in clover fields, produces seed after the first cutting. By plowing the field after the first crop is harvested, or before the plantain has had time to mature seed, the pest can be cleaned off a farm.

At the Rothamsted farm wheat grown in continuous culture for about 60 years with high yields has probably always been grown at a loss. These wheat fields have long been so foul that much hand work has been necessary to grow any wheat. The problem of weed destruction is simplified by rotation of crops.

Rotation Keeps Land Occupied.—In the case of continuous culture of most crops the land is unoccupied for a large part of the year and losses of various kinds result. For instance, nitrification takes place rapidly during the summer and nitrates would be lost on a field from which wheat was harvested in early July. Losses also take place in cornfields in September and October where wheat, rye or some catch crop is not growing to make use of the plant food made available.

Rotation Systematizes Labor.—Rotation distributes the labor over a greater period of time, thereby providing profitable work for farm help during a greater portion of the year. This balancing up of farm operations is of advantage in other ways. In the event of local natural calamities, as floods, hailstorms, drouths, frosts, etc., one crop of a rotation may escape while another perishes, thus insuring the rotation farmer against the total failure meted out to the one-crop man.

Soil Poisons May Become Harmless.—Finally, rotation may free the soil of certain poisonous elements excreted by the roots of plants. Some scientists support the opinion that crops grown on the same land several years in succession fail on this account. While the poisons injure the crop itself, they are not injurious to a different species of plant.

Profits Accrue from Crop Rotation.

As a result of all the various advantages gained for systematic rotation of crops, a much higher level of yields is both obtained and maintained. As

Crop rotation keeps up yields far better than continuous culture.

proof of this statement let us look at the effect of continuous culture as compared with rotation in the case of corn, wheat and clover.

Corn Yields Increased.—The Ohio Station furnishes an excellent example for this comparison. For more than 20 years crops have been grown continuously on the same land, and also in a 5-year rotation of corn, oats, wheat, clover and timothy, in the order named. The last two crops are seeded together in wheat, clover predominating the fourth year of the rotation and timothy the last. In order that the results may be presented most forcibly the yields will be given for 5-year periods for unfertilized land, manured land and fertilized land.

System	Treatment per acre in 5 years	Yield per 5-year period (bushels)			
		I	II	III	IV*
Continuous..............	None	26.3	16.8	10.4	8.4
Rotation.................	None	31.9	30.8	31.0	20.3
Continuous..............	Manure, 25 tons	43.1	40.1	34.6	30.2
Rotation.................	Manure, 16 tons	40.7	49.5	59.8	55.8
Continuous..............	Fertilizer, 1250 pounds	38.9	39.1	28.0	26.8
Rotation.................	Fertilizer, 985 pounds	35.8	49.5	53.9	44.1

* The decline in every case in the fourth period from the yields of the third period is due largely to seasonal conditions and insect attacks. In later years, yields are increasing decidedly.

In this rotation there are five different sections and a crop of corn is harvested every year. As many crops are averaged in one case as in another; thus any inaccuracies from seasonal complications are avoided. The outstanding fact is the greater decline in yield under continuous cropping than under rotation. This is best appreciated by comparing the first and fourth periods. When no fertilizers have been used the yield under continuous cropping has declined 67.8 percent, while under rotation it has declined 36.3 percent. When manure has been used the yield has declined nearly 30 percent under continuous cropping, even though 56 percent more manure was used than under rotation. In spite of the smaller application of manure the yield under rotation increased 37 percent. Where fertilizers have been used as indicated—the larger amount in continuous culture—the yield has declined 30 percent under continuous and increased 23 percent under rotation.

These figures show that corn grown continuously upon the same land is unprofitable. Other experiment stations show the same results. Thus, in Illinois after 30 years of experimentation corn yields varied from 27 bushels per acre in continuous culture without any treatment to 96 bushels in a 3-year rotation supplemented with lime, manure, steamed bonemeal and sulphate of potash.

Wheat Yields Increased.—The effects of a rotation in increasing wheat yields may be shown by the same Ohio rotation. The yields for a period of 20 years are averaged in the following table:

System	Treatment per acre per 5 years	Yield per acre (bushels)
Continuous........................	None	7.5
Rotation........................	None	10.7
Continuous........................	Manure, 25 tons	17.6
Rotation........................	Manure, 16 tons	22.1

The unfertilized wheat in rotation has averaged 42 percent higher in yield than that under continuous culture. Manured wheat in the same rotation averaged 25 percent higher than wheat in continuous culture with only about two-thirds as much manure.

Wheat at Rothamsted for 12 wheat rotation years (1855 to 1899) yielded 12.4 bushels per acre under continuous culture; after fallow, 18.1; and in a 4-year rotation, 28.6.

Clover in Rotation.—Continuous culture of clover, re-seeding as necessary, is unheard of in ordinary farming. It is therefore of interest to note the Rothamsted experiments in so growing it. During 29 years it was seeded 15 times, but produced a crop in only seven seasons, and of these seven only two occur in the last 18 years of the test. The hopelessness of growing clover continuously is apparent.

The problem of crop rotation is the maintenance of nitrogen and humus, so essential to the production of all crops. The growing of legumes thus sets forth one solution for this problem.

Common Rotations for the Cornbelt.

For any rotation to be practicable in maintaining the organic matter and nitrogen of the soil, it should have at least one legume, preferably one of the sod legumes. Then, too, every rotation should have one intertilled or cultivated crop, so that weeds may be destroyed and the other advantages obtained from frequent stirring of the soil. Of less importance, but still of some moment, is the number of crops grown with one breaking of the plow, as well as the time of the year these plowings are called for and the possibilities of working in catch crops for soil improvement without special harm or inconvenience to the other crops of the rotation. In order to study some of the more or less common rotations for the northern states data of this sort are arranged in the following table:

Crops in rotation	Length of rotation (years)	Crops with one plowing (average)	Legume in years
1 Corn, oats, wheat, clover, timothy........	5	1.7	5
2 Corn, oats, wheat, clover...............	4	1.3	4
3 Corn, wheat, clover....................	3	3.0	3
4 Corn, oats, clover.....................	3	1.5	3
5 Corn, corn, wheat (or oats), clover.......	4	2.0	4
6 Potatoes, wheat, clover.................	3	3.0	3
7 Corn, potatoes, wheat, clover...........	4	2.0	4
8 Corn, soy beans, wheat, clover..........	4	2.0	2
9 Soy beans, potatoes, wheat, clover.......	4	2.0	2
10 Wheat, corn, oats, clover...............	4	1.3	2
11 Corn, wheat (clover)....................	2	2.0	*2
12 Potatoes, wheat (clover).................	2	2.0	*2
13 Wheat, clover..........................	2	2.0	2
14 Alfalfa (4 years), corn, wheat, clover......	7	3.5	1.4
15 Corn (rye)†...........................	1	1.0	0
16 Potatoes (rye)†........................	1	1.0	0

† Hardly workable, except with abundance of manure.
* As a catch crop.

(1) This 5-year rotation is an old standby of the middle and central states, less in use now than 25 years ago. A difficulty with this rotation is in maintaining the nitrogen supply with a legume only once in 5 years. A legume as a catch crop between corn and oats will help, but this is not always sure.

(2) By shortening the length of the rotation a year, this objection is largely overcome. The same number of plowings are required as in the 5-year rotation. An objection to both these rotations is that the land must be plowed for wheat during the hot days of midsummer. Still, this is not as hard work as plowing a clover sod at the same season. It should be said in favor of these rotations that better crops of wheat are usually secured after oats than after corn, in large part due to the better preparation of the seed bed made possible by early plowing and thorough tillage.

(3) This is a short, useful rotation quite generally followed in Ohio and some other eastern states where wheat does well, although the yields are less than in the preceding rotation. The plowing is reduced to a minimum in this rotation and may be done at a favorable season of the year. With clover once in 3 years the nitrogen supply will be better cared for. The principal difficulty comes in getting the corn out of the way for wheat seeding. This necessitates growing earlier varieties of corn than many are now doing.

(4) In localities where wheat is too uncertain, oats should be substituted for that crop. This will make possible a catch crop between corn and oats, which was impossible in Rotation 3. In the event of its use an extra plowing will be called for. This rotation meets with considerable favor in the middle western states, as in Illinois. In the good wheat sections it is generally found that oats are a less sure crop to seed clover with than is wheat. Oats are more likely to lodge, and this condition is dangerous to clover seeding.

(5) Upon the best corn lands it is often desired to grow this profitable crop more frequently than is possible by any of the foregoing rotations. It is possible to take two crops of corn in 4 years if the fertility needs of the soil are well cared for and sure catch crops are grown to keep up the supply of organic matter. Two successive intertilled crops are excessively destructive of humus. As to sure catch crops, in case of failure in midsummer seeding of a legume in corn, rye can be seeded in September or October with almost certain success. When seeded in corn at the last cultivation it is not as sure.

(6) This rotation has all the advantages of Rotation 3 and others in addition, although with it no catch crop can be grown. Potatoes give a much better preparation for wheat than does corn, and also an earlier seed bed except where late June planting of the potatoes is practical. When potatoes are planted at such a late date, it is perhaps better to seed rye, harvesting or hogging it down, or plowing it for oats or soy beans, substituting the latter in place of wheat.

(7) There are localities adapted to both corn and potatoes. This makes

a profitable rotation where conditions favor it. There is time between the corn and potatoes to grow a valuable catch crop. If legumes are not dependable, then a crop of rye can be grown. With manure applied to the corn crop, and fertilizers used as needed, good crops can readily be secured, and likely greater profits than in any of the preceding rotations except No. 6. It is better that the potatoes should follow rather than precede corn.

(8) In the latitude of southern Ohio oats are not as successful as in the north, and soy beans are substituted for this crop in Rotation 2. Soy beans are a close rival of potatoes as a forerunner of wheat, furnishing an ideal seed bed without the expense of plowing. During a short test at the Ohio Station wheat yielded about 38 bushels per acre following oats, soy beans and potatoes, or about 10 bushels more than where this crop followed corn. Since soy beans are a legume, the drain on the soil for nitrogen would be less in any of the preceding rotations, if the beans are properly inoculated. It should be said, however, that the crop residues with soy beans are less than with clover and other sod-forming legumes. In localities where wheat after corn is rather unsatisfactory and oats are not a sure crop, soy beans should be tried in this 4-year rotation. They will be found superior to oats in both food and market value per acre.

(9) As the soy bean crop becomes better known its area is certain to be extended. It will fit well into a potato rotation, as here indicated. Rye should be seeded after the soy beans are harvested, to be plowed for potatoes. With two legumes and a surer catch crop of rye, the nitrogen content and supply of organic matter would be well taken care of.

(10) In good wheat and corn sections where trouble is experienced in getting satisfactory corn yields without interfering with wheat seeding, or where it is desired to leave the stover on the ground as a direct contribution to the stock of plant food, this rotation will prove acceptable. It permits the maximum of safe and sure catch crops. With clover seeded with the wheat and plowed the following spring for corn, it amounts practically to two clover crops in the 4 years, besides permitting a catch crop between corn and oats. The objections are the large number of plowings and the midsummer breaking of a clover sod for wheat.

(11) In localities where wheat following corn is satisfactory and it is desired to shorten the rotation, it should be possible to make the conditions so favorable for clover that without devoting a full season to it, but simply seeding it with the wheat and plowing the following spring for corn, the needs of the soil for nitrogen should be well cared for. The mineral elements can be added as needed.

(12) Here we have the same intensive system applied to the potato farm. Liberal applications of phosphorus and potassium will insure good crops of both potatoes and wheat, with a good catch crop of clover.

(13) This short rotation is in occasional use in localities where wheat is the one crop of importance. It is lacking in the requirement of a tilled crop, but considerable tillage is possible between the plowing of the clover

and the seeding of the wheat. Just how permanent this rotation will prove —in other words, how certain the clover crop—has not been fully demonstrated.

(14) While alfalfa does not readily lend itself to carefully conducted rotation cropping, it is possible to grow it in a 7- or 8-year rotation in which 3 or 4 years of alfalfa are followed by one of the common 3- or 4-crop rotations. This will likely be carried on with less precision than the ordinary rotations.

(15) and (16) Some dairymen who are farming very intensively upon limited areas, purchasing liberally of concentrated foodstuffs and consequently able to apply large amounts of rich manure annually, are succeeding in growing good crops of corn each year, using a rye catch crop to help keep up the supply of organic matter. Potatoes, with rye catch crops, when liberally manured have given good results. These systems are hardly likely to prove satisfactory except where large amounts of manure are available.

Rotations for the South.

For the South rotations must be planned to include tobacco and cotton as the main money crops. As previously stated a legume should be planned to maintain nitrogen and organic matter.

The main difference between rotations for the North and the South is with regard to the most extensive use of catch crops in winter in the warmer regions. The absence of frost in southern soils allows the manifold soil activities to proceed continuously throughout the year, resulting in a greater liberation of plant food and destruction of organic matter. Some suggested rotations follow:

	Length of rotation (years)	Crop with one plowing (average)	Legume in years
1 Tobacco, wheat, clover..................	3	3	3
2 Tobacco, wheat, grass (2 or 3 years).......	4 or 5	2 or 2.5	4 or 5
3 Grass (2 to 3 years), corn, oats, cowpeas, tobacco..............................	6 or 7	1.5 or 1.8	3 or 3.5
4 Corn and cowpeas, oats, cotton...........	3	1	*2
5 Corn (cowpeas), oats (cowpeas), cotton....	3	1	*2
6 Corn (or kafir), wheat, cotton...........	3	1.5	*3
7 Cotton, oats, cotton, corn...............	4	1.3	*2
8 Coton, oats, cowpeas, corn..............	4	1.3	4
9 Cotton, cowpeas, corn (wheat or oats).....	3	1	3

* As a catch crop.

(1) This is one of the most common tobacco rotations. It is generally followed in states from Maryland to Illinois, often with oats substituted for wheat where the latter is not as well adapted.

(2) This is an old standby in the South, but formerly was not successful; clover failed when seeded in wheat and after a year or more "resting" the land, it was put back in tobacco. Today the tobacco is liberally fertilized, usually with all three elements; wheat given little or no fertilization; and

grass land well fertilized and fitted after wheat. The objections to this rotation are that catch crops cannot be grown and the legume comes only once in the rotation. The grass mixture is made up of clover and timothy, and often redtop is added.

(3)   Corn is often handled in a separate rotation of corn, oats and grass (2 or 3 yr.). Corn is often also introduced into the tobacco rotation, as here suggested, making a 6- or 7-rotation. Crimson clover may be substituted for cowpeas after oats and with two legumes the nitrogen should be maintained.

(4)   A short rotation with cotton is here given that makes ample provision for keeping up the nitrogen and organic matter. It is common in Texas. The rows of corn and cowpeas alternate, the cowpeas being either pastured or plowed under. Cowpeas can follow oats or another catch crop can be grown after cotton. Only phosphorus is ordinarily bought and potassium for particular soils. By the adoption of this rotation, with the keeping of livestock to consume the feed on the farm, probably as much cotton could be grown as at present and meat and dairy products would not need to be bought as is the custom throughout the South today.

(5)   This is another short rotation but has the objection of requiring a plowing for each crop and no legumes are grown unless cowpeas (or soy beans) follow oats and corn as catch crops.

(6)   Where wheat is better suited' than oats, it may be substituted for oats in the foregoing rotation and kafir may replace corn. One plowing is thus eliminated.

(7)   On rich lands this rotation might be advisable, but it is very destructive of organic matter, and no regular legume is introduced. Hence leguminous catch crops should be planned. It undoubtedly is one of the most profitable rotations if fertility is properly maintained.

(8)   In order to get a legume, cowpeas may be sown instead of cotton in Rotation 6.

(9)   With this rotation the legume is grown more frequently by cutting out oats in Rotation 7. Either wheat or oats, however, may be substituted for corn. With all these cotton rotations there is the objection that a sod-forming legume is not grown. Ample provision must be made by the use of catch crops to keep up the nitrogen and organic matter.

PART TWO
FIELD CROP PRODUCTION

CHAPTER VIII.

CORN.

Historical.—While in Europe the term "corn" is construed to mean several different kinds of grain; in this country its meaning is restricted to Indian corn, or maize (**Zea mays**). The corn plant has developed by evolution from a wild form unknown to man. By some it is thought to have originated on the plateaus of Mexico, and to have been first cultivated at about the beginning of the Christian era.

The American Indians cultivated it, and Columbus sent some of it back to Europe when he first came to America. It was the main crop of the early American colonists.

World's Production.—Of the world's corn production about 78 percent is grown in North America, where more than three billion bushels are grown annually. Of this North American production 97 percent is grown in the United States, and most of our national yield is produced in 10 states, as follows: Illinois, Iowa, Missouri, Nebraska, Indiana, Kansas, Ohio, Texas, Oklahoma and Kentucky. All these states each produced more than 100,-000,000 bushels in 1915.

Distribution in United States.—Corn grows generally throughout the entire United States; in the southern states with cotton, and in the semi-arid regions often with irrigation but does best in the "cornbelt," which is restricted to the central states between the two great ranges of mountains. In this region it has optimum rainfall, temperature and sunshine. Moreover, the soils generally are of such drainage and fertility that this gross-feeding annual makes its greatest growth. This is also a great live stock-producing section of the country, as corn and fattening animals are associated together.

Kinds of Corn.

Corn belongs to the grass family and is a near relative of a coarse, grass-like tropical plant called teosinte. Wild types of corn are not to be found today, but under careful breeding and selection methods six clearly defined types have been produced. These are as follows:

Dent Corn.—Dent corn, the most common sort in the northern states, is distinguished by the depression, or dent, at the crown of the kernel. This comes from the more rapid shrinkage of the soft white starch at the crown in drying as compared with the hard, horny matter at the edges of the kernel.

There are many varieties of dent corn, varying from the little early Huron and kindred types, requiring a season of 80 to 90 days, to the mam-

The kind that wins at the Corn Show.

moth Boone County White, calling for at least 130 days. The common varieties have 12 to 24 rows of kernels to the ear.

Flint Corn.—In the flint kernel the amount of soft white starch is limited and is entirely surrounded by the horny layer. There is usually no dent in the crown, though in some varieties there is a slight crease. Dent and flint kernels may be distinguished by cutting them lengthwise; flint kernels are more shallow.

Flint corn is grown commonly in New England and northern latitudes, being earlier and smaller than the dent. It has a great tendency to throw out suckers, or tillers. Common varieties have 8 to 12 rows to the ear. Ears are long and slender, and often two are borne on a single stalk.

Sweet Corn.—This type is grown mainly to eat in the milk stage, either when freshly gathered or after being canned or dried. The mature kernel is clear and wrinkled in appearance. The smaller, earlier varieties may be eaten in 50 to 60 days; later varieties, in 90 to 110 days. Some varieties grow only 2 to 3 feet high; others 8 to 10 feet. There are 8 to 16 rows of kernels per ear.

Pop Corn.—In this type the soft white starch is almost lacking. The kernel is extremely hard, somewhat similar to flint corn in this respect. When subjected to high temperature for a short time it explodes into a white mass.

There are two groups; rice and pearl. In the rice group kernels are sharp

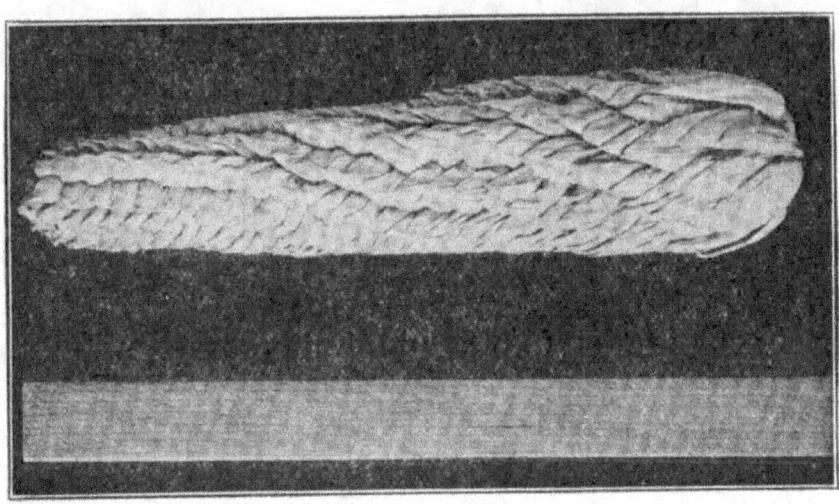

Ear of Pod Corn.

and pointed at the crown and ears are tapering. In the pearl group the crown of the kernel is smooth and rounded, the kernel more compact on the cob and the ear less tapering. The ears vary in length from 2 to 8 inches or more. Usually 12 to 16 rows are found per ear.

Soft Corn.—This type is generally confined to Central and South America. It has no hard, horny part to the kernel, and crushes into flour easily.

Pod Corn.—Pod corn has each kernel, as well as the ear, inclosed in a husk. It is grown only as a curiosity and has no commercial importance.

Selection of Seed Corn.

Variety.—Of foremost importance in selecting seed corn is the choice of a variety adapted to the conditions under which it is to be grown. While seasons vary greatly and occasional disappointments are likely to occur, they can be reduced if one chooses varieties adapted to the normal season. Variety names in themselves do not mean much, as strains of one variety may differ more than two separate varieties. The same variety that is most successful in one community may be mediocre or a failure in a different locality.

It is essential to know that the size and habit of growth of the variety will allow it to ripen in the allotted season. Very large and late varieties should be avoided for grain growing in the northern states, no matter what

Corn Investigators, U. S. D. A., Selecting the Seed Ears.

they have done for certain growers. The Ohio Station imported seed from localities farther south in that state, from Illinois and from Kansas, and compared them with a local variety. The average yield of all the foreign varieties was 64 bushels per acre, or 11 bushels below the home-grown corn. Variations of more than 30 bushels per acre have been found among varieties all grown in Ohio.

Not only season, but the soil type must be considered. Let the seed corn come from soil similar to that upon which it is to be planted. One should invest but lightly in unproved varieties. The prize ears in a corn show may not yield as much on one's farm as his own variety.

Field Selection.—Farmers today recognize that conditions are not ideal to select corn at husking time or in the crib. The largest ears thus chosen may have been produced under unusual conditions of growth, such as abundance of moisture, plant food and sunlight. Such ears were probably grown where the stand of plants was poor, one or two stalks per hill, or they were otherwise especially favored.

Selecting corn on the standing stalk is a more effective means of secur-

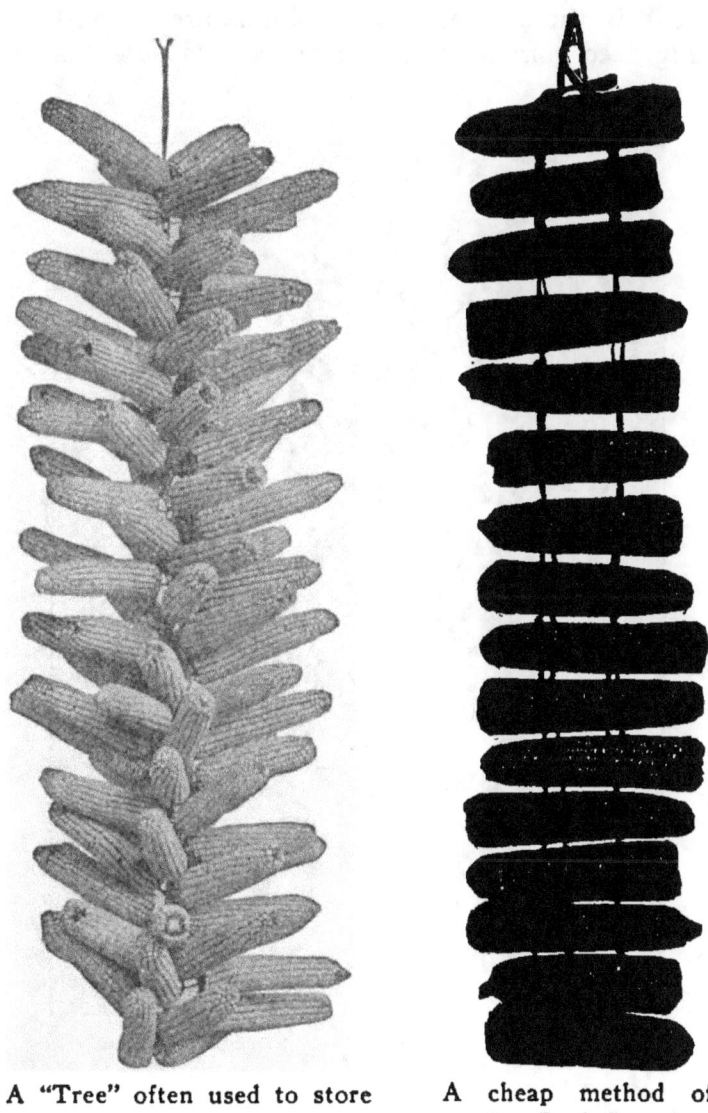

A "Tree" often used to store Seed Corn.

A cheap method of storing Seed Corn in a cellar or attic.

ing high-class seed. One can thus examine the entire plant, noting maturity, vigor, disease resistance, foliage, position of ear on stalk, etc., and particularly whether it grew under normal conditions of stand and soil. A few bushels may be added to the acre yield by selecting seed in this manner.

One wants to be sure that the excellence of an ear of corn selected for seed is due to something wrapped up in the ear (heredity) rather than to something which has happened to the ear (environment). If the same things fail to appear again the excellence will likely disappear. Moreover, by selecting in the field an advantage is offered for improvement in adaptability, or maturity. Careful selection of the earliest-maturing ears will in a few generations result in the disappearance of immature, unmarketable corn.

In selecting seed corn it is also important to note the position of the

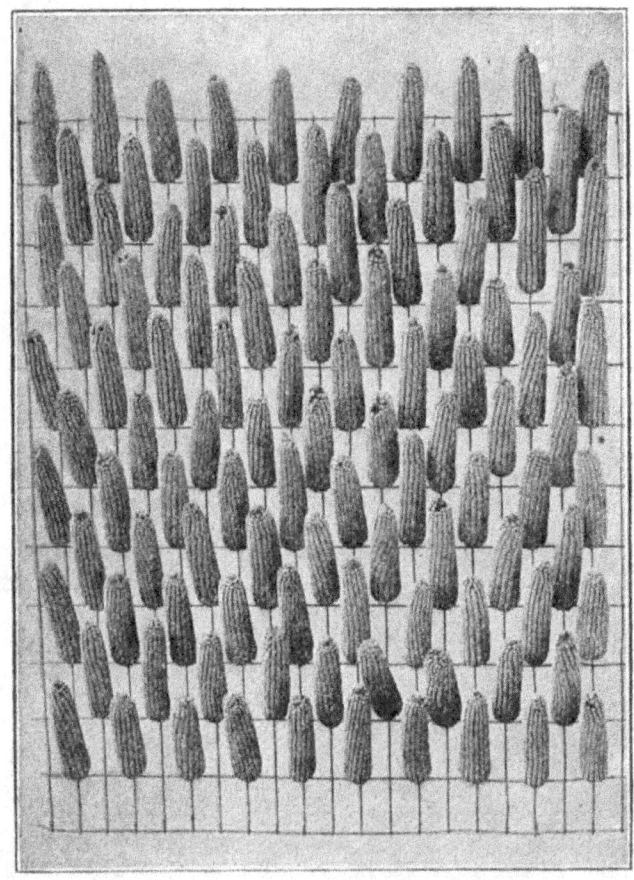

A common "Wire Fence" to store Seed Corn.

ear on the stalk. Ears at extreme height tend to pull the plant over in storms. High ears go with tall plants and late maturity. It is unwise to go to the other extreme and select continuously for extremely low ears, as this tends to reduce the size of the plant, although little difference may be noted in the yield of grain. A medium height, to be determined by each grower for his locality, should be chosen. The tip of the ear should point downward, thus shedding water.

Still another consideration is vigor. A vigorous plant has well-developed brace roots, large and strong stalk and good leaf development, and is free from disease.

The ears chosen should be marked in some way if not gathered at once, as by a little paint or string, so that they can be recognized at husking time. They should not be removed from the stalk before being well dried out. If the corn is to be husked by machinery, the chosen stalks should be shocked together separately from the main crop.

Storing Seed Corn.—Care of seed corn is fully as important as its selection, for without proper storing and handling even the best corn may be poor seed by planting time. As the corn comes from the field in the fall it contains 15 to 30 percent moisture, depending on the season. This water must be dried out before freezing weather comes, or the vitality of the corn will be lessened. A good circulation of air at moderate temperature is necessary, especially during the first few weeks after the corn is stored.

Placed on a section of wire fence or "tree" or suspended from a wire or string in a well-ventilated shed or attic, corn should be dried out well before cold weather sets in. Racks can be built at slight expense by nailing strips 2 inches by ½ inch in size on each side of 2 by 4-inch uprights, far enough apart to hold one row of ears. Such racks will last for years and pay for themselves many times. Windows should be closed only at night or on rainy days. After most of the moisture is out, experiments have shown that corn will stand almost any degree of cold without injury. If the fall of the year is wet and cold weather comes early, the seed corn should be stored where it can get artificial heat. A dry furnace room or an attic kept warm will prove satisfactory.

If mice give any trouble the corn, when well dried out, may be stored in long boxes made about a foot in width and covered at the sides with one-fourth inch wire netting. The ears should be put in crosswise of the box placed one upon another, with a tight fitting cover on the top box. These boxes may be stored in a dry furnace room, but corn should not be packed in them until it is dry or molds will develop.

Judging Seed Corn.—More seed than required for planting should be selected in the fall and stored, so that in the spring a final selection can be made for the best ears. Varieties differ with respect to length, shape, roughness, percentage of grain, number of rows, filling of butts and tips, and the like. Experiments carefully conducted for several years at the Ohio Station have shown that with all these various differences in seed corn, yields have been influenced but slightly, not more than 2 bushels per acre in any case. In other words, external appearance does not have important bearing on yield of corn. The grower should select the variety found best adapted to his community, pick out in the field in the fall the best seed from stalks that are growing under normal conditions of stand and soil, store it properly so that it is well dried out before cold weather comes, and then before planting test its viability.

Testing Seed Corn.—To insure against any mistakes in selection and storing, all seed corn should be tested ear by ear. In conducting the individual ear germination test one may follow any of a number of different methods. The kernels may be sprouted in sand or soil between moist cloths or blotters, or any way found most convenient. Many farmers have found the following method quite satisfactory: Make a shallow box 20 by 30 by 2 inches, inside measurement. Lay it off in 2-inch squares, marking the edges of the box, thus making 15 rows of 10 squares each. Number the squares from 1 to 10 across the ends of the box and from 0 to 15 along the sides. In

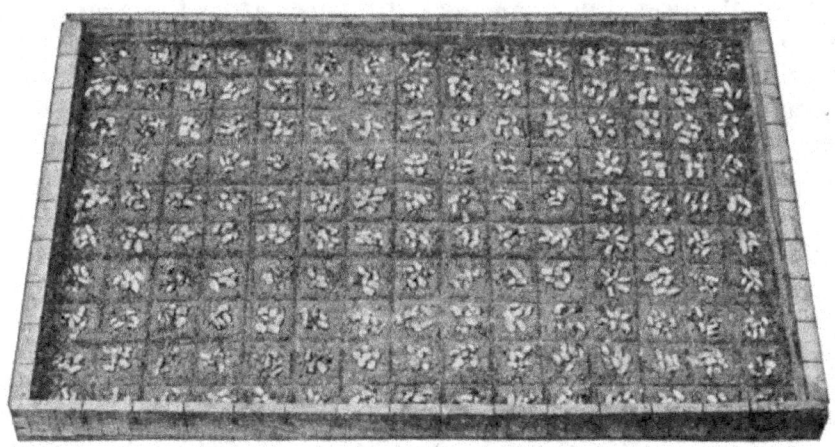

Germination Box with kernels in place.

this way it will be possible at a glance to locate any particular ear. Drive tacks in the sides and ends of the box (outside) in line with the division marks made on the edges.

Fill the box two-thirds full of fine, moist soil; then level off and firm it slightly. For convenience in planting, mark out the soil in 2-inch squares corresponding to the division marks on the edges of the box, which is then ready for planting. Number each ear, using a half-inch square of pasteboard, which may be fastened to the butt of the ear with a small wire nail, or place them after sampling in regular order where they will not be disturbed. The former method takes a little more time but it is sure. Take six kernels from each ear, making sure that the entire ear is well represented, and place them in the square corresponding to the number or position given the ear. In like manner place the kernels from the 150 ears, which the box will hold; then cover the kernels carefully with soil.

For convenience in locating the ears at the completion of the test, stretch a cord from tack to tack, crosswise of the box, both ways, thus dividing the surface into 150 two-inch squares. It is well to cover the box with a wire screen as a protection against mice.

Conditions similar to those that prevail in the field after planting time will give most reliable data. After a week or so in the germination box at a temperature of 50 to 75° F., the corn will have sprouted, and the count should then be taken when the plants are 2 to 3 inches high. For best results no ear should be saved for seed unless all six kernels are strong and vigorous; although in times of scarcity of seed corn ears may be chosen that have one or two weak kernels in such a test, care being taken always to shell out any moldy or otherwise injured areas.

For time so spent the Ohio Station has found that farmers can make more than $6 an hour, if corn is worth 50 cents a bushel. A gain of 4 bushels per acre was secured by testing, as a 5-year average. In one year, when the

Germination Box after corn has sprouted ready for count.

preceding fall was unusually wet and corn had a high percentage of moisture, a gain of 9.3 bushels per acre was secured by conducting a germination test as here described.

After the ears are tested and those showing faulty germination are discarded, the corn should be shelled and graded with respect to size of kernel. This may be done by passing the corn over a special corn sieve. By removing the small tip kernels and the ill-shaped butt kernels, more even stands can be secured.

Use of Old Seed Corn.—Corn 2 to 3 years old seems to retain full vitality, but drops off sharply at 4 years. Hence, it may be used with safety, other conditions favoring, the second or third year after its production. However, all seed corn should be tested before using it, no matter what its age may be.

Planting and Cultivating.

Fertilization.—The place of corn in the rotation, as previously discussed, is following a sod crop, preferably a legume. It was also pointed out under the subject of barnyard manure that this material is well adapted, when reinforced with phosphorus, to be applied during the winter and early spring to corn ground. From 7 to 10 tons may be spread per acre if available. The corn plant is a heavy feeder and makes good use of manure. Then it was later shown that after this material is plowed under, lime should be applied where needed, and it can then perform its full function by the time a legume comes around in the rotation. Corn yields in many cases can be increased several bushels by drainage.

With such treatment little additional fertilization need be given the

Planting Corn.

corn crop under the best systems of farming. Still, the relation of phosphorus to maturity and plumpness of grain is always emphasized; and most of our older soils, and in fact many of our newer soils, respond generously in increased yield and quality of grain to applications of this element.

Planted April 29th.

If phosphorus is not applied with manure, from 200 to 400 pounds of acid phosphate or steamed bonemeal should be used per acre. Where manure is not available, fertilizers should be applied, usually phosphorus and sometimes potassium; but nitrogen should be secured through the growing of legumes. These fertilizers should be drilled all through the soil in advance of planting, and not confined to the corn rows.

Planted May 7th.

Preparation of Seed Bed.—Early plowing at ordinary depth pays best for corn. Where manure is to be applied in the winter or soils are subject to washing, fall plowing is impossible, but early spring plowing is highly essential for maximum crops. It is an important factor in conserving moisture,

Planted May 16th.

a prime requisite for corn. Late plowing will likely result in a hastily prepared seed bed and late planting—conditions causing an inferior crop.

Thorough preparation of the seed bed is necessary for large yields. The implements to be used and the number of workings will depend upon the season and soil. A roller, disk harrow, plank drag and smoothing harrow will be serviceable and should be used as conditions demand. The ground

Planted May 26th.

should be firm, with a loose granular surface layer.

Plant Corn Early.—Where the drainage is good, early planting as an average over many years means larger yields and corn of superior quality. Late spring frosts are not as serious a menace to well-filled corn cribs as

Planted June 6th. Early planting of Corn is best—after danger of frost is past.

early fall frosts. The Ohio Station has tested plantings made at 10-day intervals for several years. The early plantings have been badly frosted in some cases; yet corn planted late in April has out-yielded that planted in early June by more than 20 bushels per acre and plantings made May 4 to 10 have yielded nearly 25 bushels more per acre than late plantings. Moreover, the quality has been better with early planted corn, differences of as much as 10 to 15 percent of moisture being noted. The best time to plant depends largely upon the locality of the grower.

How to Plant Corn.—The question of whether corn should be planted in drills or in hills is to be decided by each grower for himself, depending

Frequent shallow cultivation is best for Corn. At first it may be stirred more deeply, but later deep cultivation cuts off roots.

upon the conditions of the land and the use to be made of the crop. Early Ohio tests showed a gain of 4.5 bushels per acre in favor of drilling one kernel every 12 inches as compared with three kernels every 36 inches in the row. Where weeds are serious the advantage in cultivating corn in hills may offset this gain. Again, corn in hills is more easily harvested by hand, but drilled corn is handled better by a corn harvester.

The rate of planting in hills will depend in some measure on soil conditions. Under conditions obtaining in Ohio four kernels per hill have yielded most, but the quality of the grain was inferior to that from a three-plant per hill rate, the hills being 42 inches apart each way. Carefully conducted tests showed two plants per hill 36 inches apart each way to be most satisfactory on thin land in central Illinois, and that the stand should increase with the fertility of the soil; three plants per hill in 36-inch rows gave the largest yields on ordinary land in the northern part of that state. Three plants 39.6 inches apart each way produced the largest yields in central Illinois on reasonably good land—land that yields 50 to 65 bushels of corn per acre. If corn is fed from the shock, then smaller ears and more foliage, as produced by thick planting, are desirable.

The joys of a bounteous harvest! The results of proper tillage and treatments.

Test the Planter.—A half-day may be profitably spent trying out the planter plates to see which will give the best drop with a given grade of kernels. Then when the planter starts one can have the satisfaction of knowing that it is in condition to work properly.

Getting a Full Stand.—The last few bushels added to the yield are the most profitable. These bushels will not be in evidence without a good stand of plants. Good care of seed and ear testing will contribute much to a full stand, but the only sure way is to plant corn somewhat thicker than desired, and then thin to the proper stand. On small areas this is practical and the time required will be well paid for. On large areas it may not be possible to thin corn and dependence must then be placed on good seed and accurate planting. A full stand will range between 10,000 and 14,000 plants per acre, depending on the amount and distribution of rainfall. Three plants per hill in hills 42 inches apart each way gives 10,665 plants per acre.

Early Cultivation.—If the soil is well dried out at planting time the roller can be used to good advantage 2 or 3 days after planting, to be followed at once with a smoothing harrow. These workings should be given before the corn is up. If the ground will not bear rolling (is not sufficiently dry), then the harrow should be used alone. In rather cold weather, when the corn comes slowly, it is often possible to harrow the field twice before the plants come up. It should be harrowed once under all conditions, to smooth the surface and make a great check on the growth of actual and prospective weeds.

Early cultivation of corn is important and should be adapted to pre-

Harvesting corn for silo.

vailing conditions. If a crust has formed after a heavy rain it should be broken, so as to permit the air, so essential to germination and growth, to enter. If there is too much air in the soil and it has become too dry, it should be firmed about the seed by rolling, thus bringing about better moisture conditions. The rolling may be followed closely by a light harrowing (1 to 2 inches deep), or a weeder may be used instead of a harrow.

The weeder can be used with profit after the cultivator when corn is 8 to 12 inches high, provided it is not used early or very late in the day when the plants are brittle. Its steel fingers will work in among the corn plants, doing a work that can hardly be done by hand; and while an occasional plant may be uprooted, the good will far exceed the injury.

Numerous experiments have shown that shallow cultivation (1½ to 2 inches) is preferable to cultivating 4 inches or deeper. Weeds are killed as well; as effective a mulch is obtained, and corn roots are not disturbed by shallow stirring of the soil.

How Late to Cultivate.—Profits from late cultivation will vary with the season and soil conditions. If weeds are numerous and the soil inclined to bake, cultivating once in late July or again in early August will probably result in gains in yield sufficient to pay for the added expense.

Causes of Barren Stalks.—Barren stalks in the cornfield may result from thick planting and consequent crowding. Lack of proper feeding, attacks of such diseases as smut, attacks by such insects as white grubs or corn root aphids, and various climatic conditions contribute to the percentage of barren stalks. Excessive drouth when the ears commence to form may cause many barren plants; and extreme and continuous rains at the time of pollen distribution may interfere with the formation of some ears. Little or nothing, however, can be accomplished in the way of prevention by detasseling barren plants. Some experimenters have chosen barren plants continuously year after year to pollenize other corn plants, and the percentage of barren plants has not been increased at all, nor has the yield been affected to any noticeable extent.

Harvesting.

Topping and Stripping.—While it was the early custom, and it still prevails in many sections, to harvest the entire plant, there are parts of the country today where only the grain is harvested, the stalks being left in the field. This method is common in the West, where the labor problem is especially serious and need for rough forage small.

Again, in the South and in New England stripping of the leaves and cutting off the top above the ear while still green is the general practice. The ears are then allowed to mature on the stalk and later are "snapped" off. Such a method of harvesting adapts itself only to small acreages.

Corn Cutting.—With the increase in prices for hay and the use of corn in dairy sections, the method of harvesting the entire plant has become more general. The use of corn harvesters adapted to fairly large acreages has made the method more practicable and has supplanted in large part hand cutting, even where the corn is shocked and later husked. Machines are almost universally used to cut corn for silos.

Shred Corn Stover.—When the grain is husked from the shock, the stover should be shredded if fed to livestock. Even with large shocks (10 to 12 hills square) in which the stalks remain comparatively green during the curing process, little of the stover will be eaten if not cut or shredded. Moreover, the stover is much better to handle in feeding, while the refuse is in far better shape for bedding or manure. With hay selling at low prices, this practice might not be profitable, although it is always a great convenience.

If the stover is well dried out, one can shred a large mow or stack of it with safety, keeping it for feeding as may be needed. It makes a good feed for both horses and cattle. Many farmers are now buying small cutters of their own, and cutting up only one or two weeks' supply at a time, which obviates any danger of the stover spoiling.

Shrinkage of Ear Corn.—In storage during the winter corn loses much of its moisture and some dry matter. The Ohio Station has found that for the climate in that state the maximum shrinkage occurs in July to September after storing, the average for 6 years being 20 percent shrinkage. As the cool, moist fall weather approaches, the corn takes up moisture. The amount of shrinkage in a year depends upon the condition of the corn in the fall. Prices for corn should vary in accord with the shrinkage, corn worth $1 in November being about $1.20 to $1.30 under usual conditions.

Hogging Down Corn.—A method of corn harvesting that is exceptionally economical of labor is to turn hogs into the cornfield in late summer, allowing them to gather their own feed, some nitrogenous concentrate like tankage or skim-milk being fed in addition. Most of the organic matter is thus returned to the land and no losses in manure such as occur in barnyards are possible. The method is impracticable in the case the season is wet, and the field muddy.

Soy Beans With Corn to Hog Down.—Soy beans have proved valuable as supplementary feeds with corn for hogs at many experiment stations, and may be hogged down with corn. They may be planted with a hand planter between hills of corn one way of the field after the first cultivation, all subsequent cultivations of the corn being in the opposite direction; or they may be seeded at the rate of 2 to 3 pecks per acre ahead of the cultivator the last time the corn is cultivated. The growth of vine will not be large—in fact, this method cannot be used except for hogging down—but in seasons of average rainfall the growth will supply sufficient protein supplement for the hogs and soil fertility will be increased at the same time. Medium Green and Ito San soy beans are satisfactory varieties for this use.

Corn for the Silo.

Type of Corn to Raise.—Experiments thus far conducted point in favor of growing slightly larger and later varieties of corn for silage than one would grow for grain. Just what varieties to grow depends largely upon the locality. What interests the cornbelt farmer often is to plant as many acres and to get as much food material into the silo as possible. He can accomplish this with an early variety of field corn, and the earlier it is usually the larger the proportion of grain and the richer the silage.

Dairymen often grow no more corn than will go into the silo, and it is not unusual for them not to fill the silo. These farmers want to get the most possible food material from an acre of ground.

Experiments conducted by Dr. H. P. Armsby, of Pennsylvania, indicate that a large growing variety put into the silo yields an equal or even greater amount of total food material per acre than a smaller variety of more advanced maturity. Experiments conducted in Massachusetts showed that large varieties that bring their ears at least to the milk stage are preferable to smaller, early maturing varieties for silage purposes, although the extremely late maturing varieties yield a silage more watery, sour and of less nutritive value. By applying the net energy values of Dr. Armsby to yields of Ohio corn, it has been found that Blue Ridge silage corn yields more net energy, as well as more digestible protein, than field corn adapted to that State. It seems fair to draw the conclusion, then, that farmers wanting the

Filling the Silo.

largest possible amount of digestible food material per acre, and are not averse to purchasing some concentrates to balance up the home-grown materials, will likely find it to their advantage, as an average of all seasons, to grow a little larger and later variety than will come to complete maturity.

Time to Cut Silage Corn.—The food value of corn increases decidedly up to maturity. Experiments in New York, Pennsylvania and Indiana prove this fact conclusively. Dr. H. P. Armsby says: "The yield of total digestible food by the fully mature crop was from two to three times as great as that of the same variety in the silking stage, and 36 percent greater than at the time the ears were glazing. . . . A thoroughly mature corn crop contains about the proper proportion of grain and coarse fodder for productive feeding, and for this reason would have a decided advantage over the larger, immature corn for the farmer who depends upon his own farm for all the feed he needs. On the other hand, for the feeder who buys most of his grain, this fact would have less weight."

For this reason, an early type of silage corn should be selected so that the greatest amount of digestible nutrients may be obtained by the time early frosts come. Very late varieties, cut in a green condition before the ears and stalks are near maturity, will form a silage that is sour and watery.

If field corn is put in the silo, it should be cut about a week to 10 days before it would be cut for shocking The kernels should be well dented and the lower leaves on the stalk dry.

Effect of Frost on Corn.—A killing frost stops further growth and development of the plant. The leaves, the upper portions of the stalk, and the husks are the parts ordinarily affected. The main part of the stalk and the ears are not usually injured. The cells of the leaves are ruptured, resulting in the loss of water content, while the other contents of the cells, particularly the substances in process of formation, are likely reduced in food value as a result of the entrance of bacteria.

If the corn is allowed to stand a number of days, the portions affected dry out rapidly and many of the leaves are lost in handling. Just how great these losses are has not been definitely determined. In the early killing of immature corn there would be important losses because of the arrested development; the more immature, the greater these losses.

Such losses can be reduced to a minimum if the corn can be cut at once and put into the silo. This should be done when possible. The palatability of the silage apparently is not affected, and its food value probably but slightly. If the corn stands 2 to 3 weeks before it can be ensiled, water should be added during the filling. The amount will depend upon the juices of the stalk. Water should be added to all dry corn, either in the silo or in the blower, even if not frosted, if it cannot be cut while the stalks and leaves still contain plenty of moisture. Silage must contain considerable water, or it will not pack and then molds will develop.

If frosted corn is utilized for grain and stover, the losses would be much greater. Immaturity will mean more for one thing, and the losses in handling will be increased.

Hints on Filling Silos.—In order to get the corn to pack well into the silo and reduce losses in feeding to a minimum, corn should be cut one-half or three-fourths of an inch long; the first cost of cutting in short lengths may be greater, but the silage is much better. Have the corn distributed evenly over the entire surface of the silo; do not allow it to pack in one spot. Keep the sides a little higher than the center and pack firmly all the time. If convenient it is often advisable to refill the silo after it has settled for a few days. Some extend the height of the silo temporarily with a woven wire, with boards or, recently, with folding iron roofs.

CHAPTER IX.

WHEAT.

Historical.—Wheat is the most valuable cereal, although in production it is exceeded by corn. While corn is of comparatively recent origin, we go beyond written history to find the origin of wheat. It is supposed to have been cultivated in China about 2700 B. C.

Before the discovery of America by Columbus, wheat was not grown in this country. It never grew in a wild state, but has been dependent upon civilization for its development.

Production.—Europe produces about one-half of the world's wheat, while North America stands second. United States is the leading wheat-producing country, and here our production in 1915 was more than a billion bushels, or a yield of less than 17 bushels per acre. The five leading states in the United States are Kansas, North Dakota, Minnesota, Nebraska and South Dakota, which produce nearly one-half the wheat of this country.

Kinds of Wheat.

Like corn, wheat belongs to the grass family. It is a near relative of barley, rye and rye grasses. Wheat has the characteristic of tillering, or sending up a number of culms, depending on thickness of planting and soil conditions. The plant ordinarily varies from 2 to 5 feet in height among the various varieties. The height determines, to some extent, the resistance to winds and storms. Wheat is further characterized by being generally self-fertilized.

Spring and Winter Wheat.—In Minnesota and the Dakotas spring wheat is commonly grown because of the severe winters. The plants ripen rapidly and the average yields are less than with the winter varieties. In Kansas and neighboring states of that latitude some spring wheat, mostly durum, is grown. In nearly all the other sections of the United States winter wheats prevail. In general, winter wheat is more productive, distributes labor more evenly throughout the year, and affords opportunity for fall seedings of grass.

Bread Wheats.—The kinds of wheat ordinarily used for bread making are common wheat, which includes most of our hard and red wheats and club wheat, which is generally soft. About 95 percent of the crop belongs to bread wheats. The hard wheats are darker in color than soft wheats, and contain more gluten; hence, they are more desirable for bread making.

Club wheat has short, stiff straw and a compact, club-shaped head. The flour is used mainly for pastry and for cereal foods. Generally these varieties are heavy yielders. Their production is restricted to the states on the Pacific coast.

Durum Wheat.—Durum wheats are grown in the semi-arid sections of the West, are nearly all spring varieties, and the grains are generally hard

Smooth Heads of Wheat. Varieties: 1. Early Ripe, 2. Perfection, 3. Currell's Prolific, 4. Poole.

and flinty. The type is similar to poulard wheat. Durum wheat is a recent importation from Europe and is adapted to countries where rainfall is deficient and alkali is common. It is used mainly in the manufacture of macaroni.

Spelt Wheat.—This group includes spelt and emmer wheats. Spelt is little grown in the United States, and is used mostly as food for animals. Emmer is grown in a few sections of the United States. Its chief characteristic is drouth resistance. As a spring crop for Ohio and the northeastern states, it is not a competitor of oats.

Wheat Regions of United States.

Common Wheats.—The hard wheats are raised from Texas to North Dakota. Hard winter wheat is raised south of South Dakota while hard spring and macaroni wheat is grown north of Nebraska. Soft red wheat is grown in the more humid regions of the southern states, while semi-hard, red wheat occupies the middle section from Pennsylvania to the Missouri River. New York and the New England states comprise the white winter wheat section. Irrigated wheats are grown in the inter-mountain region. Soft wheat is produced along the Pacific coast.

Durum Wheats.—Durum wheats are adapted to dry regions, doing best

Bearded Varieties of Wheat. 1. Mediterranean, 2. Nigger, 3. Gypsy, 4. Valley.

where the rainfall is less than 25 inches. They are grown in the Great Plains regions from Texas to the Dakotas. The U. S. Department of Agriculture introduced this wheat from Europe about 1900.

Soils for Wheat.

Glaciated soils, upland, clayey and light in color raise most of the wheat. Low, dark-colored corn soils produce wheat of inferior quality. The best wheat soils lie between the typical grass and the typical corn soils. A loam or clay soil, fairly heavy and of medium water-holding capacity, is said to be best for wheat. Wheat is grown on nearly every type of land, the management of the soil depending upon its characteristics.

Moderate rainfall after seeding in the fall, followed by a fairly cold winter of even temperature and medium snowfall, and later a cool, prolonged spring, succeeded by moderately dry hot weather when the grain is maturing is an ideal condition for winter wheat. This gives quality and yield desired when other conditions are favorable.

Root development goes on in the fall and early spring; in fact, winter wheat makes its growth during the cool part of the year. Much of the world's crop is grown in cold climates. Where the climate is warm and rainfall heavy, there is produced a plump berry, low in protein, light in color, soft in texture and undesirable from the miller's standpoint.

Culture.

Place in Rotation.—Local conditions determine the place of wheat in the rotation, but it is usually after potatoes, corn or oats, or after an annual legume, like cowpeas or soy beans. It gives larger yields after potatoes or a legume than after corn.

Preparing the Seed Bed.—If wheat is grown after oats or another crop in the rotation that calls for plowing of the land, this operation should be done early. Ordinary plowing is preferable to deep plowing. When clover is on the ground, either as new seeding in wheat to be followed by wheat, or a second crop following a crop harvested for hay, something may be gained in the growth of clover by delaying the plowing. If one were sure of good rains just before plowing, which would insure proper moisture conditions for plowing and preparation of the seed bed, it would be wise to secure the extra growth of clover. More nitrogen and organic matter would thus be added. Nevertheless, as an average of many years, more will be lost as a result of poor seed beds than will be gained as a result of the extra growth of clover.

In the case of oat stubble ground, plow as early as the oat harvest will permit. One may start the plow while the oats are still in the shock. Unless the oat harvest is late, this would seldom be advisable because of harder work in hauling the oats from the field. Occasionally labor is saved by waiting for good rains, but this wait may extend for a month, and not infrequently the crop must be abandoned for the season. If the crop is not abandoned these long delays often result in poorly prepared seed beds. When early plowing is hard and the ground turns up lumpy, thorough preparation of the ground is essential, and if later rains do come they may be utilized in preparing the land.

The plow must be followed immediately by the best tillage implements. Among these are a good roller, a disk harrow, a smoothing harrow and a pulverizer or plank drag. A harrow of some sort should follow the roller. A smoothing harrow may be attached to the roller, one man with four horses doing the work of two men. It is important to roll and harrow each half-day's work at once. Even though the ground apparently is not in condition for much to be accomplished, this immediate working will ultimately be beneficial.

The work upon seed beds in rotations which do not call for plowing, as wheat after corn, potatoes, soy beans, etc., varies with the crop, the care with which the crop has been cultivated, and the character of the soil. Potato ground generally gets such a working during the process of digging, particularly when modern cultivators are used, that little work is needed to prepare an ideal seed bed. To work between the potato rows, where the digger does not disturb the ground, a disk becomes necessary. Wheat should be seeded in good season after potatoes as with other crops.

Except on some exceptionally mellow and well-cultivated loams, wheat is often seeded too hurriedly after corn. Cornfields generally are not cultivated after the middle of July. They are dry and hard, and more or less weedy by September. It is a mistake to save a few days' time for the sake of earlier seeding by disk-drilling the wheat in without working the ground, simply because the modern drill can cover the seed. Disking in advance of seeding will pay.

Summer Fallowing.—Summer fallowing usually results in excellent crops of wheat, but at great expense of the fertility of the soil. Organic matter is decomposed rapidly and a store of plant food prepared for the wheat. For instance, nitrogen will be made available; yet some of it will leach away and the total amount of nitrogen in the soil will be less because of the fallowing. To prevent this waste of time and fertility, cowpeas should be sown early in June, to be used for hay or for green manuring as most needed. From 6 to 8 pecks of seed per acre, drilled solid, will be satisfactory. Soy beans may be used instead of cowpeas, being seeded in rows wide enough to cultivate, 2 pecks of seed being used per acre.

Fertilizer.—Wheat generally receives more fertilization than any other crop. It is the crop generally sold and increases from fertilizers are speedily turned into cash. Wheat usually responds to the use of fertilizers if any crop does. The soil must first be considered in selecting a fertilizer, as has been previously explained, but usually applications of phosphorus will be most profitable for wheat soils. Potassium generally returns less, but it can be used, temporarily at least, with profit on many soils rich in total potassium when this element is not being made available fast enough for immediate needs. On soils badly deficient in nitrogen applications of this element may be profitable for wheat, though with clover or other legumes in the rotation, and with additional provision for the use of leguminous catch crops, it should be possible to avoid the purchase of nitrogen. Especially should this be true in livestock farming when the manure products of the farm are fully conserved. With an increased use of manure and legumes more potassium will be made available, so that for many soils all one will need to buy will be phosphorus, in some such form as acid phosphate.

Applying Straw to Wheatfields.—From the standpoint of insect attacks the use of straw spread on newly seeded wheatfields is not commendable. The joint worm and Hessian fly, in particular, can be spread in this manner, and the litter will make excellent winter quarters for chinch bugs. Straw can be spread thinly upon new clover seedings with safety but its best use is as bedding to make manure.

Seeding.

Manner of Seeding.—Not much wheat is seeded broadcast today. In some cases it has given as good yields as drilling, but tests at various experiment stations show gains of 2 to 8 bushels per acre in favor of drilling, as an average over many seasons. Cross-drilling is generally also unprofitable. The margin in favor of two drillings is so small that usually it does not pay to take the chances. If rains interfere before the second time over the land, further losses may occur from such delayed seeding.

Rate of Seeding.—After 17 years' tests with different rates of seeding, ranging from 3 to 10 pecks per acre, the Ohio Station has found 8 pecks per acre to be most profitable, with 9 pecks second and 6 pecks third. Farther west, in the Missouri Valley, 6 pecks is considered sufficient, and 3 to 4 pecks in the dry-farming section.

Date of Seeding.—In order to get proper root growth before winter weather comes, wheat should be seeded 6 to 8 weeks before freezing. For the latitude of the Great Lakes September 5-10 is usually satisfactory, and September 25 to October 5 as far south as the Ohio River. Spring wheat should be seeded as early as the land can be prepared. It then gets good root development and ripens earlier.

Depth of Seeding.—Depth necessarily depends upon the condition of seed beds as regards texture and moisture. Under favorable conditions, one inch or even less is ample. With a dry, lumpy soil 3 inches may not be too deep. The seed should be put in moist soil.

Rolling After Drilling.—In case heavy rains occur soon after seeding, followed closely by drying weather, it may be profitable to break the soil crust with a light harrow or weeder; otherwise poor stands may result. The rolling of wheat ground after drilling is not to be recommended. The needed firming of the seed bed should be done before drilling. A roller puts the ground in condition to lose moisture, and in dry weather this would be disastrous. If a roller is used, a light harrow should immediately follow to form a mulch that will check evaporation. The use of either implement would hardly be desirable, as the little ridges left by the drill are of benefit in the way of winter protection. They hold snow about the plants, and in freezing and thawing weather a little soil will drop in about the wheat plants.

Selection of Seed Wheat.

Variety.—The variety to select depends largely upon the locality. One should be chosen that experience has shown to be best for the section, and new varieties tried only in a limited way. Experiments at different stations have shown that seed wheat imported from other states generally yields no more, and usually less than home-grown varieties. Wheat does not "run out" if grown continuously in one locality.

In selecting a variety, winter resistance, stiffness of straw and bread-making qualities should be considered, as well as yield. Often high yielders, like Dawson's Golden Chaff and Gold Coin, have poor milling and baking properties.

Fake varieties, like Miracle and Marvelous, are often put on the market by unscrupulous men at extravagant prices, and unusual claims made for

Stiff-strawed variety of Wheat after several years of improvement
.by selection.

them. Often the ridiculous claim is made that one peck will seed an acre. Many tests have shown such varieties to require as heavy seeding as the old standard varieties.

Cleaning Seed Wheat.—It will usually pay to run seed wheat through a fanning mill for the purpose of removing weed seeds, broken kernels, badly shrunken kernels and bits of straw, which will interfere with the successful use of the drill. There is little evidence that the removal of small grains which are plump and bright is necessary, as they seem to yield as well as the larger kernels, at least when separation is made by machine. Large and small grains come from the same head and carry similar hereditary characters.

Using Old Seed.—Old seed wheat of satisfactory varieties may be used with safety the next year or two after its production. Ohio tests have shown but little dropping off in the percentage of germination with wheat until after the second or third year.

Treatment for Smut.—When wheat is affected with stinking smut and scab, it should be treated with 40 percent formaldehyde, sold at drug stores as formalin. One pint of formalin is mixed in 40 gallons of water and the solution sprinkled over the wheat spread on a tight floor or canvas. One gallon will treat a bushel of grain. After being covered over with a blanket for 2 or 3 hours or over night, the grain should be spread out to dry. In order to prevent further infection, grain sacks and drills should be disinfected with the same solution.

Harvesting and Uses.

Cutting Wheat.—Wheat should be well ripened before cutting. However, if the crop is left too long the wheat will shatter and the milling quality

Weak-strawed strain of Wheat.

will be reduced. The grain should be set up in rather small shocks allowing free circulation of air. Capping the shocks retards curing, but is desirable in rainy seasons. One should use skill and care in shocking wheat so that little damage can result from moderate rains and winds.

Threshing wheat directly from the shock is a desirable practice but is risky in rainy seasons. In threshing the grain should be fairly dry before being stored in large bins. Ordinarily wheat will shrink but little (2 to 3 per cent) within a year in storage.

Pasturing.—Wheat may sometimes be pastured with sheep or young hogs if the growth is rank. Such a practice will reduce the yield somewhat, as a general rule; but where an extremely large amount of straw is produced, as on rich soils, the yield may be increased. Seasonal conditions have much to do in determining how much to pasture wheat yields.

Wheat vs. Oats as a Nurse Crop.—Winter wheat is generally preferred to oats as a nurse crop for seeding clover for several reasons. Clover usually is seeded a few weeks earlier in wheat than in oats, which means a better root system and great resistance to summer drouths. Wheat also furnishes a firmer and more congenial seed bed, one that will supply more water than land plowed for oats in the spring. Moreover, wheat is harvested 2 to 3 weeks earlier than oats, thus removing the competing crop earlier and usually at a less critical time for the young clover. Again, wheat is not as subject to lodging as oats and consequently does not smother out the new seeding as much. Finally, wheat straw is coarser and less shade is formed than with oats.

What to Do When the Wheat Crop Fails.

When Seeding is Impossible.—At time of prolonged heavy rains at seeding time or in case a proper seed bed cannot be obtained for other reasons, wheat must sometimes be left out of the rotation. One naturally wants to disturb the rotation as little as possible, and as a clover or grass crop usually follows wheat the treatment of the land should be planned with this in view.

Accordingly, the first plan to suggest itself is to seed the field intended for wheat to some small spring cereal with which clover and timothy can be

A Wheat Field that speaks for itself.

seeded. For this purpose spring wheat, barley or oats may be suitable. Where winter wheat is generally grown, as from New England to Illinois, spring wheat is not adapted. All varieties of it have proved more or less unsatisfactory at the Ohio Experiment Station. It generally suffers severely from rust; grains are shriveled and of poor quality; and yields are only about 15 bushels per acre.

A good crop of barley of a variety like Oderbrucker or Manchuria— preferably the former—may be obtained in this section, and a good catch crop of clover or grass should be obtained in seeding with barley. Still, the oat crop is generally more satisfactory and seed more convenient to secure. Early varieties seeded lightly should permit a good clover catch, although probably not as good as when seeded with wheat.

Another scheme of utilizing ground where wheat was to be seeded is to seed it to clover and timothy alone. If seeded in time the field should yield a fair hay crop the same season and a larger crop the second. If the wheat field 'had been fitted for seeding and was in suitable condition to seed clover or grass, these can be seeded in early March when the ground is honey-combed with frost. If this condition is not found, the seeding should be delayed until the ground can be fitted with team and harrow. It will be well to harrow before seeding, and lightly after seeding, a smoothing harrow or weeder being used.

Still another course can be followed, viz, to seed oats and field peas in March or early April and then to cut them for hay in late June or early July. This should be followed by disking and seeding clover and timothy as soon as a sufficiently moist seed bed is available.

When the Seeding Fails.—When wheat fails from winter weather conditions and is not worth harvesting, spring crops should be planned to take its place. If the failure is seen in time, the ground may be disked and seeded to oats or barley, for this will make little or no change in the rotation as commonly conducted where clover and grass follow wheat. Any struggling wheat plants will add to the value of the oats as feed, and the clover and grass will come on apace, thus making no break in the rotation.

If oats cannot be seeded in time, some legume should be considered. Where clover and timothy are seeded in the fall, the wheat might be left for the sake of the seeding. A light crop of wheat is almost certain to mean an excellent catch of clover. Often a ton or more can be cut in August after the wheat is removed.

Where it seems best to plow the wheatfield several crops can be planted with profit. Corn is one of these, but it is often undesirable to double the corn acreage and change the rotation so radically. Late potatoes may be planted; yet many farmers cannot handle this crop because of labor involved. It will serve the purpose better to seed some crop requiring less work.

Sorghum may be seeded later than corn and if drilled solid the expense of intercultivation is avoided. In this event 4 to 6 pecks per acre will be sufficient if the sorghum is cut for hay. Millet, as the Hungarian and German, will make 2 to 2½ tons of acceptable hay per acre on fairly rich land when seeded in late May or in June. Three pecks should be drilled per acre. Soy beans would be a third suitable crop and they would have the added advantage of being a legume.

Whatever crop is grown, rye should be sown the next fall. This can be done at slight expense, and the crop will add considerable organic matter to make good the demands from such excessive cropping. The rye may be plowed under for soy beans the next spring, and by such a method fertility may be maintained.

CHAPTER X.

OATS.

Distribution.—While wheat came from the warmer climates of Europe and rye from the colder, oats originated in the intermediate regions. Oats were less important in early history than either of the other crops, and increased in importance only after central Europe became civilized. Oats are confined to temperate zones, and do best in cool, moist climates. The crop reaches its best development in Norway, Sweden, Canada, Germany, Great Britain and northern United States. Little of the crop is grown in southern United States.

Climate affects the quality of oats. The southern varieties are dense and less plump, have a long awn and are usually red or yellow in color. Such oats are usually winter varieties and are confined to the cotton states. North of the Ohio River fall-sown oats are winter-killed. The northern oats, grown in spring, are short, plump, smooth and dirty gray, white or black in color.

Production.—Europe generally produces about twice as much oats as North America, although the United States is the leading country in the production of this crop. Our production in 1915 exceeded one and a half billion bushels, the average yield per acre being about 37 bushels that year. Illinois, Iowa, Wisconsin, Minnesota, Nebraska and Indiana produce more than half the crop in the United States. The crop does best in the North, but it fits well into the corn-belt rotation, being suited to follow corn and producing a valuable feed for live stock.

Kinds of Oats.

Oats belong to the grass family, and differ from wheat in carrying grain in a panicle instead of a spike. The typical panicle carries whorls of branches on both sides of the head. Side oats differ from this spreading type in having all the branches on one side of the head.

The spikelet generally has two flowers, the upper one producing a lighter grain than the lower. The lower flower usually has an awn. The quality of oats depends in part upon the percentage of hull, because the hull has much crude fiber and indigestible material. The percentage of hull depends upon the variety and condition of growth, but averages about 30 per cent of the grain. The varieties best adapted to a region, maturing normally, are usually lowest in percentage of hull.

Uses of Oats.

Feeding Value of Grain.—Human consumption of oats is small compared with the amount used for feeding live stock. Oats are generally considered the best grain feed for horses, although some tests indicate that mature animals fed mixed hay do as well on corn, and the latter is much cheaper. Oats are high in crude protein and low in carbohydrates as compared with corn and wheat. Oats are also lower than corn in fat but much higher than wheat.

Since oats are a palatable feed supplying the nutrients in about the right

Regenerated Swedish Select (left). American Banner (right). Varieties of oats differ greatly in stiffness of straw.

proportion to form a balanced ration, they are considered of high feeding value for dairy cows, usually being worth at least as much as wheat bran. The oat hull in itself has little value, but it lightens up the feed by great bulk in proportion to weight. However, oats are not usually an economical feed and few dairymen use this feed in a large way on this account. Oats generally are a satisfactory feed for fattening lambs but not as good as corn. Because of their fibrous nature they are not suited for hog feeding.

Feeding Value of Straw.—Straw is poor in crude protein and fat and high in fiber, or cellulose, a carbohydrate that requires much energy for its digestion and disposal. For animals, like horses or cattle, at light work or being roughed through the winter this roughage may be an economical feed. Oat straw is the most nutritious of all straws of cereals. Wheat straw is more coarse and stiff and is lower in palatability and feeding value.

Oat hay cut when the grain is in the milk has a feeding value about equal to that of timothy, being higher in protein and a little lower in carbohydrates. Oat and pea hay is about equal in feeding value to red clover, and may replace the latter in a ration.

Oat Hulls.—Oatmeal factories have considerable by-product in the form of oat hulls, which generally contains about 30 per cent fiber. The feeding value is about the same for these hulls as for oat straw. They are used frequently in adulterating feeding stuffs, especially ground corn, the feed then selling as ground corn and oats.

Culture of the Crop.

Soils.—Oats are less dependent upon soil and cultural conditions than most cereals. The crop responds less to thorough tillage and fertilization. With favorable climate it thrives on all productive soils. A soil having great water-holding capacity is best, because oats require more water than other cereals. Oat stubble is always more dry and harder to plow than stubble of other cereals. Poorly drained soils are unsuited, as are also unusually fertile soils since they tend to produce rank growth that lodges. Weather and climate—cool temperature, plenty of moisture and slow-ripening season —are more important than soil.

Fertilization.—This fact also emphasizes the wisdom of applying fertilizers and manure to other crops in the rotation than oats. The oats can then receive some benefit from the residual effect of the application. However, when fertilizers are considered necessary, a moderate application of acid phosphate will usually suffice.

Time of Plowing.—In a rotation including oats, corn usually precedes this crop. Such a plan affords opportunity to grow a cover crop, since the corn is removed usually in good season. Soy beans, cowpeas or rape may be used, and all nitrates that form after the corn crop is made would thus be conserved. Rye is not as well suited to this use because oat seeding would be delayed if full benefit of the rye crop were obtained, and volunteer rye would spring up in the oats. Moreover, larger yields are usually secured from earlier plowing; team work is more evenly distributed during the year, and relief is afforded in the general rush of spring work if no catch crop of this kind is grown. For these reasons it is usually better not to precede oats with rye.

If oat ground is to be plowed, this work should be done in the late fall or winter. Clover and grass seeded with oats do better on fall-plowed land.

Disking Generally Better Than Plowing.—Plowing is usually unnecessary for oats except on heavy clay soils or where weeds are abundant or in cold, wet seasons. Tests made at various experiment stations have shown that plowing oat ground gives no or only slight increases in yields over disking, and the gain if any is not sufficient to pay for the extra expense involved in plowing.

Drilling vs. Broadcasting.—Results of numerous experiments indicate that better preparation of the seed bed and drilling pay more than poor preparation and broadcasting. Drilling insures a much more uniform seeding. This means that less seed is required because of better germination, and greater uniformity in growth and maturity results. Oats need to be covered only 1 to 1½ inches. West of the Missouri River a depth of 2 to 3 inches may be necessary to insure sufficient moisture.

Rate of Seeding.—The rate to seed oats depends upon the size of the grains, but generally 8 to 10 pecks per acre is sufficient. Small varieties require less seed than larger ones, and broadcast seeding more than drilling.

The amount of rainfall and the tendency to tiller influence the rate of seeding. In dry-farming regions 5 or 6 pecks per acre is sufficient. Thick planting results in less tillering; and in cold clay soils, even with thin seeding, tillering is slight.

Time of Seeding.—The date to seed oats varies with the season, but it should be as early in the spring as possible. Quality and yield are increased if the oats can grow rapidly before warm, dry weather sets in, and rust can be avoided somewhat by early seeding. In the northern region, as in Canada, early seeding is not so important.

Fall-sown oats mature 2 to 3 weeks earlier than spring varieties. In sections as far south as Alabama seeding may be done in February; in Tennessee and Kentucky, in March; and in the Dakotas, in April.

Selection of Seed.

Cleaning Seed.—Cleaning seed oats to remove small grains has been found unprofitable at the Ohio Experiment Station. Yields are increased so slightly by repeated cleanings in a fanning mill that only removal of dirt, broken straw and weed seeds in the grain is advised. Results at Kansas and Nebraska Stations support this view.

The spikelet of oats contains two grains, as a rule, but frequently three. The smaller, upper kernel is called the secondary grain. When the hull is removed from the primary and secondary grains, the kernels differ less in size than the whole grains. This is why the secondary grains weigh so much more. These grains may be used for seed, as careful hand selection of these small grains have shown they yield nearly as much as the larger grains.

Northern-grown Oats for the Cornbelt.—Seed oats brought from the North and Northwest are generally of heavier weight, but they usually take their place at once beside the home-grown varieties of the cornbelt, both as to quality and with respect to yield. There is little chance for loss to send for seed oats from the North and West in case the local crop is short or inferior. Unlike corn, the northern-grown seed oats do about as well the first year as thereafter.

Treatment for Smut.—Yields can generally be increased by treating seed oats for smut just previous to planting. A pint of 40 per cent formaldehyde, or so-called formalin, to 40 gallons of water will treat 40 to 45 bushels of grain. The oats should be spread on a tight floor or canvas and moistened thoroughly with the liquid, a sprinkling can being convenient for the purpose. After being covered with a blanket for 3 to 4 hours or over night, the grain should be spread out and shoveled over frequently until dry enough to sow. Use the same solution to disinfect bags and drills.

Testing Seed.—Testing seed oats pays in bad seasons. A convenient means is to put some of the seed between moist blotting papers or in a cotton flannel cloth between two dinner plates. The paper or cloth should be moist

Treating Oats for Smut. Shoveling and
sprinkling.

at all times, and kept at a moderate temperature. In 4 to 6 days the oats may
be counted for germination, and if faulty should be discarded. If poor oats
must be used the test will show what additional amount is required to obtain
a good stand.

Harvesting.

Cutting for Grain.—Oats should be cut when half of the leaves are still
green and the grain is in the dough stage. The composition and yield will
not be injured and loss from shelling will be avoided. When cut too ripe this
loss through cutting, shocking and hauling may be considerable. If cut rather

After treatment for Smut, Oats should be left
in a pile for 3 to 4 hours or over night and
covered with a blanket.

green the sheaves should be set up in long narrow shocks running north and south.   The half-green straw is most relished by horses.

Oats should be hauled as soon as cured, and threshing directly from the

All the equipment necessary for treating Oats
for Smut.

shock is advisable.   Rain may cause considerable injury in changing color of the grain, and in lowering the feeding value of the straw.   For the same reason care must be used in stacking oats; the center of the stack should be kept high with the outer layers slanting so that water will drain off readily.   Oats shrink less than 1 percent during the winter months.

Cutting for Hay.—In making hay of oats they are usually cut when in the milk stage, and handled like ordinary hay.   They should be cut with a mowing machine, allowed to cure about 24 hours in the swath and then finished in the cock.   The feeding value is highest when the crop is cut at the milk stage.

In choosing a variety for hay, select one carrying a large proportion of leaves and stems, provided the latter are not too coarse.

Oat and Pea Hay.—Canada field peas are frequently seeded with oats to be cut for hay, 1½ bushels of each being used per acre.   This mixture makes a palatable and nourishing hay.   The peas are drilled 4 to 5 inches deep, and the oats later at a depth of 1 to 1½ inches.   Clover will not make a good stand with these crops, because there is too much shade and the sudden removal of the crop will work hardship with the tender clover plants.   Then, too, these crops often lodge.

Cowpeas and soy beans are hot weather crops and do not fit in as well as field peas with oats.

It is best to cut the peas and oats when the pods commence to form and the oats are in bloom.   If allowed to get much riper than this there will be loss from shattering of the leaves.   The crop should be cut in the afternoon, when there is little moisture.   They should then be cured in the swath for

1 or 2 days and for a like period in the cock. The hay tedder can often be used to advantage. The hay should be well cured before hauling.

Oats and Vetch.—Winter vetch is more important and useful in this country than spring vetch. It is seeded from July to September. However,

A promising field of Oats—strain improved by years of selection.

spring vetch can be seeded with oats at the rate of 4 pecks of vetch and 6 pecks of oats per acre. The hay crop is nutritious, but not large and is usually not profitable.

Oats as a Nurse Crop.—In choosing a variety of oats to seed clover and grass, early maturity and light yield of straw are essential. The advantage in removing the competing crop a week or more earlier than the average time of ripening will be apparent in the summer and fall growth of clover, and may be the difference between success and failure in the seeding. Sixth Day and Kherson oats are among the earliest varieties and yield little straw. The former has been found most satisfactory in Ohio to seed clover and grass of all varieties now available.

As a nurse crop for clover or alfalfa, oats should be seeded at the rate of 3 to 5 pecks per acre. Usually they can ripen grain with no injury to the seeding; but in case the growth is heavy and the crop may lodge, it should be cut for hay when the oats are in bloom. Otherwise the seeding will be smothered out.

What to do When the Oat Crop Fails.

Oats are frequently delayed in seeding on account of weather or otherwise, until a fair crop is the best prospect and none at all is possible. The first thought might be to plant corn or potatoes, but often it is not desired to increase the acreage of these crops.

Soy beans and cowpeas are good substitutes, preferably the former. Either crop may also be sown in case oats are seeded and fail to grow. The total digestible nutrients in soy beans are higher than in cowpeas, and the crop generally is more profitable. For hay a mixture of the two crops would be satisfactory. Millet may also be suggested as an emergency hay crop.

CHAPTER XI.

BARLEY.

History and Production.—As with wheat, barley was cultivated before the Christian era. It was the chief bread plant of Europe until the sixteenth century. The introduction of wheat tended to cause a decline in barley production. In the United States is was used a little by the early colonists. It is now grown in all latitudes from the gulf states to Alaska.

Europe today is the main barley producer with North America second. The production in the United States is over 228,000,000 bushels a year. Russia is the largest producer, and United States is second among the countries. California, Minnesota, South Dakota, Wisconsin and North Dakota produce nearly three-fourths of the barley in the United States and rank in the order given. More than half the production in the United States comes from the north-central states west of the Mississippi.

Uses of Barley.—Barley belongs to the same family as wheat, but differs from it in structure of the grain. The outer husk of ordinary barley is tough and adheres closely to the kernel, and most varieties have long beards. Barley straw is shorter and not as stiff as that of wheat or rye. The roots are shallow and comparatively feeble.

Some species are six-rowed, others four-rowed and still others two-rowed, while there is also a hulless species. The four-rowed barley is usually called the common six-rowed variety. The hull makes up about 15 percent of the grain. Barley grain from the Pacific Coast and Rocky Mountain states is longer and plumper than that grown in the north-central and north-Atlantic states.

In general, the grain when hulled has about the same composition as wheat. With the hull it has more crude fiber than oats or wheat. The grain has more carbohydrates than oats, but less than corn and wheat; in protein content it exceeds corn and is slightly below the other two grains.

On the Pacific Coast barley is fed extensively to horses. It is often fed to cattle, but ordinarily other feeds prove cheaper. Iowa and Nebraska Stations have found it an economical supplement to corn for hogs. For hay the beardless varieties should be grown. Barley straw is nearly equivalent to oat straw in composition.

The best market barley is used for malt and the poorer grades are sold for feeding purposes. Complete, quick, uniform germination and freedom from impurities are requisites of a good malting barley. By-products of malt factories, as brewers' grains and malt sprouts, are valuable feeds, mainly for dairy cows.

Soil Conditions.—Barley seems best adapted to the north temperate zone. It requires less moisture for growth than any other cereal. A well-drained, fertile, sandy loam soil is best adapted. Cultural conditions have less influence on the quality of barley than weather.

In the rotation barley may take the place of wheat or oats. Corn, barley, oats and clover, or corn, barley, wheat, timothy and clover are common rotations. Wheat does better after barley than after oats since barley requires less water.

Culture.—Fall plowing is advisable for spring barley because of early seed-bed preparation and early seeding. Uniformity in maturing may be aided by the use of a drill in seeding. Ordinarily 2 bushels is seeded per acre, although the rate may vary from 1½ to 4 bushels per acre. Barley stools but little. On this account and also because of the small amount of foliage, it is frequently used as a nurse crop to seed clover or alfalfa.

Harvesting.—Barley is mature when the straw is yellow and heads droop over. It should be cut before the grain is dead ripe or the grain will change its color more easily when exposed to dew or rain. The bundles should be set in medium-sized, well-capped and well-aerated shocks, and hauled as soon as dry to prevent weathering. In threshing the grain should not be injured in any way; high speed and running the conclaves close together will break the kernels. Cap sheaves should be threshed separately, as a small percentage of weathered kernels will reduce the value of the grain.

Varieties.—Two-rowed varieties are all seeded in the spring; six-rowed, in spring and fall. The former are later in maturing, have weaker straw and more mealy grain and are more desirable from the malters' standpoint. Little winter barley and few hulless varieties are grown in this country. North of the Ohio River winter barley generally winter-kills.

The Scotch type, a six-rowed barley, is grown in Wisconsin, Minnesota and Iowa, but is now largely superseded by the Manchuria and Oderbrucker varieties for the states north of the Ohio River. The Bay Brewing (or California Bay) is grown along the Pacific Coast. The third common trade type is the Chevalier, a two-rowed barley originated in England. Two-rowed barley is grown in the northern climate of Canada and on higher elevations in Montana and the Dakotas. The common six-rowed barley is most generally grown in the cornbelt and the territory around Wisconsin.

CHAPTER XII.

RYE.

History and Distribution.—Rye was introduced into the Roman Empire at the beginning of the Christian era, and is supposed to have originated in northern Europe. It thrives in cool climates and often on soils too poor to support other cereals profitably. It is hardier than wheat.

Europe produces more rye than North America. European Russia produces nearly half the world's rye crop. The annual crop in the United States is about 35,000,000 bushels, and much of this is exported. The north-central states east of the Mississippi produce most of the rye crop, the leading states being Pennsylvania, Michigan, Wisconsin, Minnesota and New York. These five states produce about three-fifths of the crop. The average yield per acre is about 15 bushels, although 20 to 25 bushels per acre is a common yield.

Uses.—Rye belongs to the same family as wheat and barley, and often replaces wheat when the latter fails to thrive. Its culm is longer, more slender and tougher than that of wheat. The kernel resembles the wheat grain.

Rye is used as human food chiefly in Europe. In the United States it is commonly fed in connection with other grains to live stock, more often to hogs than to horses or dairy cows. It has more digestible protein and carbohydrates but less fat than corn, oats and wheat.

It is also often used as a soiling and pasture crop. As such it is desirable because it affords an abundance of succulent feed before other crops have entirely recovered from winter conditions. Pasturing in the fall or spring is often desirable even when the crop is to be used for grain. Pasturing should be allowed in the fall if the rye begins to joint. For fall pasture it is often seeded in late August. Unless intended for hogs, it should not be allowed to come in head, as with age it loses its palatability. It is at its best when not more than a foot in height.

Hogs frequently cause damage by trampling and rooting in early spring. Sheep may be pastured on rye earlier than either cattle or hogs. Hogging down rye is not recommended as a general practice. Hogs on rye pasture need a supplementary feed like tankage and corn for greatest gains. An acre of rye should support 10 hogs for 2 months.

Silage made from rye is not as palatable nor as nutritious as corn silage. The stems should be cut about one-half an inch long and well packed in the silo to exclude air. For such purposes the rye should be cut when the heads are well out or not later than when in bloom, an ordinary grain binder being satisfactory for cutting. Rye for this purpose may be seeded in late September, 7 or 8 pecks of seed per acre being sufficient. When the rye is cut early for silage, another crop like corn may be grown the same year. Such a practice of intensive farming requires heavy manuring and fertilization.

Rye is regarded as one of the best cover crops in the northern states, because it can withstand the winter better than any other crop used for this purpose. For plowing under in the spring it is seeded like wheat at the rate of 2 bushels per acre. As far north as the Great Lakes it may be seeded as late as the middle of October.

Rye and hairy vetch make an excellent combination to seed at the last cultivation of corn, 4 to 5 pecks per acre of the former and 20 to 30 pounds of the latter. Vetch is a legume, and thus nitrogen, as well as a large amount of organic matter, may be added to poor soils. One must watch, however, in using rye as a green manuring crop so that the growth does not become excessive, and after being plowed under act as a mulch to prevent water at lower levels from rising to the surface. It is better to lose a little in the growth of rye and plow it sooner than to take chances on trouble of this sort.

Rye is one of the most dependable Catch Crops. It supplies a great amount of green forage to plow under in the spring.

Culture.—Rye is adapted to light, sandy soils. The culture is nearly the same as that for wheat, which it may replace in the rotation. For grain the same kind of seed bed is desired for both crops. The time of seeding varies, as previously shown, with the use made of the crop. When grown for its grain the time is about the same as that for wheat; the rate, 1½ bushels per acre. It responds to fertilizers as well as other crops.

Harvesting is also nearly the same as for wheat, although the rye grain ripens about a week earlier and no caps are placed on the shocks. The straw has special uses in the manufacture of paper, baskets, hats, matting and the like.

When the straw is to be thus utilized, only the heads are thrust into the threshing machine and the bundles thrown aside are not broken. There are also special rye threshers that cause little breaking of the straw, but they are little used.

CHAPTER XIII.

BUCKWHEAT.

Distribution.—Buckwheat is not a true cereal, but is classed as such because of the use made of the seed. It was cultivated a thousand years ago

Japanese Buckwheat plant.

in China, was introduced into Europe in the Middle Ages, and early into the American colonies. Today it is grown mainly in northern Europe and Asia and northeastern United States.

The area in the United States is about 800,000 acres, or about one-tenth that of barley. New York, Pennsylvania, Michigan, Maine and West Virginia are the important producers in this country. There is little or no export, but a small import from Canada. In yield 30 bushels per acre is considered large, and 20 to 25 bushels is considered satisfactory.

Uses.—The buckwheat plant usually is about 3 feet high and produces no tillers. It has one strong, central tap-root and few branching roots. Branches are formed on the stem and considerable foliage is produced. The plant has much water when green, and becomes woody when cured. Owing to its growth on poor soils, the crop is often used for green manuring. It is often seeded with rye in July, and after harvest the rye continues to grow, producing grain or a plow-down crop the next spring. Crimson clover is

Plowing under a crop of Buckwheat. A common use of the plant for soils lacking organic matter.

substituted for rye in such a practice in some parts of the South.

The grain is triangular, gray or black in color, with a pure white, soft endosperm beneath a thick outer hull. It is lower in protein and carbohydrates than wheat and oats, but contains more fat than wheat. The grain is used for human consumption, and as feed for animals, particularly dairy cows.

Buckwheat is often grown by beekeepers, because it has a long season of blossoming and it makes an especially prized honey.

Culture.—This plant is adapted to moist, cool climates, and is quickly affected by dry, hot weather. It is the shortest-season crop, maturing in 10 weeks or less. Hot weather blasts the flowers; that is, no seed forms. A sandy, well-drained soil is best, although the crop does well on poor, acid, undrained soils if climate is favorable. Ordinarily if any fertilizers are to be applied, phosphorus should be the largest component.

The seed bed should be prepared early and thoroughly. The best time to seed is June 15 to July 15; the rate is usually 3 to 4 pecks per acre, broadcasting being a common method of seeding. Blossoms form from the time the plants are quite small until frost comes.

Harvesting is done when the first seeds are fully mature, or about the latter part of September. Ordinarily the crop is cut just before killing frost. A self-drop reaper, which leaves the green plants in small bunches, is preferred to harvest the crop. Buckwheat threshes easily, even when slightly damp, and is seldom stacked or hauled into the barn.

CHAPTER XIV.

COTTON.

Cotton is grown only in warm countries, its present production being centered in southern United States, which produces about three-fourths of the world's crop. Nearly 15 million bales are produced annually in this country, almost entirely in Texas, Georgia, Alabama, South Carolina, Mississippi, Oklahoma, Arkansas, North Carolina, Louisiana and Tennessee.

This plant was cultivated 1500 B. C. in India, and long before the Christian era the fiber was used by the Egyptians and Greeks for spinning and weaving. Later Spain and Turkey cultivated it, and it spread gradually to other European countries. Columbus found it in the West Indies, but it seems not to have been grown in the United States before Virginia colonists planted seed they brought from Europe. By 1750 it was quite generally grown in the southernmost states.

Culture.

Climate.—Cotton requires at least 180 warm days from the last spring frost to the earliest killing frost in the fall. High temperatures (70° to 75°), especially until August, and light frequent rains are ideal for its full development. Such conditions are not found north of the 37th parallel.

Soils.—Sandy loam soils are considered the best type for cotton, because they are easily drained, warm up earlier and are more easily cultivated. Clay loams are also well suited, and both types are more certain, year after year, than bottomland or heavy clay soils. Cotton does not require as fertile a soil as corn; in fact, an extremely fertile soil tends to produce heavy vegetative growth at the expense of fruit.

Preparation of Soil.—Land to be put in cotton should be plowed during the fall or winter months, unless the soil is subject to blowing or washing. If cotton precedes, the stalks should be broken down with a drag or by a special stalk cutter; they should not be burned, as they add a large amount of valuable organic matter to the soil.

Cover crops are frequently grown in the rotation before cotton, and in such a case plowing must be delayed until early spring. Level plowing not more than 8 to 10 inches deep is considered best, depending upon the type of soil. Clay soils, or those in poor physical condition, should be plowed deeper than sandy and loam soils. Fall-plowed land should be left rough during the winter, but should be disked and harrowed 2 to 3 weeks before planting. Spring-plowed land should be broken up immediately after plowing, the amount of tillage depending largely upon the type of soil, but a good firm seed bed is necessary for the best crops.

Field of Cotton.

In most sections cotton is grown on ridges several inches high and 3½ to 4 feet apart. Level culture is now coming rapidly into greater favor, especially on well-drained soils, because of the smaller amount of labor in preparing the field and because of the greater convenience in cultivation.

Fertilization.—Because it is a delicate feeder, and has a shallow rooting system like wheat, cotton requires only small amounts of readily available fertilizers. For nearly all sections phosphorus is needed most and yields greatest returns of all the fertility elements. Applications of acid phosphate hasten maturity and increase yields, and often will be the only fertilizer needed. The use of nitrogen is doubtful for most cotton soils, and this costly element should be used wisely. At the North Carolina Station the use of phosphorus and potassium returned about the same profits as the use of all three elements of plant food. The best systems of cotton farming provide for the nitrogen supply by growing legumes in the rotation. Potash is also little used in fertilizing cotton, except for soils particularly deficient in this element. From 200 to 800 pounds of fertilizer per acre may be applied, depending upon the kind and treatment of the soil. No noticeable results have been obtained from applications of lime to cotton, but succeeding crops may be benefited by such treatment.

Planting.—Early planting is always advisable, the time depending upon the latitude and season. The soil should be warm and danger from severe frosts past. In the southernmost sections April 1 may not be too early and in the northern part of the cotton area, April 15, the planting extending to May 1 to 15.

Ordinarily 3 to 4 pecks will be sufficient to plant an acre, but plenty of seed should be sown to insure a full stand. Many seeds fail to germinate and develop vigorously, and thinning 3 or 4 weeks later may be resorted to in order that extra plants may be removed.

The seed should be planted about an inch deep, and plants thinned to 15 to 20 inches apart in the row. Unlike corn, cotton requires more space on rich land, in order to allow lateral growth of its branches; and on poor land cluster types require less space.

Cultivation.—Cultivation should begin before the cotton is up, and should be repeated frequently (every 7 to 10 days) until the fruit is setting, and then later occasionally until the bolls are well formed. Weeds should be kept down and a fine mulch preserved; cultivating after each rain is important. Ordinary cultivating no deeper than 2 inches is preferable unless a hard crust has formed. A weeder and later a two-horse cultivator will be most satisfactory for this purpose.

Harvesting.

Cotton picking begins in August or September and often extends into the winter, but usually the crop is harvested before December. One person can pick from 100 to 500 pounds a day.

The seed is removed from the lint, which is then pressed into bales of about 500 pounds in weight for shipment. Most of the crop is sold locally,

Cotton Bolls.

Cotton Fibers.

but much of it is consigned to New England, Canada and Europe. About half the crop in average years has been exported.

By-products.

· Seeds are composed of about half hulls and half kernels. The separated kernels when heated and pressed form cottonseed oil and cottonseed cake. When ground, the cottonseed cake yields cottonseed meal, used mainly as a feed for live stock, but also for fertilizers, especially in the South. Cottonseed meal contains more than 40 per cent protein and is valued among the best concentrated feeding stuffs for cattle and sheep.

Cottonseed oil is used for salad and cooking oils, for preserving canned goods, as a lubricant, and as an ingredient of oleomargarine, soaps and paints. The cottonseed hulls are also used as feed for cattle, but are inferior to the meal. They are also used in paper manufacture.

CHAPTER XV.

TOBACCO.

Distribution.—Along with corn and white potatoes tobacco orginated in America. It was cultivated by the Indians and was found growing in various sections by the early explorers. While the acreage and production are not

An individual Tobacco Plant.

large, tobacco is one of the main money crops of the country. The production exceeds a billion pounds annually in the United States, which is by far the greatest production of all countries in the world. The leading tobacco states, according to acreage, are Kentucky, North Carolina, Virginia, Ohio, Tennessee, Wisconsin, Pennsylvania, South Carolina, Maryland, Indiana, West Virginia and Connecticut.

Varieties.

The U. S. Department of Agriculture has divided tobaccos into the following classes:

Cigar-wrapper Tobacco.—The Sumatra is grown in western Florida, southern Georgia and the Connecticut Valley, where sandy loam soil prevails. The Connecticut Broadleaf is adapted to sandy loam soils found in the Connecticut Valley, New Hampshire, Vermont, Pennsylvania, Ohio, Wisconsin, Minnesota and parts of Illinois and Indiana. The Connecticut Havana, one of the best general-purpose tobaccos, is adapted to light sandy soils and is grown in most of the states where the preceding type is adapted.

Cigar-filler Tobaccos.—The Cuban tobacco, adapted to sandy soils on clay subsoils, is grown in Florida, Georgia, Texas and Ohio. In the North, however, a large leaf is produced at the expense of fineness of texture and quality. The Zimmer Spanish is the most extensively grown domestic filler, and is found mainly in Ohio and Wisconsin. The Little Dutch is also grown in Ohio, along with parts of Pennsylvania, but it is a small yielder and requires much care in curing.

Pipe Tobaccos.—The North Carolina Bright Yellow occurs on the sandy soils of Virginia, Maryland and the Carolinas. Maryland Smoking tobacco is grown in Maryland, Pennsylvania and Virginia on clay loam and sandy soils.

Plug Tobaccos.—White Burley is the variety used to produce air-cured tobacco in the bluegrass region of Kentucky, Tennessee, Virginia and southern Ohio. The Orinoco and Yellow Mammoth are adapted to rich, well-drained soils, occurring in Virginia, North Carolina, Tennessee, West Virginia and Missouri. These varieties are used principally in producing dark fire-cured tobacco, or Virginia sun-cured tobacco. The Virginia types do best on sandy soil having clay subsoil.

Climate and Soil.

Suitable Soils.—Climate and soil determine the kind of tobacco best adapted to a given locality. The climate influences the quality and aroma, while the soil affects the texture of the leaves. In the extreme South, a more aromatic tobacco is grown than in the northern section, while the leaf is smaller and coarser. The best cigar fillers come from tropical climates, and cigar wrappers from tobacco grown in temperate climates. Extensive rainfall causes a thin leaf and lack of aroma.

Tobacco grows on a great variety of soils, but is easily affected by the chemical and mechanical conditions of the soil. A soil is best adapted to the growing of tobacco that is fertile and contains an abundance of organic matter, making it loose and mellow. A sod, preferably of a legume, is considered essential for the best quality and yield of tobacco.

Affect of Soil on Tobacco.—As already mentioned, the type of soil influences the quality of tobacco. Fine texture and pleasing flavor in tobacco are secured when the crop is grown on light, well-drained soils, fairly rich in humus; while the richer, heavier soils yield tobacco that is larger and coarser. The types mentioned in the classification of varieties owe their differences mainly to the soils upon which they grow.

Affect of Tobacco on Soil.—Tobacco is a heavy feeder of available plant food and draws more heavily upon the soil than most other crops do. A 1,000-pound crop, including the stalks, would contain about 46 pounds of nitrogen, 3½ pounds of phosphorus and 29 pounds of potassium. For these reasons rotations must be carefully planned which will supply organic matter and an abundance of nitrogen through the agency of leguminous crops; manure must also be carefully conserved where it is produced; and liberal application of fertilizers are necessary for maximum and most profitable crops.

Culture.

Place in the Rotation.—With the increased production of tobacco and the gradual depletion of the soil, single-crop tobacco farming has largely given way to a diversified system. In the rotation today, tobacco usually follows a sod of bluegrass or other grass or of clover, and is followed by a fall-sown cereal, like wheat or rye. Where tobacco is grown two years in succession on the same land, a catch crop, like hairy vetch, or rye and vetch, is commonly grown in order to keep up the nitrogen and humus content of the soil.

Care of Plant Beds.—Tobacco seed must first be sown in early spring in plant beds, and then when the plants are about 5 inches high they are transplanted into the field. The plant bed selected for this purpose should have a sunny southern exposure, and the soil must be warm, unusually fertile and well drained. Virgin soil chosen for such a bed may require no extra fertilization. Otherwise, manure and fertilizer must be used liberally, at the rate of 10 to 20 tons per acre of the former and 1 to 2 tons of the latter.

The seed bed should be surrounded by board walls and covered with either tent cloth or glass as protection. By having a permanent location for such beds, their value may increase from year to year, provided weeds and diseases do not become prevalent. About 200 square feet is sufficient space to grow plants for an acre. Plow or spade the ground 4 or 5 inches deep and have it finely pulverized. A liberal application of fertilizer should be applied over the bed a week or two before sowing the seed. Sow the tobacco seed at the rate of 1 to 2 tablespoonfuls per 100 square yards. In order to get an even distribution of seed, it should be mixed with wood ashes, fertilizer, cornmeal or land plaster. For the southern states, the time of seeding varies from February 1 to March 30, depending upon the locality, while in the northern states, as in Wisconsin and Minnesota, the latter part of April is more suited. After seeding the bed should be rolled, or a good plan is to cover the seed with a common garden rake.

The seed bed should be watched carefully with regard to watering, the killing of weeds and preventing attacks by insects. Water should be applied by a light spray, once or twice a day during early growth, in order that the bed may be kept moist at all times. The soil should be wet before weeds are pulled out. If the plants do not thrive and appear yellow, more fertilizer should be added. A little nitrate of soda dissolved in water may be beneficial. The temperature in the bed should not fall below 70° at night,

Tobacco beds steamed to prevent Root Rot. Small bed at corner to right was not steamed.

nor rise above 100° in the day. If the plants are too thick in the bed, thinning should be resorted to so that spindling plants do not develop. For about a week or 10 days before transplanting, remove the sash from the bed so that the plants will become hardy.

Fall steaming of plant beds is recommended to prevent root rot, while weeds and insects are incidentally killed by the process. This advantage alone is often sufficient to pay for the treatment. Where steaming is not available drenching with 40 percent formaldehyde at the rate of 4 pounds to 50 gallons of water applied over the surface will generally prevent root rot.

Preparing the Field.—On soils not subject to erosion or where a cover crop is not grown preceding tobacco, fall plowing is advisable. If manure is available, it should be applied at the rate of 8 to 15 tons per acre. The seed bed should be prepared thoroughly before transplanting. A disk harrow is often necessary to get the land in good shape, especially on clay soils. If a catch crop is grown plowing should be done as early as possible in the spring.

Transplanting.—When the plants are about 5 inches high those showing the most vigor and best-shaped leaves should be selected for transplanting. Before they are taken out the seed bed should be thoroughly moistened and care exercised in lifting out the young plants. Likewise, care must be used in the field, so that the plant is not injured if transplanting is done by hand.

Machine planting is common and profitable on large acreages. Three persons and a team can plant about 3 acres a day. These machines automatically

A field of Tobacco, mainly hybrid varieties. Tops of some are enclosed
in paper bags for seed.

water the plants after transplanting. When planting is done by hand, the soil should be watered at the same time, and it is best to transplant after a rain.

The distance to plant depends upon the variety and character of soil. The fiber of the tobacco plant is coarser if the plants are set far apart. The plants should be set farther apart on thin soils. Connecticut Havana tobacco is grown from 14 to 20 inches apart in rows about 3½ feet apart. The Connecticut Broadleaf is spaced 20 to 24 inches, while the Cuban variety is spaced 14 inches in the row.

Fertilization.—From the composition of the tobacco plant, it will be seen that it is not a heavy user of phosphorus, but since this element is commonly lacking in soils applications of it usually pay. Tobacco is a heavy user of potassium, and more of this is taken from the soil than by most other crops; the sulphate should be used instead of muriate of potash. Nitrogen influences the yield more than either of the other elements, and it must be well supplied to the crop. Barnyard manure tends to produce a coarser tobacco, but wherever

Transplanting Tobacco.

available it should be plowed under for this crop. Where this is done, or where legumes are grown in the rotation, the need for commercial forms of nitrogen will be lessened. Lime benefits the entire rotation, particularly the legumes, but usually has little effect on tobacco, although the Virginia Station has found that by it the yield may be increased slightly and the color of the tobacco darkened. Many farmers add 400 pounds of a fertilizer analyzing 3-8-3, which has about sufficient plant food in it to supply a 250-pound tobacco crop. Heavy applications pay with this crop if the season is normal. In Ohio 1,100 pounds per acre of a 4-9-8 fertilizer has been most profitable, while the Virginia Station found that applications of 1,400 pounds of a 3-8-3 fertilizer paid $19.58 more than 800 pounds per acre.

Cultivation.—It is particularly important that all weeds be kept down during the early stages of growth in the field and that a mulch be maintained. Weeds should be hoed out, and shallow cultivation at frequent intervals until the plants are too large should be practiced.

Topping.—When the plants begin to head all except those kept for seed should be topped. The aim should be to have all plants mature at the same time; therefore, the grower must judge the individual plant and the soil in which it grows to determine how many leaves to cut off. The top should be broken off below the first seed sucker, but the height of topping varies with the variety, climate and soil. Removing suckers may be necessary two or three times during the season, so that the growth and quality of the leaves are not interfered with by them.

Harvesting.

Cutting the Crop.—The time of harvesting depends upon the season, but it can usually be started when irregular, light yellow patches appear on the leaves and the leaves commence to crumple and thicken. The plants are usually cut with a knife or hatchet, and allowed to remain in the field until they

A Tobacco Shed for storing and curing the crop.

wilt. A common method in Kentucky and other states of that section is to split the stalk to within 6 inches of the ground and place it astride a stick. Another method in the same section is "spudding," by which the stalks are forced on a stick; such a method does not permit as rapid curing of the plants.

The tobacco plants are either hauled directly to the barn or left on scaffolds in the field for a few days to cure. Special wagons are most satisfactory to haul from the field.

Curing.—Tobacco sheds should be closed and have windows and ventilators. These should be opened in the morning and closed at night or on rainy days during the curing process, which usually takes 4 to 6 weeks. Too rapid curing lowers the quality of the tobacco. In damp weather fires should be built in the shed, coke, charcoal or pine wood being used, and some hardwood in case a different flavor is to be imparted to the leaves. The midrib should be cured before the tobacco is taken to the packing house.

Seed Selection.—No farm crop is better adapted to improvement by selection and breeding than tobacco. Growers in testing new varieties should sow only a small amount of the seed at first, retaining the rest of the seed, as it keeps its vitality for at least 5 years with proper care. If the new variety is satisfactory more can be seeded the succeeding year.

In obtaining seed select the best plants in the field; that is, with respect to size, color, maturity, resistance to disease, etc. Put a 12-pound paper bag over the flower to prevent cross-fertilization. For the first week or two raise the bag a little every few days to prevent crowding by rapid growth. When the seed pods have turned brown, cut off the top of the plant; pick out the large, heavy seed pods, and then after replacing the bag store in a shed or attic where air has free circulation.

A seed separator should be used to remove light seed, hulls and chaff. The Kentucky Station found about 10 days difference in maturity by selecting large seeds, and various tests have shown the yields to be better from the heavier seed.

CHAPTER XVI.

SUGAR CROPS.

Sugar Beets.

Production.—About one-half the world's sugar comes from sugar beets, of which about one-sixth are grown in the United States. Colorado, California and Michigan lead in the acreage devoted to this crop. Sugar beets do best in cool climates. Mangels also belong to the beet family, their chief use being for live stock feeding. The two types require practically the same culture; although in harvesting, mangels, which grow about one-half out of ground, can be pulled by hand, but sugar beets, being deeper in the soil, are usually pulled with a beet puller.

Sugar Beets and Soil Fertility.—Sugar beets cause an unusual drain upon the fertility of the soil. A 15-ton crop removes more nitrogen and phos-

The Sugar Beet.

phorus than 65 bushels of corn or more than both 25 bushels of wheat and 50 bushels of oats. Such a crop likewise removes four times as much potassium as 65 bushels of corn, 25 bushels of wheat and 50 bushels of oats combined. For these reasons, sugar beets require land in a high state of fertility for the best crops, and careful attention must be given to the maintenance of this high condition.

The sugar beet calls for large amounts of potash, as do all crops which store large quantities of starch and sugar. While phosphorus is not called for in large quantities, yet the tendencies of older farming have been to deplete the soils of this element. Considerable available nitrogen is also essential for this crop. Large applications (400 to 700 pounds per acre) of fertilizers are usually most profitable.

Culture.—For best results beets require well-drained, deep, rich loam,

141

although they may be grown in various soils. Deep plowing is advisable, and fall-plowing is recommended where conditions permit. Manure, if applied, should be used the fall before planting to be plowed under. The ground should be put in the best of tilth, much the same as for corn.

The beets or mangels may be planted a little before corn planting time, April or May. From 6 to 10 pounds per acre should be seeded in rows 28 to 34 inches apart for mangels and 18 to 20 inches apart for sugar beets. Regular beet drills or an ordinary grain drill may be used for planting, the usual depth being about 1 inch.

Cultivation should begin a few days after planting in the use of the weeder, and interculture as soon as the rows can be followed. Special beet cultivators adapted to the narrow rows are most satisfactory. This cultivation should be kept up at frequent intervals until the tops meet in the row. Considerable hand working is usually necessary. When the plants have four leaves they should be thinned to a single plant every 5 inches. Often they are "bunched" with a hoe, small blocks being left the proper distance apart. Soon afterward, the plants should be thinned by hand to one at a place.

Sugar Cane.

Adaptation.—Sugar cane is said to have been first grown in China and India. It was introduced into United States by New Orleans planters and is grown in Louisiana most extensively today. Japanese sugar cane is cultivated as a forage crop in latitudes south of the 33d parallel.

Sugar cane is a gross feeder and requires abundant rainfall, besides a soil well supplied with organic matter and plant-food elements. It occupies the land often 2 or 3 years from a single planting.

Culture.—Seed cane should be allowed to grow as late as possible in the fall and planted either in the fall or early spring. Where there is no danger from winter-killing fall planting is preferable, because there is less rush of work and the canes sprout quicker the following spring. The land should be thoroughly prepared and rows opened 4½ to 5 feet apart, into which the seed canes are dropped and covered 3 to 4 inches deep. A rolling afterward is advisable in dry weather. For spring planting the seed cane should be cut in the fall and covered with moist earth till spring. The Mississippi Experiment Station recommends half lapping the canes in the drills and using about 3,500 stalks per acre. The Louisiana Station has shown that cutting the stalks is inadvisable when they are straight enough to lie flat in the furrow. The top five or six joints may be used for seed and the bottom part of the stalk for grinding. Frequent shallow cultivation is essential during the growing season. Manure used directly on this crop gives a disagreeable taste to the syrup.

Harvesting.—The blades and tops are removed from the stalks in the fall, and the stalks then cut near the ground. This operation is usually done by hand, as harvesters cannot get crooked canes. Yields of 20 tons of plant cane or 15 tons of stubble cane per acre are common on rich land and in good seasons. A ton yields about 20 gallons of syrup.

CHAPTER XVII.

FIELD BEANS.

Production.—Common field beans were cultivated by the American Indians before the discovery of America, and were first probably grown in New York for commercial uses. Their production in United States exceeds 11,-000,000 bushels annually, or an average of about 14 bushels per acre. Michigan is by far the largest producer, and it with California and New York produces 80 percent of the national crop.

Characteristics.—The field bean is a legume. It is rather exacting in moisture and temperature conditions in the soil. Conservation of soil moisture by tillage and early preparation of the seed bed are therefore essential.

The bean is used mainly for human consumption, and is one of the cheapest foods to supply protein. Cull beans are used for feeding livestock, especially sheep and hogs. They are also fed to some extent ground for cattle. They contain, according to Cornell University analyses, 21.6 percent protein, 47.5 percent nitrogen-free extract, 1.2 percent fat and 3.7 percent fiber. This analysis for protein is twice as high as that of corn or oats. Bean straw contains about three times as much protein as oat straw and slightly more of the other components. As a feed for sheep, horses and cattle, it is but slightly inferior to timothy.

Culture of the Crop.

Soil.—Well-drained loam soils of medium fertility are best adapted to field beans. Heavy clay and sandy soils are less desirable, while soils loaded with organic matter produce rank growth of vine and beans of inferior quality. Since this is a leguminous crop, it does best on soils well supplied with lime.

Fertilizers.—Being a legume, the field bean requires little nitrogen, particularly if grown on soils rich in this element or preceding a legume, as is common in such rotations as clover, beans and wheat. In most tests phosphorus in readily available form has been found most profitable. Often from 200 to 400 pounds of acid phosphate per acre will be the only fertilizer needed.

Selection of a Variety.—This group of beans includes the pea bean (often called the navy or soup bean), the kidney bean and the marrow bean. Soil conditions and the locality should be considered in choosing a variety.

Planting.—The ground should be plowed early in the spring and dragged thoroughly before planting time so that a firm, mellow seed bed is obtained. Planting should be done when the soil is warm and all danger of frost past. In New York and states west to Indiana planting may be done in late May or early June, although under favorable conditions beans planted late in June will mature a crop. Experiments in South Dakota and West Virginia showed planting in drills preferable to hill planting.

The beans should be planted 1 or 2 inches deep in drills 28 inches apart, a two-row corn planter or grain drill being used when a special bean planter is not available. With the marrow and small pea bean 2 or 3 pecks will seed an acre, while 4 or 5 pecks of the kidney beans will be needed. More seed is necessary under unfavorable soil conditions.

Cultivation.—The beans should be cultivated frequently to keep down weeds and conserve moisture. Do not cultivate deep, or when foliage is wet with dew or rain, as cultivation then tends to spread diseases.

Harvesting.—The time of harvesting depends on the variety, season, time of planting and soil conditions. The crop should be mature before harvesting, but weather conditions in general largely determine this time. In large bean-growing districts the crop is harvested with horse pullers, which are far less expensive than hand labor. Men should follow this harvester with forks and throw two or three rows into windrows or small piles. Five or six men are needed to keep up with one team. A side-delivery rake will serve the same purpose, taking two rows at a time. This, along with the hay loader, is coming into common use in Michigan and New York with the increasing high prices and scarcity of farm help.

The vines should be turned two or three times in rainy weather so that they dry thoroughly and do not discolor. Do not leave them in rows as left by the harvester. They should not be put into the barn until thoroughly cured, as they heat and color easily if the work is rushed too much. Beans are always considered a most risky crop, because chances for loss in harvesting during rainy seasons are very great.

In order to prevent shelling, the beans should be loaded on the wagon without tramping, or if a hay loader is used the vines should be tramped as little as possible. Slings are an advantage in that there is little or no tramping in unloading.

Except for small acreages the old use of the flail and subsequent cleaning of the seeds in a fanning mill is now being replaced by machine separation. In the large districts special bean hullers are common. By adjusting the conclave and using suitable sieves an ordinary threshing machine may be satisfactory.

CHAPTER XVIII.

POTATOES.

Production.—Irish potatoes constitute the largest of all crops in the United States, the annual production being more than 400,000,000 bushels, or an average of about 110 bushels per acre. This, however, is only about one-sixteenth of the world's production, Europe producing most of the crop.

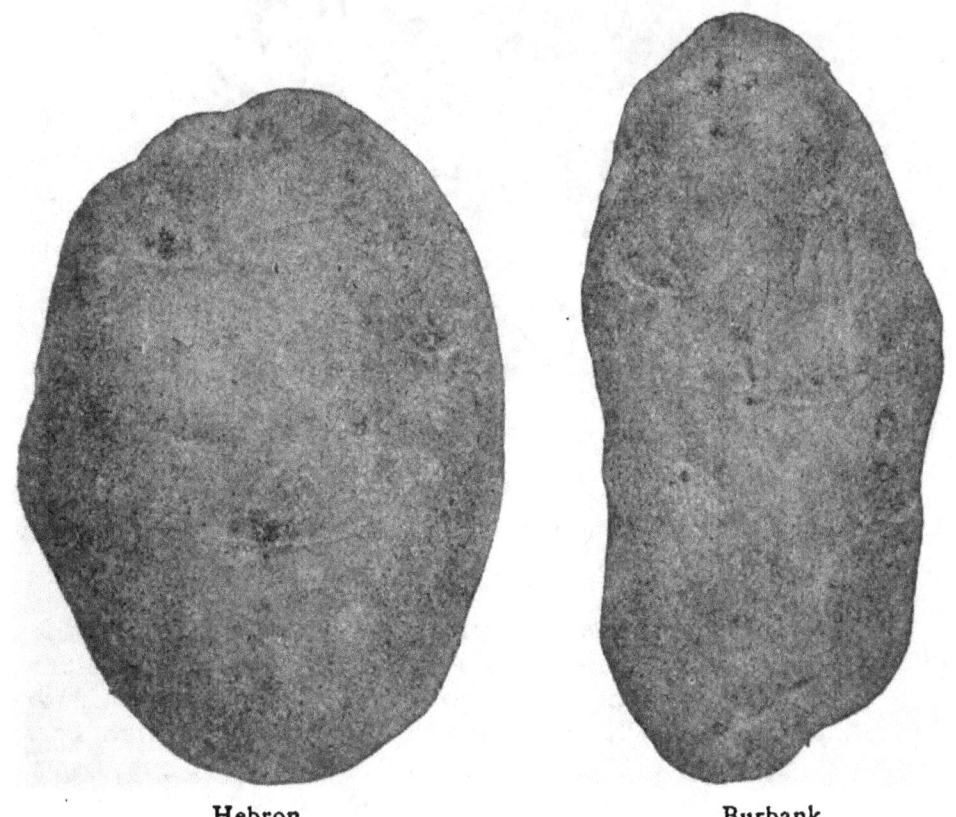

Hebron.   Burbank.

The potato is a native of America, having been cultivated first probably in South America, although Columbus found the plant growing here. From the intensive culture of the crop later in Ireland, it got the name "Irish potato." In this country New York, Michigan, Pennsylvania, Wisconsin, Maine, Minnesota and Ohio are the leading potato-producing states and rank in the order given.

Distribution of Groups.—The U. S. Department of Agriculture has divided potatoes into 11 groups according to shape, color and maturity. In the northeastern states Irish Cobbler and Triumph are grown for the early crop, although the latter group is less productive. For the late crop Rural and Green Mountain types prevail; the former mainly in New York and the latter in cooler and higher elevations of that section.

In the southeastern states early varieties are extensively grown for northern markets. The Rose group, one of the oldest and a pink-colored type, predominates, with Triumph and Cobbler also used.

Rural.

In the north central states Rural and Burbank groups are grown chiefly, but in the corn belt Early Ohio is the most important. Preference is given on the market for the Rural type. Rural, Pearl and Burbank groups are the main types in the West.

The varieties of the type of which the Rural New Yorker No. 2 is perhaps the original, are grown more generally in the best potato sections than others. These are late varieties and are quite well adapted to comparatively late planting. The list includes Carman No. 3, Sir Walter Raleigh, Prosperity, White Giant and others. Of the early varieties Early Ohio and Irish Cobbler are favorites. As far as possible growers should plant varieties that have proved most profitable in the locality and on their type of soil.

Climate and Soil.

Climate.—A cool, moist climate is ideal for potatoes. Scotland and northern Europe are especially suited, while in United States the largest acre yields are found in northern states or those of high altitudes, particularly Maine, Colorado and the northwestern states. In warmer climates they do not thrive as well, and in the South winter planting is resorted to. In warm climates the potatoes degenerate and seed from the North must be secured each year.

Maine, New York, Michigan and Wisconsin are important as seed-growing states.

Soils.—The best potato soils are light, loose loams which carry good percentages of sand and silt, and are well supplied with humus. Some soils

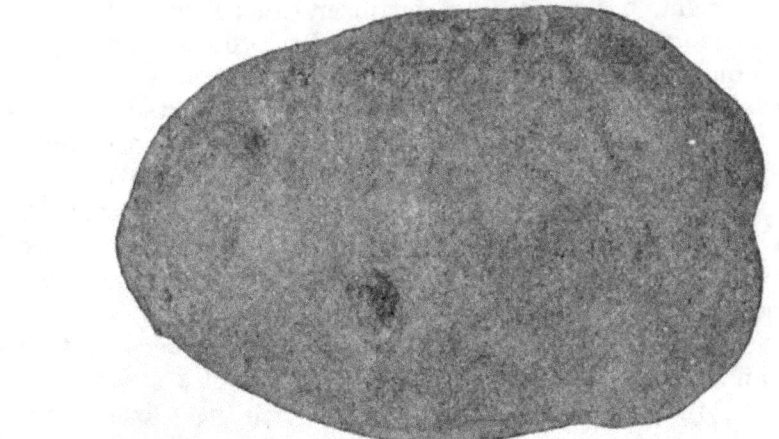

Early Rose.

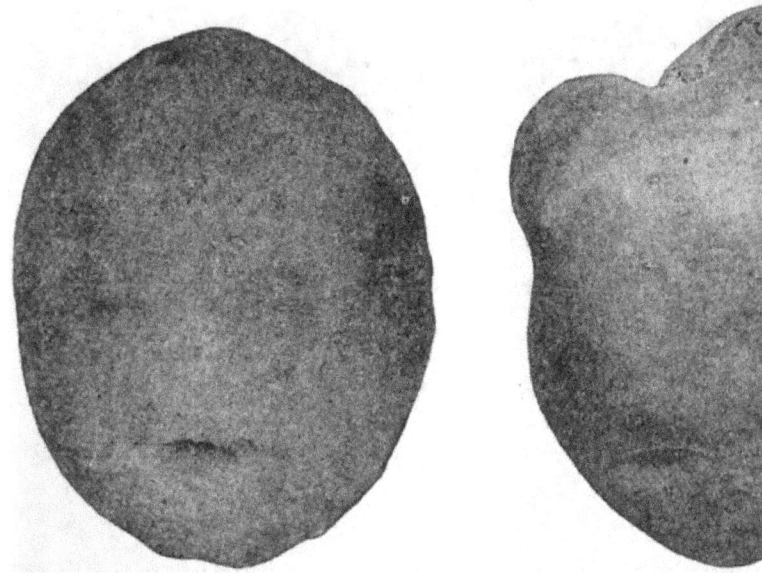

Early Ohio.   Irish Cobbler.

classed as clay loams fill these requirements. Potato soils should be deep, mellow and easily worked. Some of the requirements can be put into soils by good treatment on the part of the farmer, but not all for certain types of soil are more favorable to potato growing than others. New soil usually produces **good** potatoes, and a clover sod is ideal to plow under for this crop.

Manures and Fertilizers.—In the central states manure is extensively used for the potato crop. At the Ohio Experiment Station it has produced larger total yields and larger increases over untreated land than an equivalent amount of fertilizer. The most profitable fertilizer for this section would probably be one analyzing high in phosphorus and potassium and rather low in nitrogen and applied at the rate of 400 to 800 pounds per acre. One analyzing 3-10-7 might be about right, but no definite fertilizer can be given for all soils. If clover precedes potatoes nitrogen might be dropped out entirely. Sulphate is preferred to muriate of potash.

In the eastern states fertilizers are used far more extensively, as manure is often not available and potatoes are frequently grown 2 years in succession on the same land. On the sandy soils of this section a fertilizer analyzing about 3-10-10 is common. Here applications of 1,000 to 2,000 pounds per acre are common. In the West on new land fertilization has as yet received but slight attention. The amount to apply can best be determined by field tests by each farmer.

Where potatoes are grown in rotation, as in a 3-year rotation of potatoes, wheat and clover, or a 4-year rotation of potatoes, soy beans, wheat and clover, or corn, potatoes, wheat and clover, the fertilizer should be distributed all through the soil in advance of planting. The New York (Gen-

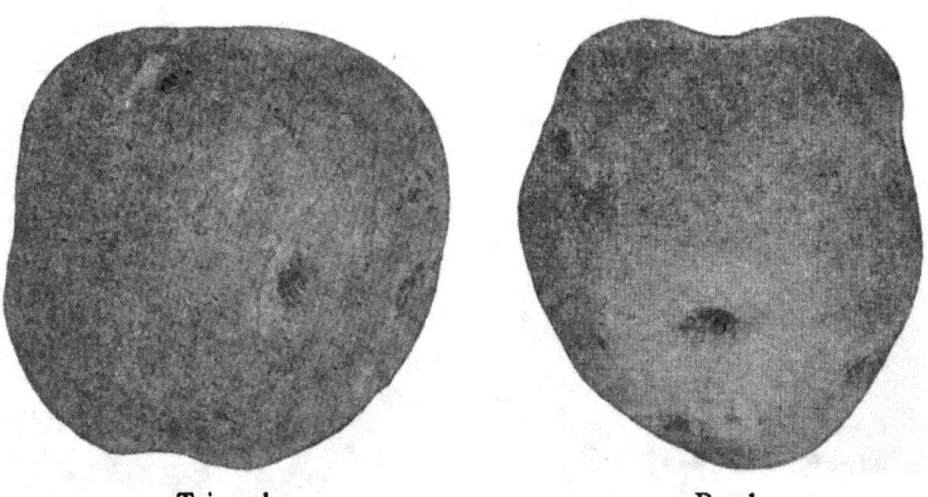

Triumph.   Pearl.

eva) Station found a slight gain in potatoes from drilling fertilizer in the row, but in a rotation, succeeding crops should be considered, as they also are benefited by the fertilizer.

A catch crop of rye is possible in the second 4-year rotation above, or in the soy bean rotation when this crop preceded potatoes. Rye is suitable to plow under for potatoes, but excessive spring growth is undesirable as it may cut off water from the surface soil when turned under.

Lime.—With such rotations liming is necessary for acid soils, and when potatoes are grown only every third or fourth year they are not injured by lime, if it is applied following potatoes rather than just before. However, diseases, especially scab, become common on potatoes grown continuously on limed soil. The best potatoes from the eastern states, where definite rotations are not practiced, are grown on acid soils.

Culture.

Handling Seed.—Tests made at several experiment stations indicate that about 12 to 20 bushels of seed cut in halves to plant an acre give most profitable returns, although some find whole potatoes more profitable than halves. The time to cut seed potatoes is just before the planter starts. The work of some experiment stations shows that gains of several bushels per

Green Mountain.

acre have been secured by deferring cutting until planting rather than 4 to 6 days before. In case potatoes must be cut a few days before planting some air-slaked lime or land plaster should be sprinkled over them, as it seems to dry the wet surfaces; and the potatoes should be kept from heating. Numerous experiments have been conducted to determine whether there is any difference in value of the seed end and the stem end of the potato for seed. The two ends may be said to have substantially the same value.

The manner of growth and care of seed have more to do with its value than has its source. Southern growers will find it profitable to secure new seed each year from northern states, although they may grow a second crop, planted in August and harvested after frost. Such immature seed is often poor in germinating the next season.

When planting is deferred till late in May or June, it is difficult to keep seed potatoes in good condition. It is undoubtedly an advantage to sprout the seed in the sun. A few weeks before planting, and before the potatoes have started to sprout to any extent in the cellar or pit, they should be spread out in a thin layer on a barn floor or placed in racks arranged one above another, where the sun will shine on them a portion of the day. The pota-

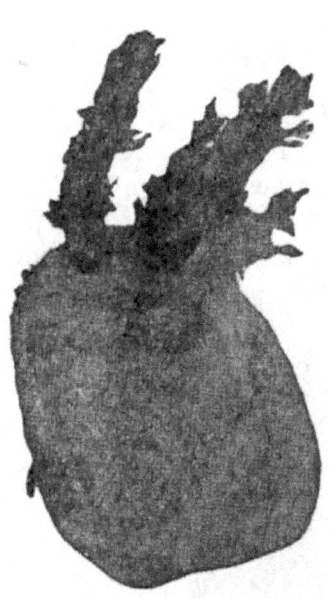

Potato Sun Sprouted Seed.

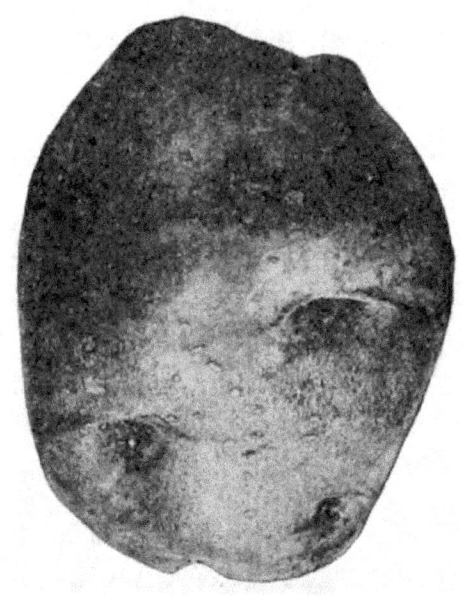

Peachblow.
Examples of Different Varieties and Types of Potatoes.

Early Michigan.

toes will become green and hard, and develop a stubby, green sprout that will start off with a rush when planted. At Cornell University gains of 19 to 37 percent in yield were reported from sprouting 36 days; at Rhode Island Station 32 to 54 bushels per acre from sprouting 4 to 6 weeks **at a** temperature of 60° to 75° F.

Potato scab and potato rosette are held in check by formaldehyde (formalin) treatment. This consists in soaking the seed for 2 hours in a solution of 1 pint of formalin to 30 gallons of water or for 1½ hours in a solution of 1 ounce of corrosive sublimate in 8 gallons of water. Potatoes should be treated before cutting, and if they are not planted at once they should be spread out and dried. The same solution should not be used more than two or three times. If carefully done this treatment will not injure the germination of the seed.

Method of Sun Sprouting Seed Potatoes.

Preparing the Seed Bed.—A sod, whether manured or not, should be plowed in winter or early spring. Early plowing is also essential if a catch crop of rye precedes potatoes. A seed bed for corn will be suitable for potatoes.

Planting.—Potato rows should be 30 to 33 inches apart, with seed dropped 12 to 14 inches apart in the row. The planting should be 3 to 5 inches deep below the level surface for best yields and highest quality. Planting less than 3 inches is likely to result in crowded, exposed and consequently sunburned tubers and a large proportion of small ones. When level culture is practiced a little deeper planting is necessary to prevent sunburning than when potatoes are ridged somewhat.

The time of planting varies with the variety and locality. Early planting is necessary in the corn belt and southern states in order to get much growth before hot weather comes, and most of the potatoes are planted in May. In New England the main crop is planted about June 1; in Florida, from December 15 to March 15 for shipping to northern states, and in August for home use.

Cultivation.—Cultivation of the potato crop should begin a few days after planting in the use of a smoothing harrow. Fields should be harrowed both ways, two or three times before the potatoes come up. Before the planting rows are lost sight of it will be an advantage to cultivate deeply between the rows to loosen the ground packed by the planter wheels and the tramping of horses.

Keep the cultivator going at intervals of a week to 10 days until the vines fill the rows. The weeder can be used to advantage to keep the rows clean. Comparatively level cultivation should be given. Hilling or ridging is objectionable in that it increases the temperature of the soil, causes the land to lose more moisture and permits more water to run away.

Harvesting.—Much of the potato crop is still dug by hand, but horse diggers are more common in the large sections. In storage potatoes should be kept dry, cool and well ventilated. During the first 6 months of storage potatoes shrink 6 to 8 percent, and often 4 percent a month under cellar conditions in the spring. Montgomery, of New York, recommends that seed stock be kept at temperature of 32° to 40° F., and for table use, at 40° to 50° F.

CHAPTER XIX.

ALFALFA.

Distribution.—The original home of alfalfa is thought to be in southwestern Asia, and from that continent it was introduced into Mexico and South America. From the latter country it was brought to southern California in 1854.

Its culture spread eastward, and the crop was found adapted to irrigated and dry-farming areas. In Kansas and Nebraska it found a congenial home and here are the largest acreages in United States today. Colorado, California, Idaho, Utah, Montana, Oklahoma and Wyoming follow in the order given. In the eastern states, in competition with clover and timothy, it has not been so extensively grown.

Hot, dry climates are most favorable for alfalfa. North of a line from South Dakota to central New York most varieties winter-kill, a few being able to stand the rigor of cold weather.

Culture of the Crop.

Soil Requirements.—Alfalfa first of all requires a well-drained soil. Soils not having natural underdrainage—that is, porous subsoil—should be tiled. While good surface drainage is helpful, trouble may be expected with a heavy, retentive subsoil unless tile drainage is resorted to. Alfalfa does not thrive where the water level is permanently within 3 feet or less of the surface. It will stand flood water for considerable time during the dormant season, unless the water freezes, but during the growing season flooding for 2 days or more is disastrous.

Furthermore, no farm crop is more sensitive to soil acidity. Alfalfa should not be seeded where red clover fails to thrive and sorrel is growing in its place. Uplands usually become acid before bottomlands. As a rule the first thing to do with all soils long in cultivation, unless they are naturally well stocked with carbonate of lime, is to apply lime in some form. Even limestone soils frequently give better results if limed, as the lime in them is gradually leached out of the surface soil and also used by farm crops. At least two tons per acre of limestone is recommended.

While alfalfa, like other legumes, may take care of its own nitrogen problem when once it gets well established with the proper nitrogen-gathering bacteria aiding it, yet a soil rich in all plant-food constituents is an important factor in getting it established. For enriching the soil for alfalfa there is nothing better than a liberal application of stable manure. This should not be applied to the surface soil just preceding the seeding of alfalfa, for the weed seeds in it will cause considerable trouble. If applied the same season as alfalfa is seeded, it should be plowed under. However, manure is best applied to a cultivated crop a year ahead of alfalfa.

In the absence of manure it is well to make use of such leguminous catch crops as field peas, cowpeas, soy beans and the like, to fill the soil with humus-forming material. Organic matter is a highly essential food for alfalfa. With an application of acid phosphate (or steamed bonemeal) and muriate of potash the fertility conditions should be well cared for in any soil, and often only phosphorus is needed to give alfalfa a good start. Alfalfa is an unusually deep-rooted plant and is thereby enabled to draw upon greater depths than

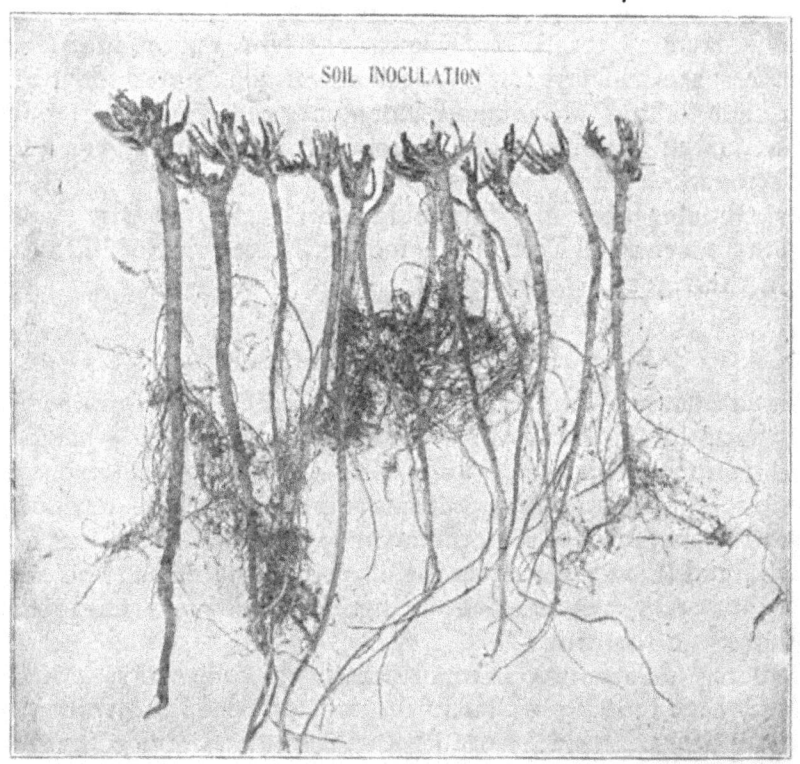

SOIL INOCULATION

Alfalfa Roots.

most crops for phosphorus and potassium, and yet these elements must be supplied at regular intervals if good alfalfa crops are to be grown year after year.

Probably more alfalfa failures are due to competition with weeds than to any other cause. It is a mistake to seed alfalfa in soils filled with weed seeds until this problem is solved. This may be accomplished in large part by growing an intertilled crop, as corn, potatoes, tobacco, etc., for a year or two previous to seeding alfalfa, and giving these crops extremely clean cultivation. A cultivated crop like early potatoes may also be grown the season of seeding, or a crop like thickly seeded field peas, or peas and oats that smothers weeds will serve the same purpose. When the situation is so critical as to justify the practice, the problem can best be solved perhaps by early plowing and giving clean fallow cultivation for 6 to 10 weeks before seeding alfalfa. While this is more expensive, it is pretty sure to result in a good clean stand of alfalfa if other conditions are favorable and in the long run may prove most economical.

Still another requirement for alfalfa is an inoculated soil. Early alfalfa growing was a failure in this country until the ground gradually became inoculated with the micro-organisms that thrive in the nodules on its roots. Healthfulness and vigor of plant and yields are dependent on these nitrogen-fixing bacteria. At the Ohio Station the difference between inoculated and uninoculated plots has been discernible half a mile away, simply by the difference in color of the alfalfa—the inoculated plots being dark green and the others a pale, greenish yellow. Soil from old alfalfa or sweet clover fields is best to inoculate new fields.

Preparing for Alfalfa.—It must be remembered that alfalfa is no ordinary crop, but requires special care and lives for 4 years or longer if successful. Hence, mistakes made with this crop mean more than with those planted each year. One should plan for alfalfa at least a year in advance, filling the soil with organic matter and plant food and cleaning it of weeds. It is not possible to plow and prepare a good seed bed for alfalfa in a week's time, as is often done with other crops. Sufficient time should be taken to prepare a firm, fine, moist, clean seed bed. For spring seeding, as in western Kansas, fall plowing is advisable.

In order to avoid losing the use of the land the season alfalfa is seeded, a number of different crops have been grown in advance of seeding, among them early potatoes, oats, soybeans (for hay), field peas and wheat. In potato sections there is no better crop to precede alfalfa than early potatoes. This crop may be out of the way for late July or early August seeding, and when it has been kept clean a good seed bed is speedily available. Next to potatoes likely stand field peas, as they are taken off early and leave a clean seed bed. Neither potato nor pea stubble ground should be plowed for alfalfa. Oats are almost too late for sure seeding of alfalfa unless planted early and cut for hay. Wheat, while earlier, is more or less uncertain, much depending on moisture conditions. Alfalfa should never be preceded by a bluegrass sod, for the bluegrass will crowd out the alfalfa.

Nurse Crop vs. Seeding Alone.—Under some conditions light seedings of oats and barley have been successful. If alfalfa is seeded in early spring (April or May) a nurse crop is better than seeding alone, but it should be planted more thinly than when it is grown alone. Oats cut for hay may be satisfactory, but barley is usually preferable because it takes less moisture from the soil. The nurse crop should always be cut early in dry seasons, giving the alfalfa a better chance to grow. Mid- and late-summer seedings should be made alone, a nurse crop being unnecessary and injurious. Seeding with corn at the last cultivation is occasionally successful but always uncertain.

Time of Seeding.—Midsummer seeding (July 15 to August 15) without a nurse crop, upon land that has been thoroughly prepared, probably furnishes the most favorable conditions for securing a stand of alfalfa in humid sections. This preparation may consist of 4 to 8 weeks of fallow cultivation, or of a preparatory crop, as early potatoes or field peas, which is out of the way in good season.   In western Kansas and the intermountain states, spring seedings have been most satisfactory, because moisture is less abundant in the late summer or fall and fall seedings winter-kill.   Still, even in this section under favorable conditions of rainfall, midsummer seedings are now becoming more common. Under irrigation in the Northwest April seeding is advised by the Montana Experiment Station, and also early spring seeding after a year of fallow under dry-farming conditions.

Seed Selection.—Alfalfa varies widely in hardiness.   Seed from the South or that grown where irrigation is necessary generally cannot withstand northern climates.   Northern- or northwestern-grown seed is uniformly more satisfactory in the corn belt.   One should not buy alfalfa seed without knowing its source, and first examining a sample for adulteration and impurities.   Yellow trefoil, sweet clover and bur clover are sometimes used as adulterants, while dodder, bind weed, Russian thistle and narrow-leafed plantain seeds are also to be guarded against.   The Kansas Station has found home-grown standard varieties more satisfactory than new varieties from other sections.   Home-grown seed is preferred to imported if other conditions are equal.   In states as far north as Wisconsin better profits are usually obtained by the use of seed from western states, as Kansas and Nebraska.

Varieties.—Since about 1900, various varieties and strains of alfalfa have been developed to meet different climatic conditions, some being adapted to the cold North and Northwest, while others succeed in the warm South and Southwest.   The U. S. Department of Agriculture has divided alfalfa into five groups, namely:

(1)   Common or ordinary smooth, purple-flowered alfalfa, including numerous regional strains generally grown for seed throughout the West.   This is the most common group, except in the extreme northern states, Canada and the states along the Gulf of Mexico and the Pacific.   It is now being recognized by the states or conditions under which it grows, as Montana-grown, Kansas-grown, irrigated, nonirrigated, dry-land alfalfa, etc.

(2   Yellow-flowered alfalfa, often referred to as Siberian, of itself is not profitable.

(3)   Turkestan varieties are shorter and more spreading in growth than common alfalfa and are almost always inferior in yield and hardiness.

(4)   The variegated group includes Grimm, Baltic, Canadian variegated and sand lucern, all, with the exception of the last, being more resistant to winter conditions than are ordinary strains.   They are grown in New England and states of that latitude west to the Rocky Mountains.   Where common alfalfa will succeed the U. S. Department recommends it above variegated alfalfa because of greater yields and lower cost of seed.   The sand lucern cannot compete with others of the group.

(5)   The non-hardy group is adapted to southern climates where a long growing season is possible. The Peruvian and Arabian varieties are most common.   The Peruvian is unusually hairy and rapid growing in mild climates,

as in the Gulf and Pacific Coast states, and more particularly in Arizona, California and New Mexico. The Arabian is undesirable because short lived.

Since alfalfa produces little and inferior seed in the eastern states, it is useless to try to grow it there for this purpose.

Manner of Seeding.—It makes little difference how alfalfa seed is sown so long as it is evenly distributed and well covered. The seed may be distributed through the grass-seeding attachment of an ordinary grain drill, the seed dropping in front of the hoes, which then cover it. This method is

A special Alfalfa and Clover Drill. Here used to seed clover on wheat in the Spring.

usually better than drilling the seed—that is, distributing it through the drill hoes the same as wheat—for there is danger of covering it too deep. The comparatively new clover and alfalfa disk drill is an excellent implement for seeding. Alfalfa may also be seeded with a common grass seeder, or broadcasted by hand, but the last-named method requires more seed. Where it may possibly be seeded unevenly, it is well to divide the seed and cross-seed. Except when covered with drill hoes, the seed should be harrowed in.

Depth of Seeding.—The depth to seed alfalfa depends on soil and moisture conditions. It should not be covered too deeply, one inch usually being sufficient. Heavy soils require more shallow seeding than loose sandy soils.

Rate of Seeding.—It has been found at the Ohio and Wisconsin Stations that for ground plowed about 2 months previous to seeding and harrowed frequently in this time 10 to 15 pounds of seed per acre is sufficient, but for the average soils, 15 to 20 pounds will insure more satisfactory stands. Seeding more than 20 pounds per acre has not been found desirable. Weeds are not held in check by seeding at the rate of 5 pounds per acre in humid regions. In the semiarid sections of the West, particularly in states west of Kansas, 8 to 12 pounds per acre will be ample, and even less on uplands. Well-prepared seed beds do not demand as much seed as those poorly prepared, and localities of slight rainfall demand less seed than those of abundant rainfall.

Cultivating Alfalfa.—Weedy alfalfa fields may be cleaned up effectively by the use of the disk or spring-tooth harrow. The latter implement is to be preferred. By using it both ways of the field all weeds and grasses may be exterminated with little or no injury to the alfalfa. Being deeper rooted, alfalfa is not injured by such cultivation as will uproot its competitors. Alfalfa should not be given severe cultivation until a year or two old, and such cultivations should be made immediately after a crop has been harvested. It will make little difference which cutting is followed if weather conditions are favorable. It is useless to try to kill weeds during wet weather.

Alfalfa is often grown in rows, especially for seed, in western states, and cultivated. It thus succeeds under somewhat drier conditions than when seeded solid and responds favorably to cultivation. This method is usually best adapted to alfalfa culture on the uplands of semiarid regions.

Clipping Young Alfalfa.—Spring seedings of alfalfa will likely need one clipping; midsummer seedings may or may not. Alfalfa should not be clipped much later than the first of September, or in an unfavorable season it is not likely to go into the winter with sufficient growth for protection. It should have 6 to 8 inches of growth for winter protection.

There is little reason for clipping midsummer or late seedings of alfalfa the season of seeding, unless disease is prevalent.

Harvesting.

Time of Cutting.—The time of cutting varies under different conditions. If one waits until a certain percentage of the bloom appears, he may sometimes wait in vain, finally realizing that his alfalfa is rapidly drooping its leaves and becoming less valuable every hour. Generally the crop is cut for hay when about one-tenth in bloom, as it has its best quality for hay for cattle and hogs at this stage, but for horses it may stand a little longer. As a rule, the crop should be cut when new shoots are 1 to 2 inches high, as early cutting aids the next crop. The cutter bar should be adjusted so that about 2 inches of stubble is left.

Curing.—In order to make the best hay in humid regions alfalfa should be cured as much as possible in the cock. When mown in the middle of the day when weather promises fair, alfalfa should be ready to cock in 24 hours and should then be cured for 2 days or more in the cock, which should be opened for a little while before hauling. In this way the leaves can be saved to best advantage, and they carry 70 to 80 percent of the crude protein and more than 50 percent of the nitrogen-free extract.

Hay caps greatly simplify the matter of making good hay in bad weather. Good duck covers cost 25 to 50 cents, and heavy unbleached muslin is often used with good results. The nutrients of alfalfa hay are lost in considerable

Cutting Alfalfa.

-amount by leaching, a third of the protein and a seventh of the nitrogen-free extract being lost by 1.75 inches of rain in one instance reported by the Colorado Station.

In the great alfalfa sections in Kansas, with favorable weather the crop is cut in the morning, raked in windrows in the afternoon, bunched in the evening or the next morning, and hauled the second afternoon. In this section cock curing is unprofitable except in rainy weather. Alfalfa is too valuable a feed to stack; provision should be made to store it under cover.

Should Every Farmer Grow Alfalfa?—Thousands of dollars have been lost upon unsuccessful alfalfa fields because of wet, acid, impoverished soils, weeds, lack of inoculation, poor seed beds, over-nursing, ill-timed seeding, poor seed, excessive clipping and other causes. Some farmers find the crop ill fitted to their rotation, and prefer clover instead, which often is a far better crop for their conditions. Whenever a farmer decides to try alfalfa on land unacquainted with this crop, he should try it first in a limited way and failures will then mean less to him.

Feeding Value.

Hay.—Alfalfa is noted for its excellent feeding qualities. It contains more protein than any other hay crop, and holds a unique position with respect to mineral content. Well-cured alfalfa hay contains about 10.5 percent digestible crude protein, 40.5 percent carbohydrates and 0.9 percent fat, which is about the same analysis as for soy bean hay. Alfalfa hay is higher in protein than most grain feeds and red clover hay, although the latter has twice as much fat as alfalfa has. Because of its greater yields it has a much higher acre-value than clover or timothy.

Alfalfa hay leads all roughages for dairy cows and sheep, and may be used to a limited extent with fattening hogs and is largely employed for breeding hogs. For horses it is too laxative and should be fed but sparingly, and then largely to those at slow steady work.

Green Feed.—Dairy cows do well on alfalfa pasture, but should not be turned on it for the first time until the dew is off, or bloat may result. It is an excellent forage crop for hogs and sheep, although the latter graze it too closely. As a soiling crop for dairy cows it has high value, but should be used for silage only when it cannot be cured for hay.

It is better if alfalfa fields are not grazed the first year or just after a crop of hay is harvested. At no time should a field be overstocked, or the plants will die out.

CHAPTER XX.

THE CLOVERS.

Red Clover.

Distribution.—Little is known of the early history of the clovers, but they were cultivated two or three centuries ago in European countries. In the United States the culture of red clover is from the Ohio River north, and westward to the Great Plains region. The crop is comparatively hardy and kills out

Red Clover.

less than alfalfa in severe winters. Hot and dry weather limits its culture in southern and far western states, although it is seeded in the South in the fall, dying the following summer.

Red clover is thus seen to differ from alfalfa in preferring humid regions

161

and moist soil. With respect to drainage and lime requirement it is much like alfalfa, but can grow on less fertile soil and in competition with more weeds. Since it can grow under conditions unfavorable to alfalfa, it is better adapted to many farms in the northcentral and eastern states. Because of its wide distribution inoculation of the soil is seldom necessary.

Value and Uses.—There is no crop that can take the place of red clover. Alfalfa, other clovers and still other legumes are valuable in certain places, but none of them can replace red clover. This one crop occupies a unique place in our agriculture. Labor incident to seeding, including cost of preparing a seed bed, is slight; the cost of the seed is generally less than that of other legumes; and as a sod-forming and hay-making legume for a short rotation, all seasons considered, it has no near competitor, although alsike clover compares favorably. Red clover should have a regular place in the rotation, and other legumes should be supplementary.

As compared with alfalfa and sweet clover, red clover is lower in digestible nutrients, especially in protein. It is far superior to timothy as a feed for dairy cows, but not for horses. With equal care in harvesting it is a close rival to alfalfa as a dairy feed. Its chief use is to supply protein and furnish a highly palatable roughage. Moreover, one should not overlook the fertilizing value returned when clover is fed to livestock, nor its advantages in soil improvement.

Seeding.

Nurse Crops.—Clover is usually sown with a nurse crop, the preparation for which is all that is necessary for seeding clover. The most common method is to seed wheat fields in early spring when the ground is honey-combed with frost. Then after the wheat is removed the clover comes on, covering the ground by winter, and yields two crops the next season.

Clover is also frequently seeded in the spring with barley, oats, or oats and peas cut for hay. Where these crops do not grow thickly and the ground is moist, clover may succeed. Such a practice is common in the Middle West, but in Ohio and the East such seedings are usually more or less uncertain. Generally such nurse crops grow so dense that the sunlight and moisture supply is too scanty for the clover and grass, and these generally perish before the crop is harvested. If they do survive the harvest, the change in conditions is so great and abrupt that the spindling plants usually die. There is also danger of lodging, and this means destruction to the seeding.

Seeding in Corn.—Upon rich lands well filled with humus many farmers are getting good results from seeding clover at the last cultivation of corn. For this purpose 10 pounds of clover and 4 pounds of timothy seed per acre should be sufficient.

Seeding Alone in Spring.—Under favorable conditions it is possible to harvest a fair though not a maximum crop of clover hay from early spring seeding of clover alone. Ground should be free from weeds and the seed bed well compacted. Some fertilizer (mainly phosphorus and sometimes a little potassium) will likely give the clover a good start. Dropping the seed in front of the drill hoes by the use of the grass-seeding attachment of an ordinary grain drill and allowing the hoes to run at a moderate depth to cover the seed will be the most satisfactory means of seeding. Seeding clover in the spring in this way means more competition with weeds and harvesting about 6 to 8 weeks later than the usual clover crop.

Seeding Alone in Midsummer.—Because clover rarely succeeds when seeded with spring crops in regions of rather heavy rainfall when the latter are planted at the usual rate, seeding in July and early August is frequently resorted to. In this way a crop of oat and pea hay or early potatoes may be harvested,

Field of Clover.

and a good seed bed can be prepared by disking, if one takes advantage of occasional rains; and a timely seeding of clover and timothy will result in a good stand. Seed just after a good rain, and cover the seed with a light smoothing harrow. A full crop of hay may be expected the next season.

Some farmers may be tempted to put in soy beans or cowpeas after wheat is removed and then seed clover in the forage crop. In the average season the supply of moisture will be pretty well exhausted by the wheat crop, and later rains will be necessary to start the forage crop. With an unusual amount of rainfall late in the summer, a catch of clover might be secured; but with dry weather plus the shade of the soy beans or cowpeas, the chances for a stand would be small. It would be better either to seed the clover alone or to follow the forage crop with wheat or rye and seed then. Where seeding fails in wheat and then clover is seeded after its removal, a thorough disking and a light harrowing will be all that is necessary to cover the seed.

Why Seed Clover and Timothy Together?—In general farm practice clover and timothy are seeded together, whether with wheat or rye as a nurse crop or alone. The reason is that the life of the clover plant is usually ended when it produces a crop of seed, and the next year the timothy, which is little in evidence the first season, takes possession of the ground. When clover is clipped so that seed formation is prevented, or when pastured in the fall, it usually dies out after one full season of growth, largely on account of the many insects

Curing Clover in the Cock under cover.

and diseases that prey on it. In case the clover fails, the timothy will come on the next season after seeding.

Red Clover alone is not easy to cure, and makes a dark, often dusty hay if weather conditions are not the best.

Rate of Seeding.—The conditions of seeding, including the fertility and sweetness of the soil, have much to do with the rate of seeding. A common mixture of timothy and red clover is 6 to 8 pounds of the former and 8 to 10 pounds of the latter per acre; or if alsike is also included, use 2 pounds of this and reduce the red clover to 6 or 7 pounds. For seeding alone in the spring about 12 pounds of the red clover would be needed for a good stand.

Clover as a Catch Crop.—Red clover is seldom used as a catch crop, and yet it can be used with profit for this purpose when seed can be obtained cheap. Mammoth clover, or crimson where adapted, is better suited for such a purpose.

Harvesting.

Clipping New Seedings.—When clover is seeded alone in the spring, a crop of hay may be cut in late summer. When it is seeded with a nurse crop, the best plan is to clip it after the cereal crop is removed and the clover has made fair growth. This insures a cleaner crop of clover the following year, and usually a thicker stand. Moreover, if the clover makes a large growth some heads may form seed, and this should be avoided. If the growth is too rank in the fall and is not cut, field mice may become numerous, while if left in the swath it may smother out the clover. A late summer cutting should not injure the next year's crop.

The time to clip will depend largely upon the season. In seasons of frequent rains when both clover and weeds start off promptly, it is advisable to clip in 2 to 3 weeks after the wheat or other cereal is removed, but in some seasons it may be delayed longer.

Occasionally it is necessary to clip a second time in order to prevent seed from maturing. This will occur in seasons of frequent fall rains and high temperatures. This second clipping should not be made so late that the clover cannot make a good growth for winter protection.

When to Cut for Hay.—Clover hay is more difficult to make than timothy because the ground usually is not so thoroughly dried out and also because green clover carries more moisture and the leaves shatter more easily. And finally, clover hay does not shed water as readily as timothy. The total yield of hay and the amount of protein, nitrogen-free extract, fat and ash are higher when the clover plant is in full bloom than at any later period. The crude fiber is the only constituent which increases with maturity. A large acreage should be harvested in due season so that the quality will not deteriorate seriously.

Methods of Making Hay.—Before clover hay can safely be stored in mow or stack its percentage of water must be reduced from about 75 percent to 20 percent or less in as little time and with as little effort as possible, if the best hay is to be secured. It is hardly possible early in the season to cut, make and store clover all in one day, though later in the season this can be done with excellent results where the clover is not extremely heavy.

The best clover hay is made by way of the hay cock. This is hardly practicable on large areas, but on small or medium-sized farms it is much in vogue. The clover is cut in the forenoon as soon as the dew is off; if heavy, the tedder should be started in 2 hours; early in the afternoon it can be raked into loose windrows and put in large, well-made cocks during the afternoon, where it may be left for 2 or 3 days. If hay caps are available they can be used to advantage. Before hauling, the cock should be opened a little for an hour or two.

Cured in this way the leaves and blossoms are preserved almost entirely and the color and flavor are at their best. The objection to this method is the great amount of hand labor required. Upon large farms, therefore, it is more common to mow the clover late in the afternoon, and then the next morning the tedder can be used once or twice to advantage. Early in the afternoon it should be possible to start the side-delivery rake, and soon afterward the hauling, depending upon the sun and air and also the dryness of the ground. With everything favorable the clover may be stored within 24 hours after cutting.

As the season advances and clover becomes riper it is possible to hurry matters a little, and under favorable conditions put the hay in the mow the

Side Delivery Rake used on large acreages especially with hay loader.

same day it is cut. This will usually involve frequent tedding and turning with the side delivery. Whichever method is followed it is not advisable to store or stack half-cured hay. With too much moisture in the hay one may expect a musty, fire-fanged product.

Save the Leaves.—With the improved hay-making machinery of recent years, hand labor has been largely eliminated, except on the smaller farms of the East, where the slower curing methods are still followed. There is danger in this machine handling of the hay of losing much of the more valuable portions of the plant. The pulling and shaking of the tedder and the sifting of the loader tend to leave behind the finer and richer part, provided the plant is overdried rather than cured. The leaves carry fully three times as much protein and fat as the stem, and the heads more than twice as much. Moreover, the stems have much woody material and are largely indigestible.

Obtaining Clover Seed.—For the production of clover seed the first crop should be cut early. The conditions which make a good seed crop, however, are mainly a matter of weather. If the early cutting of the first crop is followed by moderately wet weather until the second crop gets started, thus insuring a fairly thick stand on the ground, and then the weather turns dry until seed harvest, the conditions are ideal. A wet August will result in a tall, rank-growing second crop which is quite likely to be worthless for seed.

Red Clover as a Forage Crop.—Red clover is a close second to alfalfa as a forage crop for swine. It will not endure drouth quite as well, but if not pastured until it has made a fair growth, and a portion of the field is

Hogs on Clover and Bluegrass.

cut at about the usual time for harvest, it will come on fresh and do better than if allowed to become woody. With corn at 56 cents a bushel an acre of red clover has had a value in replacing grain of $33.67 at the Ohio Station and $36.88 at the Iowa Station.

Clover Silage.—Clover may be put satisfactorily in the silo, although the silage is likely to be a little soft and acid. For such a purpose it should be cut when just past full bloom, with about one-fourth of the heads turning a little brown. It should go into the silo as fast as possible after being cut. It should be run through a silage cutter both for the sake of cutting it in short lengths and for the sake of elevation. It needs more tramping than corn. Ordinarily it is not economical to put clover in the silo, for the reason that an enormous amount of water must be handled. Then, too, clover hay is consumed with great relish and profit, and where a succulent feed is desired corn silage will prove more practical.

Other Varieties of Clover.

Alsike Clover.—Alsike clover is medium in size and appearance, is not as deep-rooted as red clover, nor does it produce as large a top growth. Since it has a creeping habit of growth it is often seeded with red clover, timothy or both. It will do well on wetter ground than will red clover, and is not so exacting in lime requirements. It differs from red clover in seeding habit, in that the first crop usually seeds freely except when conditions favor a rank growth and the crop reseeds itself from year to year.

White Clover.

Mammoth Clover.—Mammoth clover is a large variety and under ordinary condition is a biennial. It makes a coarser hay than red clover, matures about 3 weeks later and produces seed in the first crop. It finds its chief use as a soil improver and to a limited extent as a soiling crop. For hay it is inferior to red clover, and when the second crop of the latter is considered, it has but slight advantage from a fertility standpoint.

Sweet Clover.—Sweet clover (Melilotus) is widely distributed over the United States, growing freely along roadsides and waste places. Under these conditions it is hardy and persistent. In certain sections of Wyoming, Kansas and Iowa, in Kentucky, in Mississippi and in other southern states it is grown more or less extensively. It resembles alfalfa in several characteristics, and has much the same soil requirements, although it will grow on poorer soils with less drainage than alfalfa. In the West it is more resistant to alkali. Like alfalfa it requires inoculation, and the bacteria of one will grow upon the other. While its greatest value is as a soil improver, adding large quantities of organic matter and nitrogen wherever grown successfully, yet it may prove valuable as a forage crop, or as a catch or cover crop. It is fed to some extent as a hay, its analysis being similar to that of alfalfa. It may be seeded much like alfalfa as to preparation of seed bed, time and rate of seeding. The germination of the seed is usually low, and hence all seeds should be tested for germination before buying. An implement has been recently put on the market for the purpose of improving the germination of hard sweet clover and alfalfa seed by scratching, or scarifying it. It is a safe plan to buy only scarified seed.

Crimson Clover.—Crimson clover is quite generally grown in the South, and it does not do well north of the 37th parallel. It is grown largely for hay, but is the most admirably adapted to catch-crop uses of all the clovers. When seeded in corn at the last cultivation it makes a rapid growth in the early fall, completes its growth in early spring, and is off the ground in time to plant regular crops. North of the Ohio River it does not do well when seeded in the spring, as it ceases to grow when hot weather comes. When seeded in the fall this far north it may come through the winter all right and make excellent spring growth, but commonly it is almost entirely winter-killed. It is usually seeded at the rate of 15 to 20 pounds per acre.

Japanese Clover.—Japanese clover (Lespedeza) is an annual, adapted to the South, where it has a long season for growth. North of Virginia it is not a profitable legume, and other varieties of clover should be used in its stead. It should not be seeded until all danger of frost has past, about 15 pounds per acre being sufficient. In the South it reseeds itself persistently, often dominating over all pasture grasses. It ordinarily grows 4 to 6 inches high and is used chiefly for pasturing.

White Clover.—White clover is a perennial that spreads by creeping stems. It is best adapted for pasture uses, as it reseeds itself and is unusually persistent. It grows in warmer climates, and in poorer, wetter, more acid soils than red clover. For lawns or permanent pastures it is well adapted to be seeded with blue grass, 2 to 5 pounds per acre being used.

CHAPTER XXI.

OTHER LEGUMES.

Soy Bean.

Distribution.—The soy bean is a native of China and Japan, where it is said to have been grown for centuries. Within recent years it has become of considerable importance in the United States. The wide variations in different varieties regarding time of maturity permit it to be generally distributed throughout the United States from states just north of the gulf states to the northern limit of corn growing. It is slowly but steadily finding a place for itself from Massachusetts to Kansas.

Uses.

Grain.—In the United States soy beans are not used extensively for human consumption, as in China and Japan. The beans have a high feeding value for all classes of livestock, especially for fattening lambs and hogs and for dairy cows. Because of the high protein and oil content they should be fed with a starchy food like corn. They are in general a profitable substitute for oats, particularly in regions where oats are somewhat uncertain. The Ohio Station found them more satisfactory than oats in a 4-year rotation of corn (or potatoes), oats, wheat and clover, and the two legumes keep up fertility better than one.

Hay.—Soy beans can hardly claim a permanent place in a rotation, but as a substitute crop where others have failed they may take an important place. In case of a failure of clover seeding, or of any spring crop, there is plenty of time to grow a good crop of soy bean hay. The hay is equal to alfalfa and superior to clover hay in nutrients contained, although the stems are usually coarse and woody. The leaves are particularly valuable and should not be lost in harvesting.

Silage.—Soy beans alone do not make satisfactory silage, but when mixed with twice the weight of corn or sorghum a good silage may be produced. Many farmers have found it profitable to grow soy beans and corn together for this purpose, though with large varieties of corn the beans make little growth.

Soiling Crop.—Soy beans make satisfactory soiling crop for late summer and early fall. They are not as palatable as peas and oats, but they delight in midsummer heat and hence may be used after the other crop is gone. Yields of 5 to 10 tons green forage per acre are common. Because of the richness in protein, this forage is fed to advantage with corn, millet or sorghum.

Pasture.—Young hogs do well on soy bean pasture, although it is usually inferior to rape. The crop is ready in 6 to 8 weeks after seeding. In case of loss of beans from shelling in harvesting hogs should be turned in to utilize them.

Culture.

The use of soy beans as soil improvers has previously been discussed. The crop works in well in case of failure of spring crops, at the last cultivation of corn, or as a midsummer catch crop.

Seed Selection.—Soy beans are usually classified according to color of seed and time of maturity. Early varieties usually yield more in proportion to size of plant, and are inferior for hay or silage on this account. The best-known and most widely grown varieties are the Ito San, Medium Green, Medium Yellow,

Field of Soy Beans

Hollybrook, Wilson and Mammoth. The Medium Green is a high yielder of both grain and hay, but it shatters easily. The Ito San is an early, well-known variety of great merit. The Mammoth probably has a larger acreage than any other variety. Many selections have been developed at the experiment stations from these varieties. The grower should select a variety recommended for his particular section and uses.

Soils.—Soy beans require a soil like that for field beans. On fertile land, especially mucks, they produce most forage, though the seed is less than on land of medium fertility. Good drainage, lime and inoculation are necessary for the best crops, though soy beans do better on more acid soils and with either excessive moisture or drouth than red clover. Stable manure, or a fertilizer carrying only phosphorus and potassium, may be all that is needed to improve fertility when the crop is once started, and usually, as when they follow corn, no fertilizer is applied.

Seed bed.—Soy beans require a seed bed as for corn and when planted in rows the cultivation is similar for both crops. Foxtail is a common weed enemy, giving most trouble in fields drilled solid for hay. It may be controlled by running a weeder over the field before the beans come up, and again when they are about 3 inches high.

Seeding—Soy beans are not seeded until all danger of frost is past and the soil is warm. North of the Ohio River conditions are usually favorable in May and June, or 2 or 3 weeks later than for corn.

At the Ohio Station, when drilled in rows 28 inches apart, 3 pecks per acre has given the highest yield in the case of Medium Green. On account of variation in size of seed, results may differ with varieties. For hay a 3- or 4-peck rate is satisfactory, although a finer hay is produced with twice this quantity drilled solid. For silage 3 pecks per acre in rows is recommended.

The oat feed of an ordinary grain drill is satisfactory to seed soy beans. By stopping a part of the runs, the distance between rows can be varied to suit conditions. The drill should be tested before seeding by lifting one wheel and turning off a small part of an acre, noting the rate the beans are fed through. Under ordinary conditions soy beans should be planted about an inch deep. It is a waste of seed to broadcast it and cover with a harrow.

Planted in rows soy beans should be cultivated two to four times depending on the season. One-horse cultivators having 12 to 15 small teeth, are preferred for 28- to 30-inch rows, and for wider rows two-horse cultivators may be used. Shallow, level cultivation should be practiced.

Harvesting.

Hay.—The harvesting of soy beans may be done with the usual implements of the harvest field—the ordinary mowing machine and the rake. They should be mowed when the dew is off and raked when nicely wilted.

Soy Bean Pods on Plant.

For hay soy beans are best cut when pods are well formed, as then few leaves will be lost and the crop is at its best regarding palatability and nutrients. Afterward the stems get woody and coarse. The same rules pertain to curing as for clover hay. Since the leaves shatter easily most of the curing should be done in the cock.

Silage.—Soy beans should be cut for silage after the beans are well formed, but before the leaves begin to drop. This will be several days later than the time for cutting for hay. Corn should be ready to cut at the same time. An ordinary grain binder is most satisfactory to cut soy beans for silage.

Seed.—For seed most varieties should be allowed to stand until nearly matured. Season and the variety determine the proper time; they should be cut earlier in a hot, dry fall than in one that is cool and moist. The pods should be brownish or black in color and about half of the leaves fallen in case of varieties that shatter easily. Some varieties may stand until practically bare of leaves.

An ordinary mowing machine with side-delivery attachment or self-raking reaper is satisfactory for harvesting. Much the same care is required as when the crop is cut for hay, but in case of rain at this season the cocks should be turned from time to time to prevent beans lying on the ground from molding.

At the time of threshing soy beans should be slightly tough to prevent cracking of the beans. An ordinary thresher may be used if run with blank concaves and with the cylinder at moderate speed by means of special pulleys which allow the rest of the machine to run normally. Small areas can readily be threshed with a flail. After threshing the beans should be spread out to dry and shoveled over occasionally.

Cowpeas.

Adaptation.—The cowpea is a summer annual resembling somewhat the garden bean and field bean. It is said to be a native of the Orient. In the United States it is grown mainly in the South, but within recent years its culture has been extended northward, until it is grown in Ohio, Michigan and other states of this latitude. Heavy frost is always fatal to it.

Cowpeas will stand considerable drouth if well tilled, but growth of vines will be less than when moisture is abundant. They do best on fairly rich, sandy loams, and for forage purposes make most growth on rich soils. They are not as exacting in drainage and lime requirements as soy beans, and hence find wider adaptation for green manuring. Wherever the soy bean can be raised it is to be preferred.

Uses.—Like soy beans, cowpeas are grown for food and soil improvement. For livestock they may be used for grain, hay, silage, soiling and pasture. As pasture they are most useful for hogs, because in cattle and sheep they often produce bloat.

Culture.—With few exceptions the same rules obtain regarding seeding, cultivation and harvesting of cowpeas as for soy beans. On account of their having a larger, more vigorous root growth, the Kansas Station recommends seeding cowpeas a little thinner than soy beans. The Oklahoma Station recommends 1 to 3 pecks per acre in rows 36 to 42 inches apart, planted any time from late April to July 15.   The Mississippi Station recommends 1¾ bushels drilled or 2½ bushels broadcasted per acre for hay.

When harvested for hay cowpeas should be cut when the first pods turn yellow; for seed, when half the pods are ripe. Cowpea hay compares favorably with clover hay in digestible nutrients, but is a little lower in protein. Black, Iron, New Era and Whippoorwill varieties are most generally grown and Buckeye, Red Ripper and Early Buff are also common.

Field Peas.

Adaptation.—Field peas, commonly called Canada field peas, are all summer annuals, grown north of the Ohio River and chiefly in Canada, Wisconsin and Michigan. They grow on nearly all kinds of soil, and like other legumes need lime for best development. The plant resembles the common garden pea, but is taller, from 2 to 5 feet, and has such weak stems that it falls to the ground. For this reason it is usually seeded with a crop like

White Canada, Scotch Beauty, Blackeyed Marrowfat Field Peas.

oats. Field peas are often mistaken for cowpeas, which are much different. The latter is a hot-weather plant and is not adapted to seeding with oats.

Oat and Pea Hay.—Field peas find their best use as a forage crop when seeded with oats. The two make a palatable, nutritious hay, comparing favorably in composition with clover hay.

Field peas and oats should be seeded as early as the ground can be put in condition—in March or early April. About 3 bushels of seed of the two crops should be used. This is generally equally divided, though sometimes 1 bushel of peas and 2 bushels of oats are used. They should be drilled separately, the peas much deeper than the oats. The proper depth for drilling peas will vary with the soil, ranging from 2 to 4 inches, and the oats 1 inch. The oats are often broadcasted in advance of drilling the peas with good results, the use of the drill in putting in the peas covering them sufficiently. Any midseason variety of oats (like Big Four or Silvermine) may be used. and the Golden Vine, often called White Canada, is the most common variety of peas. Other good varieties of field peas are the Marrowfat, Arthur and Prussian Blue.

These crops are at their best when cut for hay when the oats are in milk and the peas well podded. Do not let the oats get riper than this even if the peas must be cut a little green. They should be cured about like

Peas and Oats make an excellent combination to feed green to dairy cows.

clover, and should be expected to yield about 2 to 4 tons of cured hay per acre. In Canada the grain is threshed from such a mixture and makes excellent feed for livestock.

For feeding green to dairy cows oat and pea forage is excellent. Seedings should be made at intervals of 2 weeks apart, just enough to be sown that it can all be utilized in 2 weeks, as when allowed to stand longer than this the forage gets woody and unpalatable. In Colorado peas are sown for pasturing sheep and hogs.

Peas for Green Manuring.—Field peas, if not cut for hay, may be used to advantage as a green manuring crop. About 2½ bushels of seed per acre should be used, and drilled 4 inches deep. The crop is not adapted to midsummer seeding, but is sometimes seeded late for a fall growth.

Vetches.

Hairy Vetch.—The winter or hairy vetch, is a winter annual, often acting as a biennial, adapted to the colder climates of northern United States and Canada. It makes but little growth the first year, and reaches maturity about July the second year. It is considered drouth-resistant, and does well on nearly all productive soils. It is especially advantageous in growing in sandy soils. Like other legumes it needs lime and inoculation when introduced in a new locality.

It is best planted in July or August and is usually sown with rye, wheat or timothy, 4 or 5 pecks of the cereals or 10 to 12 pounds of timothy with 20 to 30 pounds of vetch being used per acre. It is difficult to harvest alone

Rye and Vetch to plow under.

because of its twining vines. If allowed to mature seed the plant dies when cut, but there is often sufficient seed shattered to reseed the field. Spring seedings are seldom advisable, though somtimes a moderate crop of hay is produced the same year.

If cut for hay when in bloom the fall seeding will furnish a second cutting the same season under favorable conditions. It makes an excellent hay, the richest of all legumes, and yields 1½ to 3 tons per acre.

Rye and hairy vetch are commonly grown together for seed. They should be cut when the lower pods of the vetch are ripe. Vetch does not ripen all at once, but more seed will be saved at this stage than later. It is difficult to separate the two seeds; in fact, it is almost impossible with an ordinary fanning mill and grader. On this account many prefer timothy. The range in yield of seed varies from 2½ to 12 bushels per acre. Spring seedings of hairy vetch seldom produce any seed the year of seeding. Most of the seed is imported from Russia and eastern Germany.

In common with other legumes vetch is a valuable soil improver. Seeded with rye an excellent combination for a cover crop is afforded which may be plowed under early in the spring. Vetch is not to be regarded as a weed, for it is in fact less persistent than the clovers. The chief objection to it as a weed is when it turns up in the wheatfield; it is well to time seedings to remove vetch as far as possible from the wheat crop if this is grown in the rotation.

Spring Vetch.—Common or spring vetch is an annual adapted to cool climates. In the northern states it is grown in the spring; in the South, in the fall. Hairy vetch is much preferred in the North. In the South common vetch is sown from September to November, from 1 to 2 bushels per acre being used when it is seeded alone, and from 30 to 60 pounds when seeded with oats at a like rate. In culture and harvesting it resembles hairy vetch.

Peanuts.

Adaptation.—Peanut culture in United States has developed rapidly since 1870, and today they are cultivated north of the great cotton belt, notably in North Carolina and eastern Virginia, with Georgia, Florida, Alabama and Texas following. Here they have a long growing season with no frost, warm weather and light rainfall during the growing season.

Peanuts grow on nearly all types of soil, but well-drained, light, sandy soils are preferred. Such soil affords opportunity for a good mulch and pods of clean appearance. Lime is necessary to correct any acidity in the soil. From 300 to 500 pounds per acre is generally used of a fertilizer high in both phosphorus and potassium, and usually without nitrogen. Manure is not to be applied to the peanut crop, as it produces too much vine.

Culture.—Peanuts usually follow an intertilled crop, like corn or cotton, in order that weeds will be overcome. The peanuts are sown in rows 28 to 30 inches apart, or as much as 36 inches in case of the large varieties; and the plants 6 to 8 inches apart in the row for small varieties; and 12 inches for the larger varieties. About a half bushel of shelled peas, or 2 bushels with pods, is required per acre. Planting is deferred till warm weather, in May in Virginia and a little earlier farther south. Cultivation as for corn or cotton is recommended.

Harvesting.—At maturity the leaves turn yellow and the pods tend to shed. Potato diggers, or the special machines, are used satisfactorily in harvesting the crop. From 4 to 6 weeks is necessary to cure the crop. A stake about 7 feet high and sharpened at both ends is set firmly in the ground and the peanuts are stacked around it loosely. After curing in this manner the pods should be picked during dry weather. The straw may then be used for feeding.

Velvet Beans.

Adaptation.—Velvet beans require a hot growing season of about 200 days. The crop is grown almost entirely in the Gulf states. The vines are very large, often growing 30 to 40 feet long.

Culture.—When planted alone the crop produces little seed. Hence, it is usually seeded with corn or sorghum, one plant every 5 feet in the row, or in alternate rows. Because of the large growth the crop is best harvested by grazing.

CHAPTER XXII.

PERENNIAL GRASSES.

Timothy.

Adaptation.—Timothy is a native of Europe, and was brought to this country by the early colonists. It is also known as herd's grass in New England, and is more generally cultivated in the United States than any other grass. The acreage is chiefly north of Tennessee and east of the Missouri River, with small acreages in the Rocky Mountain states.

It is adapted to a great variety of soils, though it does best on rich, moist loams and clays. On light sandy soils it is not as successful. It is a hardy perennial, with a bulbous enlargement at the base of the stem. It is distinctly a grass for hay rather than for pasture. It may be used in pasture mixtures for early or temporary pasturages, but it will soon give place to the better sod-forming grasses, as it does not take kindly to close grazing and tramping. As a meadow grass, however, it stands at the head.

Culture.—Timothy is usually grown in a 4- or 5-year rotation, being seeded with clover in wheat or oats. The first year the clover is the principal crop, timothy predominating the next year. Near good city markets, it is often grown year after year on the same land, good yields being maintained by annual dressings of manure and fertilizer.

When seeded with wheat in the fall, it may occupy the ground the next year in case the cereal does poorly. Fall seedings are generally preferable to spring seedings. The grass may also be seeded alone in July, August or early September, and a full crop of hay may be expected the following season. For such purposes 12 to 15 pounds of

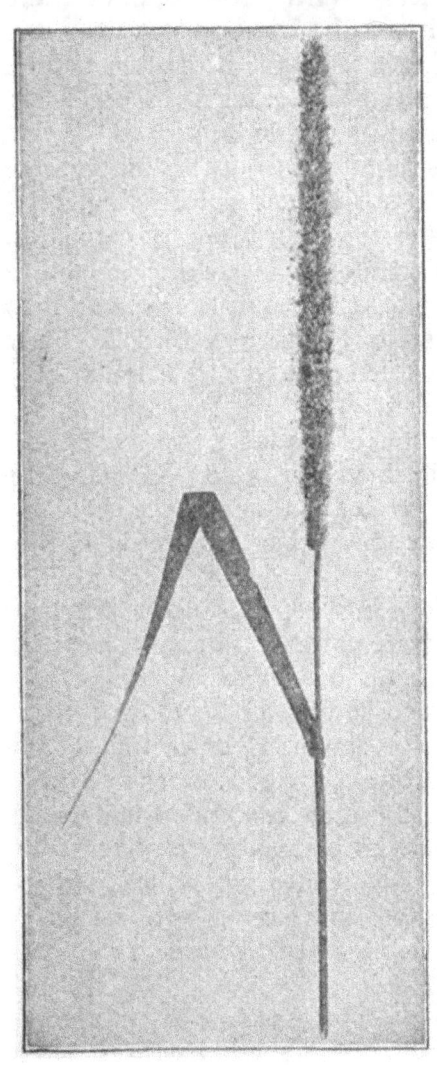

Timothy Head.

timothy is a full seeding; or 6 to 8 pounds is sufficient, along with 7 pounds of red clover and 3 pounds of alsike clover. A clean, moist, compact seed bed is essential, and a light covering with a harrow or weeder is all that is needed afterward. When seeded with wheat or oats the timothy can be distributed through the grass-seeding attachment of the drill and dropped in front of the hoes.

Harvesting.—The total dry matter of timothy increases until the seed is close to maturity; the total protein and fat, as well as the digestibility of the different nutrients, and nitrogen-free extract increase during this period. It is best to defer timothy harvest until the blossoms have fallen, then to rush it through with all speed, endeavoring to complete the harvest before the seed is in the dough. If this cannot be done with the labor available the harvest may begin a little earlier. With good weather there is little difficulty in curing and storing timothy in the barn or stack the day it is cut. Some time may be saved by mowing late in the afternoon before.

At the Ohio Station timothy has led all other grasses in yield per acre. From 3 to 4 tons per acre may be expected on fertile land.

The popularity of timothy is due to its high yields, palatability, ease and cheapness of harvesting, high price, moderate expense of seeding, relative certainty of securing a stand and its longevity when properly cared for.

Disadvantages of Crop.—Like all grasses timothy is lacking in ability to utilize nitrogen of the air, and, hence, is more exhaustive on land than clover is. Still, it does leave considerable residue in the way of sod. Only in exceptional seasons, as of unusual rainfall during July and August, may any second crop be expected.

Seed Production.—Timothy can usually be depended upon as a seed producer. The yields vary from 5 to 12 bushels per acre. Usually the crop is cut with a grain binder, shocked without caps, allowed to cure about a week and then threshed later with a common grain separator having special sieves.

Improving Meadows.—Success in hay farming depends upon the land adapted to the growing of grass and the systematic fertilization of such land. Sandy soils ordinarily yield but one crop of clover or mixed hay, while clays and clay loams are adapted to a course of 2 or 3 years of grass.

Land for a second crop of grass should be fertilized or manured, and this treatment can be made to pay a good profit, provided one has a good stand of grass. The stand may be lacking in vigor and only moderately thick, but if uniform will still be worth improving by use of fertilizers; but poor, uneven stands should be plowed up and seeded anew or, perhaps better, be passed through the regular rotation, for heavy fertilization of such meadows will result in luxuriant growth of weeds where grasses are wanting.

It will not be profitable to thicken up an old meadow by reseeding without plowing. While this may be done with permanent pastures where plowing is impracticable, it is rarely satisfactory with meadows.

A 3-ton crop of timothy hay removes as much nitrogen as a 75-bushel crop of corn, a few pounds more of phosphorus, and about four times as much potassium. For this reason plant food must be added as for other crops. The lime content of the soil is also important, not only to insure success with clover, but also because timothy does better on a sweet than an acid soil. Nitrogen is the important element for the production of hay, and the nitrogen must be in soluble form if the immediate crop is to be greatly benefited. Where clover precedes timothy additions of nitrogen for this crop are unnecessary, but for timothy the next year a moderate application of nitrate of soda (75 to 100 pounds), combined with acid phosphate and a little muriate of potash, will generally be profitable. This fertilizer for meadows may best be applied in the spring as the grass is starting to grow. Top-dressings with stable manure are always satisfactory when applied in the fall or early winter.

Redtop.

Adaptation.—Redtop is quite generally distributed, and in northcentral and northeastern states it is second to timothy as a grass for hay, and to bluegrass for pasture. It is a hardy, long-lived grass, making a heavy tough sod that stands tramping well. It does best on moist clay loams and loams. It will do well on soils too acid for clover or timothy to thrive. It should always be included in mixtures for meadows or pastures on wet, heavy bottomland.

Seeding.—Redtop may be seeded as recommended for timothy and blue grass, 12 to 15 pounds of the best recleaned seed being used per acre when it is seeded alone. It can be seeded when the ground is honey-combed with frost in February or March. Thin or recently cleared woodlands may be set in grass in this way, and also wet lands with inferior herbage. For the sake of variety and season, redtop should be included in all pasture mixtures. For hay redtop yields from 2 to 3 tons per acre.

Bluegrass.

Adaptation.—Kentucky bluegrass is a hardy, persistent perennial, adapted to temperate regions of relatively high humidity, but in arid regions succeeds under irrigation. It holds the same position as a pasture grass that timothy holds for hay, making the best sod of all grasses, and tending to crowd out other grasses, particularly in permanent pastures. Its area extends farther south than that of timothy, as it survives hot summer weather and grows late in the fall.

It is at its best on limestone soils, although it does fairly well on a wide range of soils in the timothy and clover region. It does better on clay than on sandy loams. On poor soils and when subject to drouth it gives way to other grasses like redtop and Canada bluegrass.

It is not well adapted to rotation farming, and is seldom used for hay, since it yields but a small crop and unless harvested promptly inclines to be wiry and unpalatable.

Seeding.—Bluegrass may be seeded with wheat in the fall, or in early spring along with other grasses and clover; but probably the best time is to seed it in late summer or early fall either alone or with other grasses. When seeded with timothy in the fall, and clover added the next spring, the meadow may be cut for 2 years, and the land then pastured when bluegrass will occupy the land. When seeded alone 20 to 25 pounds should be used per acre, and much larger quantities are advisable for lawns. It is important that the seed be tested for germination before any purchases are made, as the seed is generally poor.

For hay it yields from 1 to 3 tons per acre. Seed production is largely confined to a small area in Kentucky.

For Lawns.—Bluegrass is the basis of all grass mixtures. It stands frequent clipping well, although it turns brown in dry weather. A mixture of 30 pounds of bluegrass, 15 pounds of redtop and 3 pounds of white clover per acre makes a good seeding. Unless the soil is rich it should be made so by the use of manure or commercial fertilizers. The fertilizers should be added before seeding and after lime has been worked into the soil if this is needed. After seeding the ground should be raked over to cover the seed and to make a smooth surface. Sprinkling is advisable for dry seasons.

Improving Pastures.—The failure of most pastures is due to four causes; pasturing is started too early in the spring. A pasture allowed to grow 4 inches will grow twice as fast and furnish twice as much feed as one 2 inches high. Most pastures are kept too closely grazed, and the grass plants are weakened, gradually giving way to weeds. Fertile conditions are not properly maintained. Even if all the manure from livestock feeding upon it should be evenly distributed over the pasture, this contribution would not meet the requirements, for mineral elements are carried off in dairy products and in the bones of animals. There have been no determined and systematic efforts to destroy weeds.

Where the land will permit plowing, often the best thing is to plow, manure and fertilize and grow an intertilled crop for a year or two before reseeding in pasture. Cultivation must be thorough, and the soil must be properly enriched if good results are to be obtained.

Where plowing is not possible, a harrowing may give the grass a foothold where reseeding is desired, and it will break up and distribute masses of manure. The regular mowing of weeds, practiced once or twice a year and timed so as to catch the worst offenders before they go to seed will result in great improvement in the pastures. Mowing machines should be set to run as high as possible. For reseeding a mixture of grasses is desirable, of which bluegrass and redtop should be the principal constituents with a little timothy for early results, and some alsike and white clover.

Manure can be used with profit, especially on the thinner portions, and should be supplemented with acid phosphate. Ground limestone where needed should be applied in late summer or fall, and all manure the following spring.

Orchard Grass.

Adaptation.—Orchard grass is a native of Europe, and in the United States is grown mainly in the South. It grows earlier in the spring than most grasses and is persistent in partly shaded places. It does best on rich, well-drained loams. It calls for drier soil than redtop, and it is somewhat more resistant to drouth than is timothy or bluegrass.

It is equally valuable for hay or pasture. It makes a more rapid growth after mowing than most grasses, furnishing considerable aftermath for pasturage, and in some seasons a fair second cutting. Unless kept closely cropped in pastures it is likely to be avoided by livestock, as it seems to lose in palatability with size and age. Unless cut for hay promptly (when in blossom) it deteriorates rapidly, becoming very woody and is then not relished by livestock. The tendency to grow in tufts is much greater than with other grasses. It is difficult, if not impossible, to secure an even sod. It costs about three times as much to seed an acre in orchard grass as in timothy.

Seeding.—Orchard grass may be seeded on wheat ground in early spring when the ground is honeycombed with frost, but a better way is to wait until the ground is dry enough to put a smoothing harrow on it, thus insuring a better covering. Like other perennial grasses, it is satisfactory if seeded in midsummer or early fall, alone or with other grasses. It rarely yields a crop of hay the first season, but can be pastured.

When seeded alone about 30 pounds per acre of seed is necessary; for seed crop, 12 to 20 pounds. The Illinois Station found a mixture of 9 pounds of timothy and 6 pounds of clover to yield 18 percent more hay than 17½ pounds of orchard grass and 6 pounds of clover. If orchard grass has a place in the northern states, it is in localities where timothy does poorly.

Brome Grass.

Adaptation.—Brome grass is a long-lived perennial, a recent introduction from Europe. It is proving valuable in the West and Northwest, but in recent years its cultivation has slowly diminished. It is of doubtful value where timothy and bluegrass can be grown successfully. Brome grass has a deep rooting system and makes a good drouth-resistant pasture grass. The Dakotas and Manitoba are the regions of its greatest culture today.

Seeding.—A common rate of seeding for this grass is 20 pounds per acre. In regions where is is chiefly grown, it is usually seeded in early spring, a nurse crop being used if rainfall is deficient.

Brome hay is nutritious and palatable, and yields of 1 to 4 tons per acre are secured. Under favorable conditions of moisture two cuttings are possible.

Tall Oat Grass.

Tall oat grass is a European perennial, little cultivated in this country. In the United States it is adapted to about the same climate as orchard grass, and is grown mainly near the southern limit of timothy production. It is a deep-rooted grass and stands drouth well.

It ripens a little earlier than orchard grass and deteriorates rapidly after it is ready for harvest. It starts growth early in the spring and produces a great deal of aftermath when cut for hay. At the Ohio Station it was found to be less palatable to horses than mixed hay, timothy, blue-grass or redtop.

Rye Grasses.

Perennial Rye Grass.—Perennial rye grass, the most important grass of Europe both for hay and for pasture, has never gained à prominent posi-tion in this country. It is frequently recommended for pasture, but cannot compete with bluegrass, redtop, orchard grass and timothy. It is best adapted to rich moist soils. When seeded alone 30 pounds of seed per acre is needed. It may be sown either in the fall or in the spring, the former being preferable. At the Ohio Station it has stood lowest in yield of hay in a test of 10 grasses.

Italian Rye Grass.—Italian rye grass is a biennial, a rapid grower, having taller and coarser stems than perennial rye grass. It is adapted to moist regions with mild winter temperatures, such as obtained on the At-lantic and Pacific Coasts. Its greatest success seems to be under irriga-tion, and it is frequently used for a quick covering of green in lawns. For field conditions 35 pounds of seed per acre is required. It yields compara-tively little as a hay crop under ordinary field conditions, although under most favorable conditions it is remarkable for the number of cuttings in a season and the large total yield.

Bermuda Grass.

Bermuda grass is a long-lived perennial grown from Pennsylvania west to Kansas, and south to the Gulf, and also in Arizona, New Mexico and California. In general it is adapted to the same areas as cotton and in this region it is relatively as important as Kentucky bluegrass in the North. It grows best on rich, well-drained bottomland, and has a marked ability to withstand close grazing. It is well adapted to shade and is an excellent soil binder on sandy soils and on eroded slopes. The seed is usually sown in the spring, about 5 pounds per acre being sufficient if an excellent seed bed is prepared. It is also planted by tearing the sod into small pieces and dropping them in furrows 2 to 3 feet apart each way. The average yield is not much more than a ton per acre. When seeded alone or in mixtures it makes excellent pasturage, but should be closely grazed, as the stems get tough and unpalatable with age.

Johnson Grass.

Johnson grass is adapted to the cotton section and also to New Mex-ico, Arizona and California. It usually winter-kills north of the 37th par-allel. Rich, moist soil is preferred for its growth. It is used for hay and for pasture, and when once established it is hard to eradicate. It is com-monly seeded with oats or oats and vetch, two crops of Johnson grass hay being possible the same season in the South. From Kentucky north it may be sown in the spring and will produce a fair hay crop.

CHAPTER XXIII.

ANNUAL GRASSES.

Sorghum.

Adaptation.—Sorghums have been grown in the United States since colonial times, but the last half-century has seen most of their development. They are adapted to regions having a warm summer climate, and they are injured by light frosts. In general the soil adaptations are identical with those of corn. Since sorghum is a shallow-rooted plant having a comparatively short growing season, it seems to exhaust, temporarily at least, the available plant food in the surface soil, and therefore has a reputation of being hard on land.

Three types are usually recognized: saccharine, nonsaccharine and broom corn. They have been developed for sugar, grain and broom straw, all three groups yielding forage as a by-product. Saccharine sorghum, on account of the sweet juices in its stalk, was formerly grown for syrup; and agriculturally the term sorghum is restricted to this division. Broom corn is grown almost exclusively for seed and brush for brooms. The nonsaccharine sorghum is grown chiefly for grain, almost entirely in semiarid regions.

The principal sorghums are Amber, Orange, Sumac, Gooseneck, Honey and Planter. Nonsaccharine sorghums embrace kafir and the durra group consisting of milo (often called milo maize) and feterita. Teosinte is a southern plant similar to sorghum and Indian corn; it is a heavy producer with abundant rainfall in the South, but is inferior to corn and sorghum in the northern states.

Culture.—Sorghum requires a seed bed similar to that for corn, but greater care must be taken in cultivation to keep out weeds. Seeding should be deferred until the ground is thoroughly warm. The seed may be broadcasted or drilled solid, about 1 bushel per acre being used; or it may be drilled in rows 36 to 42 inches apart, 8 to 15 pounds being used per acre. The former method is better for hay and the latter for soiling or silage.

Harvesting.—For silage sorghum is best harvested with an ordinary grain binder when the seed is in the dough stage. For soiling it may be cut at any time after it is large enough to handle well, though to best advantage from the time it blooms until toward maturity. Sorghum is consumed more closely than corn, and its drouth resistance makes it a valuable crop for the production of succulent feed during hot, dry summer months. For hay it may be cut at any time after heading, although for the best quality it should be harvested soon after blooming. It may be cut with a mowing machine, allowed to lie in the swath for several days, and then left in cocks until thoroughly dry. Yields of 4 to 8 tons per acre of dry forage may be expected. For pasture sorghum sometimes causes bloat in cattle and sheep, and is often poisonous when the growth is checked by drouth or frost.

Millet.

Millets are rapid-growing annuals, thriving best in midsummer and useful as green or dry forage. This crop usually has no definite place in any rotation but is used as substitute or emergency crop where others fail, or where two crops a year are desired.

A common grouping of millets is foxtail, broom corn and barnyard. Of the foxtail millets the Hungarian and German varieties are most highly prized. The Hungarian should be used on uplands of moderate fertility and where earliness is important. The German is more adapted to long seasons and to rich bottomland; it is rather coarse, vigorous growing and usually producing but one stem for each seed. The Siberian variety is somewhat earlier than the German. Broom-corn millet is grown but little in this country, mainly in the Dakotas and Manitoba, though it is well adapted to the Northwest. It is shorter and coarser and seeds are larger and of more varied color than those of the foxtail millet. The forage is lower in yield and more woody and less leafy than the foxtail millet. Barnyard, introduced from Japan, is often called billion-dollar grass. It is coarser and yields more than foxtail millet.

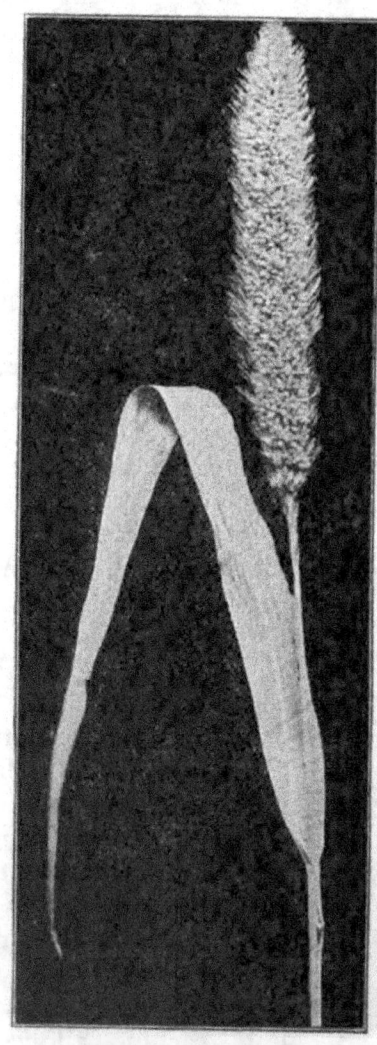

Hungarian Millet.

Culture.—Millet does best on rich, mellow soils, although it grows on almost any kind of land. Like sorghum it bears the reputation of being hard on land, and manure or fertilizer should be applied liberally on poor soils.

A clean, firm, fine, moist seed bed should be prepared and millet seeded after all danger of frost is past. In the North early varieties, such as the Hungarian, may be seeded as late as the latter part of July. Millet may be broadcasted and harrowed in or drilled 1 to 1½ inches deep. When drilled solid for hay, soiling or pasture, from 3 to 4 pecks per acre is usually sown. If grown for either seed or silage it is usually seeded in rows 24 to 30 inches apart, 1½ to 2 pecks per acre being used. Barnyard millet drilled solid is usually seeded at the rate of about 2 pecks per acre.

Harvesting.—For hay, millet should never be allowed to stand until the seeds are ripe. The hay is handled much the same as timothy, although it requires a longer time to cure. The hay is slightly superior to timothy, but it should be fed moderately and not too exclusively, as some danger accompanies its   feeding. For soiling, millet may be harvested at any time

Sudan Grass in Bloom.

after it is large enough to handle. It is most suitable for silage when the seed is in the dough stage. When grown for seed it may be cut and threshed with the same implements used in harvesting wheat and oats.

Sudan Grass.

Sudan grass was introducd in the United States in Texas in 1909. It is strictly an annual, closely related to sorghum and is much like Johnson grass in appearance.

Adaptation.—In the opinion of its introducers Sudan grass promises to be best adapted to Texas and the western part of Oklahoma, Kansas, Nebraska and south-central Dakotas. Where timothy, alfalfa and red clover do well there is little need for the annual grasses. Sudan grass is not especially exacting as to soil, though it is at its best on rich, well-drained loams.

Culture.—A firm, well-prepared seed bed is desirable, and seeding should be deferred until the soil is warm. For hay it is best drilled solid or broadcasted, 25 to 30 pounds of seed being used per acre. For seed production it should be drilled in rows 30 to 36 inches apart, 3 to 6 pounds being used per acre. For the best quality of hay Sudan grass should be cut soon after it is in full bloom. It is handled much like millet, cut with a mower and cured in the swath or hay cock. While it often yields three or four cuttings in the South and Southwest, north of the Ohio River one or two cuttings will probably be the limit. The forage is coarser than German millet. For seed Sudan grass should be cut when the first heads are fully ripe. An ordinary grain binder may be used and the bundles cured in shocks like grain.

CHAPTER XXIV.

ROOT CROPS AND RELATED FORAGE CROPS.

The root crops are chiefly biennials, having similar uses and adaptations, and usually characterized by an enlargement of the primary root and stem. They are an excellent source of succulent food for nearly all classes of livestock. The term includes beets, or mangels, rutabagas, turnips and carrots. Of these, beets or mangels have already been discussed in connection with sugar crops.

Field of Rape.

They are grown mainly in Canada, the Pacific Coast states, Colorado, Utah, New Mexico, Wisconsin, Michigan and New York. They have no regular place in a rotation and cannot compete with corn as a source of palatable succulent feed.

Turnips.

Turnips thrive best in silt loams, stiff clay and sandy soils not being well suited. A cool, damp climate is most favorable.

With respect to seed bed, cultivation, harvesting and storing nearly the same operations employed in the culture of sugar beets apply in the growing of turnips. However, in the early stages of growth, turnips grow more rapidly and less difficulty is experienced in keeping the seed bed clean while the plants are young. They may be seeded from May to August, either with a hand garden drill in rows 18 to 30 inches apart, usually 3 pounds of seed per acre being used, or seeded broadcast at the rate of 4 to 6 pounds. Turnips are often seeded at the last cultivation of corn for fall feed. Yields of 10 to 20 tons are considered satisfactory.

White-fleshed varieties are best suited to fall and early winter feeding and are usually seeded broadcast. The yellow-fleshed varieties grow less rapidly, are firmer, have higher feeding value, are more resistant to frost and may be kept sound for longer periods.

Rape.

Adaptation.—The soil and climatic adaptation of rape is similar to that of turnips. It is partial to rich loams and, like corn, responds readily to liberal applications of manure and fertilizers. Rape is especially valuable for furnishing

Rape for Hog Pasture.

green feed in autumn and early winter, thus conserving the stock of hay and silage for winter use. It is a most valuable source of green feed for sheep and swine. If not pastured too closely it will furnish green forage through the season, the amount varying with the fertility of the soil. Several small pasture fields seeded at different times will be found desirable.

Culture.—Rape calls for the preparation of an excellent seed bed. It may be seeded from April to the last of July. As a catch crop it may be seeded at the last cultivation of corn. Fall sowing is most satisfactory in the South. It may be sown solid (broadcasted or drilled) or in rows 24 to 28 inches apart. The seed is best distributed from the grass-seeding attachment of the drill. If broadcasted, 5 or 6 pounds of seed is required per acre; if drilled in rows 2 to 3 pounds.

While some varieties are annual (or summer), others are biennial (or winter.) The biennial rapes, of which the Dwarf Essex is leading variety, are most commonly grown in this country. It produces seed only where it can withstand the winter.

CHAPTER XXV.

WEEDS AND THEIR ERADICATION.

Annuals.—In this chapter only the more common weeds are listed, and recommendations given for their control. If the weed is an annual that reproduces by seed, the prevention of seed formation may keep it in check. A systematic crop rotation is most effective in controlling annual weeds. Most weed seeds retain their vitality for several years, and hence continued prevention of the production of seed is necessary to check their spread entirely. For permanent pastures and along roadsides, this is the most practical method, and when persistently followed is effective. Allowing a full crop to seed, however, may do more harm than the good resulting from a few years of consistent cutting.

Cutting weeds in late summer does not prevent seed ripening of many weeds in meadows, such as whitetop, docks, prickly lettuce, mustard, pigweed and giant hawkweed. Still, such weeds will be carried off the meadow in the hay crop. New seedings are often overrun with ragweed and foxtail grasses, especially in wet weather. Clipping is therefore advisable in such cases.

On tilled areas cultivation largely controls the weed problem. Summer fallowing may be practicable on foul land for crops, like alfalfa, that cannot stand competition with weeds. A mistake is often made when land is broken and left unoccupied during the growing season. For land that is foul with weed seeds a catch crop in corn at the last cultivation or in potatoes will aid materially in keeping down weeds. Harrowing before a crop is up checks weeds in early growth. A little extra work before planting time may be extremely profitable later in reducing the number of weeds.

Biennials.—Biennials, like burdock, wild carrot, and bull thistle, store nourishment in the root the first year and produce seed the second year, or if seed formation is prevented they live longer, often sending up several stalks from the base. Cutting such weeds below the crown, as with a small spade or mattock, usually kills them, and decay is almost certain if a little oil or salt is dropped on the freshly cut surface. Weeds in this class are most abundant in old pastures, along roadsides and in waste places where seldom disturbed. Early cutting is desirable, fall cutting the first year of growth being a favorable time to check their development; then in the second year they should be cut before seed is formed and later as necessary to prevent seeding.

Perennials.—Perennials reproduce both by seed and by underground stems, as Canada thistle and wild onion. Seed formation must be prevented by mowing when the first flowers appear and the underground portion killed. To kill the rootstocks they may be dug up if only a small area is infested; or a material like salt, oil or strong acid on the freshly cut root destroys it; or they may be starved by keeping green leaves cut off or smothered by dense grasses or other crops. Plowing out such rootstocks to expose them to the direct action of the sun during a dry spell in summer is effective, but in wet seasons is of no avail. Milkweed and horse nettle are similar in habit of growth to perennials in their long root. Smothering them out with a crop like buckwheat or rape is efficacious. Horse nettle is usually controlled by thorough cultivation and hoeing in an intertilled crop.

Importance of Clean Seed.—Manure and agricultural seed are both important means of spreading weeds. Manure well decomposed has few weed seeds capable of germinating. Clover and alfalfa seed in particular often carry weed seeds, such as those of mustard and dodder; a careful examination should be made before buying such seed. Agricultural seed is perhaps the most important means whereby weeds are introduced upon a farm. With a small magnifying or reading glass, a farmer, after a little study, can soon recognize the common weeds in seed.

Treatment of Pastures.—Over-grazing will result in predominance of weeds over pasture grasses; hence, the removal of live stock for a short time in such cases may be beneficial. Mowing weeds once or twice before seeds form is advisable. Fencerows and roadsides should be handled in like manner. Clipping pastures after harvest was found by the Ohio Experiment Station to result in more and better grasses, clover and bluegrass occupying ground where briars and weeds formerly predominated.

Oil Spraying—The Ohio and Indiana Stations have found spraying with crude oil efficient to control wild garlic infesting meadows and grainfields. Early spring treatment before the garlic forms heads is necessary. Spraying in April and May is most effective, depending upon the latitude. About 75 gallons of oil per acre should be applied, a power sprayer that furnishes pressure sufficient to cause a mistlike spray being used.

Destroying Weeds With Chemicals.—Canada thistle, dandelion, mustard, corn cockle, shepherd's purse, bindweed. pigweed, ragweed, cocklebur and some other weeds have been successfully sprayed by such solutions as copper sulphate (blue vitriol), iron sulphate, common salt and corrosive sublimate. Carbolic acid (one part to four of water) will kill Canada thistle when placed in contact with the roots, but has not been effective against quack grass. Common salt is not entirely effective against Canada thistle, quack grass, morning glory and milkweed, but can be used to kill orange hawkweed and mustard, about one-third of a barrel to 52 gallons of water being necessary. Copper sulphate was found satisfactory in Iowa to kill burdock, prickly lettuce, common mustard, pigweed and goosefoot, when used at the rate of 12 pounds in 52 gallons of water, but was not effective on morning glory, knotgrass and foxtail. Slaked lime is sometimes partly effective, but is not generally recommended, and the same is true with formalin. Because of its poisonous character corrosive sublimate is not recommended. Sodium arsenite and sodium arsenate have both proved successful in exterminating certain weeds. The arsenite (1.5 pounds to 52 gallons of water) is considered excellent to kill Canada thistle, burdock and other weeds having similar root systems. At the Iowa Experiment Station iron sulphate was found effective in exterminating dandelion, knotgrass, purslane, yarrow, ragweed, sorrel, sour dock, mustard, velvet leaf, lamb's quarter, sow thistle, bull thistle, wild carrot, shepherd's purse, pigweed and a few other weeds, when used at the rate of 100 pounds to a barrel of water and sprayed over the leaves on a quiet day.

Some Special Weeds.

Canada Thistle.—Simply cutting Canada thistles once a year will do little or nothing toward exterminating them. All growth above ground should be prevented for an entire season. The patch should be gone over once a week during the growing season with a sharp hoe, and every plant should be clipped off just below the ground. The next year the fight must begin again. If the field is plowed the young plants can be easily seen and killed. In pastures it will help if salt is placed upon each cut stem and live stock, like sheep, allowed to eat and trample over the patch. Rootstocks of thistles carried on plows or other implements to different parts of the farm spread infestation. Sodium arsenite is effective and carbolic acid partly so.

A badly infested meadow may be disked after the hay is cut early in June, and then fallowed until late October. Most thistles will be killed by this treatment, and the field may then be seeded to rye to be plowed early the next spring for an intertilled crop like corn. Frequent cultivations and hoeing in this crop should then practically clear infested land of Canada thistles.

Quack Grass.—Quack, or couch grass, has a persistent rootstock under ground, sending up new stems from its many joints. Treatment which involves a shallow (2 or 3 inches) plowing in midsummer during drouthy weather is effective if the weather continues dry. Being a shallow-rooted plant it receives a severe shock when the rootstocks are thus exposed to the sun. Any stray survivors can be taken care of the next year with a hoe and some salt. If the weed appears in corn or potatoes clean cultivation with hand hoeing should be effective.

Morning Glory.—Patches should be handled separately as roots dragged on implements to other land will extend the infested area. Absolutely clean cultivation will starve this weed out if continued throughout the summer and fall. Some farmers report this weed to be exterminated in a few years if the field is pastured. Hogs and sheep will eat the young shoots.

Milkweed.—The same treatment as for wild morning glory will destroy milkweed. If the field is pastured it seldom becomes serious. Hand pulling is not practical.

Foxtail.—Plowing grainfields as soon as the crop is removed prevents seed formation of foxtail grasses. Clean cultivation, including hand hoeing, will kill out these grasses. The same treatment will apply to wild oats, wild barley, smartweed, pigweed, ragweed and similar weeds.

TABLE OF TROUBLESOME WEEDS

Common Name	Where Found	Dura-tion	Time of Seeding	Root Stocks	Method of Eradication
Barnyard grass	Minn.–Mont.	A	July–Sept.		Prevent seeding.
Bracted plantain	Ohio, Iowa.	A	May–Oct.		Prevent seeding
Buffalo burr	Iowa, Cal.	A	July–Nov.		Prevent seeding; cut in fall.
Bull thistle, common thistle	Everywhere.	B	July, Nov.		Prevent seeding; cut in fall.
Burdock	New Eng., Wis.	B	Aug.–Oct.		Prevent seeding; grub.
Button weed	Md.–Ala.	A	July–Nov.		Prevent seeding; cultivation.
Canada thistle	New Eng., Mich.	P	June–Sept.	*	See text.
Chess, cheat	New Eng., Wash.	A	Aug.–Oct.		Clean seed.
Clover dodder	N.Y. and N.C., west.	A	June–Nov.		Clean seed.
Cockle, corn cockle	New Eng., Wash.	A	July–Sept.		Clean seed.
Cocklebur, small burdock	Everywhere.	A	Aug.–Nov.		Prevent seeding; cultivation.
Couch grass, quack grass	New Eng., Minn.	P	Aug.–Sept.	*	See text.
Crab grass	N. J., Mo.	P	July–Oct.		Prevent seeding; cultivation.
Dandelion	Nearly everywhere.	B	May–Nov.		Cultivation.
Devil's weed, golden hawk-weed	N.Y., Ohio.	P	Aug.–Oct.	*	Sheep pasturing; cultivation; heavy cropping.
Dewberry	Md., N.C.	P	June–Aug.	*	Cultivation; smothering crops
Dog fennel	Everywhere.	A	July–Sept.		Prevent seeding.
False flax, wild flax	Mich., Minn.	A	June–Aug.		Prevent seeding.
Fleabane, whitetop	Maine, Minn.	A	July–Sept.		Prevent seeding.
Horse nettle	N.J., Iowa and South.	P	Aug.–Nov.	*	Alternate cultivation and heavy cropping.
Johnson grass	N.C., Cal.	P	Aug.–Sept.	*	Alternate cultivation and heavy cropping.
Live forever	N.Y., Pa.	P	Aug.–Sept.	*	Close cultivation.
Milkweed	N.Y., Neb.	P	Aug.–Sept.	*	Prevent seeding; cultivation; heavy cropping.
Morning glory	Del., Cal.	A	Aug.–Dec.		Prevent seeding; cultivation; heavy cropping.
Narrow-leafed stickweed, beg-gar tick	Everywhere.	A	July–Oct.	*	Clean seed; cultivation.
Nut sedge, nut grass	Md., Texas.	P	Aug.–Nov.		Alternate cultivation and smothering crops.
Orange Hawkweed	N. Y.,	P	Aug.–Oct.	*	Prevent seeding; cultivation.
Ox-eye daisy, white daisy	Maine, Va., and Ohio.	P	July–Oct.	*	Prevent seeding; cultivation.
Pigeon grass, foxtail	Everywhere.	A	July–Nov.		Burning; thorough cultivation
Pigweed	Everywhere.	A	Aug.–Nov.		Prevent seeding; thorough cultivation.
Poison ivy	Everywhere.	P	July–Aug.	*	Cultivation; repeated grubbing.
Poverty weed	Mont., N. Mex.	P	July–Sept.	*	Close cultivation; smothering crops.
Prickly lettuce	Ohio, Ia.—Utah, Ore.	A	July–Nov.		Prevent seeding; burning.
Purslane, pursley	Everywhere.	A	June–Dec.		Close cultivation.
Ragweed	Everywhere.	A	Aug.–Nov.		Prevent seeding; burning.
Rib grass, plantain	Nearly everywhere.	P	July–Nov.	*	Clean seed; cultivation.
Russian thistle	Minn., Colo.	A	Aug.–Nov.		Burning; cultivation.
Shepherd's purse	Everywhere.	A	May–Dec.		Cultivation.
Smartweed	Ohio, Neb.	A	Aug.–Sept.		Cultivation; prevent seeding.
Snap dragon	New Eng., Wis.	P	Aug., Nov.	*	Cultivation; heavy cropping.
Sorrel	Nearly everywhere.	P	June–Nov.	*	Cultivation; smothering crops
Sour dock	S.C., Ga.	P	June–Nov.	*	Thorough cultivation.
Sow thistle	New Eng., Wis.	P	Aug.–Nov.	*	Thorough cultivation; smothering crops.
Teasle	Ohio, Tenn.	B	Aug.–Oct.		Prevent seeding; cultivation.
Wild buckwheat	Mich., N.D.	A	July–Oct.		Clean seed; cultivation.
Wild carrot	New Eng., Va.	B	July–Nov.		Grubbing in fall; cultivation.
Wild mustard	New Eng., Ore.	A	July–Oct.		Prevent seeding; cultivation.
Wild onion, wild garlic	Pa., S.C.	P	Aug.–Sept.		See oil spraying, in text.
Wild parsnip	New Eng., Wis.	B	July–Oct.		Prevent seeding; cultivation.
Yellow daisy	New Eng., Ohio.	B	July–Sept.		Prevent seeding; cultivation.
Yellow dock, bitter dock	New Eng., Wis.	P	July–Sept.		Prevent seeding; cultivation.

A—Annual.   B—Biennial.   P—Perennial.

* Adapted from U. S. Dept. Agrl. Farmers' Bul. 28 (1895), 24 ff.

CHAPTER XXVI.

THE HOME ORCHARD.

In this chapter it has been the aim of the authors to confine their remarks to the planting and care of the small farm orchard—one of from 1 to 5 acres or a little more—such as may be found on nearly every farm. No attempt has been made to go into detail regarding common practices or to present the more recent untried methods of culture. Rather it has been the purpose to give the farmer practical suggestions for the care of his orchard to supply fruit for his family and the local market. The subject of commercial orcharding is beyond the scope of this work.

In general, the more common dependable varieties of orchard fruits and small fruits are given. The reader is reminded here to study such phases of fertility maintenance as manure, legumes, fertilizers and tillage, as they apply to orcharding as well as the field-crop production.

Planting the Orchard.

Buying Nursery Stock.—The first essential in buying young trees is to get nursery stock free from diseases and insects. Crown gall is especially common and hairy root is also to be watched for. San Jose scale, woolly aphis and other insects are often carried on nursery stock, when close inspection is not maintained.

Nursery stock is best shipped in boxes, lined with building paper, with moist moss around the roots and straw around the tops. The trees should be unpacked immediately upon arrival unless frozen, in which case they should be left in the box to thaw out slowly in a cellar.

When the trees are not to be planted at once, they should be "heeled in." A well-drained spot having a gentle slope (about 15°) is selected and a trench dug 12 to 18 inches deep, the dirt being shoveled to the south side. The roots should be covered with soil and kept shaded and protected from wind, care being taken that varieties are not mixed at this time. Roots, body and nearly all the top should be covered with 8 to 10 inches of soil, especially for peach and apple trees. Bundles should not be heeled in without untying or the roots will dry out.

In buying nursery stock it is well to choose young trees to facilitate transplanting. For peaches and cherries and generally for plums 1-year-old trees are preferred. In some cases 1-year-old apple and pear trees are preferred but in the hands of a novice 2-year-old trees will likely be more successful. At all times only reliable nurserymen should be patronized. Good, straight, strong trees, true to name, typical of the variety, and free from disease and insects are best. It is advisable to get trees from one's own latitude; that is, a grower in the

North should not buy in the fall from southern nurseries, although spring purchases from the South may be as satisfactory as those from northern nurseries.

Location of Orchard.—Orchards are best protected when near bodies of water. Hill sites are less likely to be visited by frosts than level areas. Good

Yellow Transparent.

natural drainage or tile drainage is necessary for orchards. The slope does not count for much, although southern slopes allow varieties of apples to ripen sooner and give better color to the fruit. Windbreaks are an advantage to protect the orchard if they are not planted so close to the orchard as to prevent air currents and to shade the fruit trees and if they do not harbor insects and diseases.

A sandy soil is best for peaches and cherries, while apples grow on almost any soil that is fertile and well drained. Pears and plums require a heavy soil. A potato soil well drained, limed and well stocked with plant food is best for the farm orchard.

Preparation of the Land.—An ideal way to prepare a field for orcharding is to plow under a crop of clover and plant an intertilled crop to which manure is

A neglected Apple Tree before pruning.

added liberally. Trees should not be planted in sod except in extreme cases where plowing is impracticable. If the orchard is on a hill a catch crop should be grown after a cultivated crop is removed.

Time of Planting.—For northern states spring planting is preferable because the danger is eliminated of young trees winter-killing. Early planting is advisable so that growth may be well started before hot weather comes. Trees should be ordered in the fall as nurserymen have their best stock then and spring orders may be delayed too late for the best planting.

Space Between Trees.—The space between trees depends on soil and varieties. On good soils apple trees need about 40 feet each way. Peach and sour cherry trees need a distance of 20 feet; and standard pears and sweet cherries should be spaced 25 feet, and dwarf pears only about 15 feet. During the early years, filler trees, which mature quickly and yield a profit before the permanent trees are developed may be planted. Extra provision must be made for maintenance of fertility and the fillers must be removed when they crowd the permanent trees. It is advisable not to mix species in the orchard. Peach and other vigorous-growing trees are unsuited as fillers in apple orchards when

Same tree after being pruned of superfluous branches and twigs.

planted in the rows with apple trees. Small fruits, like currants, gooseberries, raspberries and blackberries, may be profitable for a few years when set in rows of young fruit trees. Yellow Transparent, Duchess, Wealthy (all summer varieties) and Wagener (a winter variety) are commonly used for fillers in apple orchards. Fillers are not generally used with stone fruits.

Manner of Planting.—Various methods of arranging trees in the orchard have been devised, but the square plan is most simple and practical. The trees should stand at least an inch deeper than they did in the nursery. Holes should be dug sufficiently large to accommodate all the roots in a normal position. The soil should then be tramped firmly about them with the surface soil loose and a mulch on top.

Pruning.—Pruning is done to train the tree in its early stages, to check transpiration of water, to open the center to sunlight, and in old age to prevent decay and remedy errors. The young tree at transplanting time is generally cut to a whip. The second year from three to five scaffold branches should be selected so that an evenly balanced, symmetrical tree develops without any crotches. Future pruning will be a thinning of superfluous branches and those which form

Orchard and Field Crops cannot be grown with profit simultaneously on the same land.

crotches or rub, and to head back vigorous growth on some part of the tree that causes it to be unbalanced. Pruning is best done in early spring, but it may be done in moderate winter weather or even in summer.

A general fault in pruning is to give too little attention to the outer rim of branches midway between the top and bottom and to spend too much time cutting out the tops. Large branches should not be removed if better results can be secured by cutting a number of small ones. Heavy pruning, especially of large branches, lets in too much sunlight and may cause sunscald.

Thinning.—Thinning of fruit may be done at pruning time by removing crowded small twigs. After the fruit has set and is yet small it may often be thinned advantageously, so that a higher quality and larger size may be obtained in the remaining fruit.

Care of Old Trees.—Old trees should have the dead limbs removed; the other branches frequently should be headed back by moderate pruning. Cutting back the top is commonly necessary to facilitate spraying and picking operations. And diseased limbs or branches which cross should be removed. If the trees are planted too closely it may be necessary to remove some of the tops in order to facilitate orchard operations. It is well to remember at all times that excessive pruning may let in too much sunlight upon the bare bark and branches, and therefore two years or more should be taken to complete the work. The first season the topmost branches should be cut out, all healthy side branches being left. The next year these horizontal branches may have their extremities cut back so that a uniform, well-rounded, symmetrical top is secured. Large branches should be sawed on the under side first and then on the upper side a few inches farther out on the branch so that it does not split in breaking off. All wounds should be dressed with a thick paint, as white lead and boiled linseed oil. Such trees should then be pruned and cared for in future years or the benefits of rejuvenation will be lost. Mulching and fertilization may work wonders in renewing old orchards.

Cultivation.—Cultivation of the orchard is as necessary as for other farm crops. There are in general two different systems; namely, clean cultivation with cover crops, and sod mulch. The choice of these methods depends largely upon local conditions. Tillage would be impractical on hillsides subject to washing and a sod mulch would be better. Liberal use of manure and fertilizers along with straw or other mulching material is necessary on poor soils to promote the growth of grass, and applications of lime may be beneficial. Clean culture is disastrous to the soil, and some provision for cover crops must be made. With

Four season's fertilization with 5 lb. nitrate of soda, 5 lb. acid phosphate and 2½ lb. muriate of potash, along with a mulch of a bale of straw per tree, produced this difference in vigor on trees formerly uniform in condition.

such a system the land is plowed early in the season and a mulch is maintained until time to plant a cover crop, such as clover, soy beans, cowpeas, rye, vetch or other plants. If other crops are grown and removed the humus supply will soon be depleted, and later it will be almost impossible to secure a satisfactory growth to plow under.

The chief advantages of the mulch system are that the trees bear earlier, fruit is more highly colored and erosion is prevented. It is well to start the young orchard with cover crops to be plowed under. When bearing time comes it is necessary to check growth by limiting food and moisture. The sod mulch plan includes the use of all herbage in the orchard and occasionally hauling in additional organic matter.

Fertilizing.—Fertilizing apple orchards with 10 pounds each of nitrate

The results of Proper Fertilization. Row on left, fertilized with
5 lbs. acid phosphate, 2½ lbs. muriate of potash, yielded 46 barrels of apples. Row on right, unfertilized, yielded 9 barrels.

of soda and acid phosphate and five pounds of muriate of potash per tree all over the ground paid about $20 an acre more than using half this quantity with a mulch of one bale of straw about the trees, in experiments conducted by the Ohio Station. This larger amount of fertilizer was most beneficial in encouraging the growth of grasses for mulch. Where acid phosphate was used alone or combined with muriate of potash and applied at the rate of 350 pounds per acre, the mulch in the orchard consisted of red clover. Where the same amount of nitrate of soda was used the poverty grass and weeds formerly occupying the orchard land were replaced by timothy, redtop, and bluegrass, no seed being sown in either case. Acid phosphate should be applied evenly over the ground to produce a mulch, and nitrate of soda is necessary only when the trees lack vigor and need prompt fertilization. As previously explained each crop on the farm needs separate fertilization, and experimental tests with different fertilizers are needed to determine which are most profitable for any one orchard.

Spraying.—The three essentials of successful spraying are timeliness, proper form and strength of material and thoroughness of application. The life history of insects and diseases determine the proper time to spray; thus, San Jose scale is effectively controlled by spraying when the tree is practically dormant. Likewise, apple scab is spread when the blossom clusters unfold, and applications of spray after this time are wasted. Blight is most easily and rapidly carried about by insects at blossoming time, and infection should be checked as much as possible at this time.

Definite formulas should be used for particular diseases and insects. In order to become acquainted with the details of mixtures and the process of compounding various sprays, the farmer should write to his state experiment station for directions. Often fungicides and insecticides may be combined and two or more pests removed by one spraying.

Insecticides are used differently for sucking and biting species; the former being killed by a contact poison, like lime-sulphur, nicotine sulphate or oil. that corrodes the flesh or prevents their breathing, while biting insects are

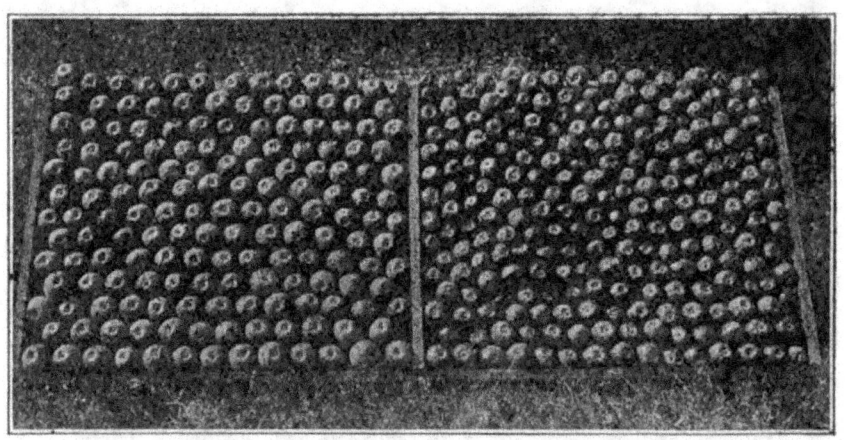

The results of fertilization. Apples on the left from fertilized trees. Apples on the right from unfertilized trees. Yield of fertilized trees, 3 barrels. Yield of unfertilized trees, 3 pecks.

killed with a stomach poison, like arsenate of lead, paris green, etc. Plant lice therefore need different treatment from that given the codling moth.

The year's program should start with the dormant spray applied wherever scale insects occur. Early spraying to kill scale insects before any leaves appear may be the difference between success and failure in the fruit crop. About 90 per cent of all the good from spraying comes from that done in March, April and May. All fruit trees, except sour cherries since they are seldom attacked by San Jose scale, should be sprayed with either commercial concentrated lime-sulphur solution diluted with seven parts of water, or a good miscible oil diluted with 15 parts of water, during March or April before the leaves expand. It is better to spray even as late as when the blossoms are showing pink than to omit this spray for trees infested with scale. The material may be applied with either a hand or a power sprayer.

For apple trees the calyx-cup spray should never be omitted, commercial concentrated lime-sulphur in 40 parts of water, along with 3 pounds of arsenate of lead paste to 50 gallons of spray, being generally recommended for application just after the blossoms fall. One part nicotine sulphate to 700 of spray is effective if plant lice (aphids) are present. The same strength of lime-sulphur and arsenate of lead—2-2-50 Bordeaux mixture where blotch or bitter rot occurs—should be applied in July or August. This Bordeaux mixture is made by dissolving 2 pounds of copper sulphate in hot water, and pouring the solution in about 20 to 25 gallons of water; to this is added the

Proper Spraying is a prime requisite in Profitable Orcharding.

milk of lime obtained by slaking 2 pounds of quicklime; and water is then added to make 50 gallons.

Peach trees require the dormant spray; and then often the petals drop. 3 pounds of arsenate of lead paste in 50 gallons of water should be used to prevent wormy fruit. About 10 days later the same amount of arsenate of lead in self-boiled lime-sulphur solution should be applied. Self-boiled lime-sulphur applied again 3 or 4 weeks later is commonly recommended to prevent brown rot.

Plums need a spray of 4-4-50 Bordeaux and 3 pounds of arsenate of lead to 50 gallons of solution when the buds are swelling. For European varieties the same material after the calyx drops and later as necessary is recommended to prevent fungus diseases and curculio attacks. For American and Japanese varieties self-boiled lime-sulphur must be substituted for Bordeaux.

Wormy cherries and attacks by plant lice may be prevented by spraying just after the blossoms fall with 3 pounds of arsenate of lead paste in 50 gallons of water plus one part of nicotine sulphate to 500 of spray, plus 2 pounds of dissolved soap in each 50 gallons.

Culture of Small Fruits.

Among small fruits are included blackberries, raspberries, strawberries, gooseberries and currants. Of these the raspberry is the most important and the gooseberry of least significance. In quantity of fruit produced, the blackberry about equals the raspberry, and is gathered wild more than any of the other species named.

All the bush fruits do fairly well on nearly every fertile soil that is well drained, although extremely rich soil is objectionable. A leguminous crop plowed under in the fall or early spring is excellent to precede the planting of small fruits.

Blackberries.—Blackberries are usually planted in the fall, plants being set 2 by 8 feet apart. Shallow cultivation to keep down weeds is necessary.

Raspberries.—Raspberries are divided into blackcap, purple-cap and red varieties. The red raspberries are generally planted in the fall in hills 5 feet apart each way, or in rows 6 feet apart with the plants spaced 2 feet apart in the row. The young sprouts coming up along the row should be cut off with a hoe or cultivator. If too many sprouts grow the berries are small and have a reduced market value.

The black and purple varieties are planted in the fall or spring on loose sandy soil, but always in the spring on clayey soils. These varieties may be propagated by planting the ends of canes of the current season's growth. The rows should be about 7 feet apart with the plants about 2½ feet apart in the row. After the first year these varieties may be grown without cultivation by mulching with leaves or straw so as to prevent weed growth.

The Cuthbert is the leading red raspberry, and the Turner is another hardy, prolific bearer. Early King is also recommended. Cumberland and Kansas are excellent blackcap raspberries and Conrath is a satisfactory early sort. The Columbian is a good purple-cap raspberry.

Raspberries and blackberries are trained much alike. Little pruning is done the first year. During the second and succeeding years the canes are cut back at the tips when they become more than 3 feet long. This summer pruning makes the canes more stocky and with more side branches. Red raspberries and blackberries are better if tied to wires than if allowed to stand alone. These bush fruits should be pruned in the spring, laterals being kept back to within 18 inches of the central cane and the dead canes being removed.

Currants.—Currants are usually planted in the fall 3 feet apart in rows 6 feet apart. Plants 1 or 2 years old may be used. Wilder, Red Cross and Perfection are dependable varieties, while Victoria, Red Dutch and White Grape are also common.

Pruning consists in cutting out old branches in the center of the bushes and in shortening the new growth as in the case of raspberries and blackberries. Gooseberries require practically the same training.

Strawberries.—Strawberries do best on a well-drained, fertile clay loam that is retentive of moisture. An intertilled crop followed by a cover crop plowed under in early spring is an excellent preparation for strawberries.

The bed should be prepared much as for wheat and well fertilized. The hill method of planting is adapted to small patches where large berries are desired. By this method plants are set a foot apart in rows 30 to 36 inches apart and no runners are allowed to start and set roots. By the single-row plan the plants are set 15 to 18 inches apart in rows 30 to 36 inches apart and two runners are allowed from each plant. By the double-row system about six runners are allowed to start from each plant. By the matted-row plan runners are allowed to root around the plants, forming matted rows from 18 to 24 inches wide. Larger yields are obtained by growing strawberries in matted rows, but the size of the berry is not as great as that of berries grown in hills.

Cultivation of the strawberry patch is essential, the small spike-tooth and sometimes a five-tooth cultivator being used. Spring planting is most common and summer or fall planting is rarely followed. In the fall the vines should be mulched with a thin layer of straw to kill weeds and to obtain clean fruit. Cheap hay, shredded corn fodder and brome sedge may also be used as a mulch if available. Sometimes an additional spring mulch may be profitable.

Strawberry beds are generally profitable for 2 years. After this they should be mowed off, allowed to dry and to be burned. If the straw is well loosened up with a hay tedder before burning no injury will be done to the crown of the strawberry plants. Attention may then be directed to the cultivation, fertilization, spraying and other care of the new vigorous plants.

RULES FOR MEASURING FEED.

Small Grains.—A bushel holds 2150.4 cubic inches. To find the capacity of a bin, multiply the number of cubic feet by 4/5.

Ear Corn.—Two bushels of ear corn are usually considered equal to a bushel of shelled corn or small grain. To find the capacity of a crib multiply the number of cubic feet by 2/5.

Straw.—Usually a ton of settled wheat straw occupies about 1200 cubic feet. This figure varies with the time of settling and the depth of the mow.

Hay.—A ton of settled hay usually occupies about 500 cubic feet, but some have found that with mows 16 feet high from 565 to 615 cubic feet of space are necessary to make a ton 5 months from filling. The Washington Experiment Station gives 512 cubic feet for a ton of settled hay in a stack; the U. S. Department of Agriculture, 515 feet for timothy or mixed hay in a stack after settling for 74 to 115 days.

Silage.—The following table (prepared by the Dairy Department of the Ohio Experiment Station) gives approximately the capacity of silos of various sizes, and the number of acres required to grow the crops specified, when averaging 15 tons per acre:

Diameter	Depth	Capacity	Area of crop required
Feet	Feet	Tons	Acres
10	20	26	2.0
12	20	38	3.0
12	24	49	3.4
12	28	60	4.0
14	22	61	4.5
14	24	67	4.7
14	28	83	5.7
14	30	93	6.0
16	26	97	7.0
16	30	119	8.0
18	30	151	10.2
18	36	189	12.3

SEEDING CARD FOR FARM CROPS*

CROP	Usual date of seeding	Distance apart of drills	Amount of seed per acre	Usual weight per bushel
		Inches	*Pounds*	*Pounds*
Alfalfa...............	Apr.–Aug.	Broadcast	10–20	60
Barley...............	Mar.–May	8	72–120	48
Beans, field...........	May–June	24–32	30–75	60
Buckwheat...........	June–July	8	35–60	52
Clover, red.............	Feb.–Aug.	Broadcast	8–12	60
Clover, alsike.........	Feb.–Aug.	Broadcast	7–10	60
Clover, white.........	Feb.–Aug.	Broadcast	6–8	60
Clover, mammoth......	Feb.–Aug.	Broadcast	8–12	60
Clover, crimson.......	July–Aug.	Broadcast	10–15	60
Clover, sweet........	Mar.–Aug.	Broadcast	‡10–20	60
Corn, hills or drills......	May	36–44	6–12	56
Corn, solid...........	May–June	8	100–125	
Cotton...............	Mar.–June	28–36	33–66	33
Cowpeas, solid........	May–July	8	60–120	60
Cowpeas, in rows.......	May–July	28–32	30–45	
Emmer...............	Mar.–Apr.	8	80–100	40
Flax.................	May	8	28–56	56
Grass, blue...........	Mar.–Apr. or July–Aug.	Broadcast	15–20	†14–32
Grass, brome.........	Mar.–Apr. or July–Aug.	Broadcast	15–25	14
Grass, Italian rye......	Mar.–Apr. or July–Aug.	Broadcast	30–35	†12–24
Grass, meadow fescue...	Mar.–Apr. or July–Aug.	Broadcast	15–20	†12–30
Grass, orchard........	Mar.–Apr. or July–Aug.	Broadcast	15–25	†14–24
Grass, perennial rye.....	Mar.–Apr. or July–Aug.	Broadcast	20–30	†18–30
Grass, redtop.........	Mar.–Apr. or July–Aug.	Broadcast	10–15	†14–40
Grass, tall oat........	Mar.–Apr. or July–Aug.	Broadcast	30–35	†10–16
Grass, timothy........	Mar. or July–Sept.	Broadcast	10–15	45
Kafir, for grain........	May–June	36	3–5	56
Mangels..............	Apr.–May	28–36	6–12	..
Millet, Hungarian......	May–July	8	25–50	50
Millet, German........	May–June	8	25–50	50
Millet, barnyard.......	May–June	8	20–30	50
Millet, pearl, thick.....	May–June	8	15–25	50
Millet, pearl, in rows....	May–June	24–30	4–8	50
Oats.................	Mar.–Apr.	8	64–96	32
Peanuts..............	May–June	28–36	44 (in pods)	22
Peas, field...........	Mar.–Apr.	8	120–180	60
Peas and oats.........	Mar.–Apr.	8	{60–120 (peas) / 32–64 (oats)	
Potatoes.............	Mar.–June	30–36	600–1000	60
Pumpkins, in hills.....	May–June	100	3–5	..
Rape, solid...........	Apr.–July	Broadcast	5–8	50
Rape, in rows.........	Apr.–July	24–30	2–3	
Rye.................	Sept.–Nov.	8	85–115	56
Sorghum, solid........	May–June	8	50–100	50
Sorghum, in rows......	May–June	36–42	8–15	
Soybeans, solid........	May–June	8	60–120	60
Soybeans, in rows.....	May–June	24–36	30–45	
Squash, in hills.......	May–June	60–80	3–5	..
Sugarbeets...........	May	18–28	12–20	..
Tobacco..............	Mar.–Apr.	Broadcast	1 tablespoonful per 100 sq. yd. (for 6 A.)	..
Turnips, solid.........	May–Aug.	Broadcast	3–5	..
Turnips, in rows.......	May–Aug.	18–28	1–2	..
Vetch, hairy..........	July–Sept.	8	30–60	60
Vetch, spring.........	Apr.–May	8	30–60	
Wheat, winter.........	Sept.–Oct.	8	75–135	60
Wheat, spring........	Apr.	8	90–120	

* For the southern states, dates of planting may vary a month earlier than here given.

† Formerly the lower figure prevailed, but owing to improved methods of cleaning the seed, it is possible to reach the higher figure.

‡ Of hulled seed; if unhulled, 2 to 3 pecks per acre.

CHAPTER XXVII.

VEGETABLES IN THE HOME GARDEN.

The small vegetable garden may be a prominent source of profit on any farm if properly managed. A large area is not necessary for an ordinary family, and when some common farm practices are applied to gardening, allowance being made in most cases for the more intensive culture, a small plot may furnish an abundance of green vegetables and others for winter storage at slight expense.

A well kept and orderly garden. Overhead system of irrigation shown. In foreground are hot bed and cold frames.

Principles of soil management and fertilization apply to vegetable gardening as to field crop production. The reader should therefore study the preceding chapters on these subjects, as only comparatively brief mention is made of them here. Where two crops or more are grown in a year on the same land, provision must be made for such intensive culture. Likewise, the thicker plantings of vegetables than of common field crops call for extra plant food and tillage.

Preparing the Soil.

Selecting a Site.—The garden soil should first of all be easily worked. A rich loam or sandy loam soil is ideal, although a clay soil by proper care can be fitted as a home for vegetables. Natural or artificial drainage is essential and must be considered before any planting is made.

When the soil is of proper physical condition, it breaks up fine and crumbly without any clods. It allows the best development for plant roots. The site must permit an abundance of sunlight.

Fitting the Seed Bed.—After the garden is plowed, it should be harrowed and raked fine. A condition suitable for seeding a lawn should be obtained for the garden. To secure such a mellow condition plowing should be delayed until

A heavy crop of Tomatoes saved by tying to stakes.

the frost is out of the ground and the soil is sufficiently dry. If the soil crumbles when squeezed in the hand, it is ready to work.

Fertilizers.—Well-rotted or composted manure is the best fertilizer for the garden. Sheep manure and poultry droppings are especially valuable. Manure may be applied liberally, often 3 to 5 inches deep if it is available.

Many garden soils may be benefited by an application of lime. Twenty-five pounds of hydrated lime is sufficient for a square rod. Peas and beans need lime if it is deficient in the soil.

Commercial fertilizers have their place in the garden, and should be carefully selected and used. Acid phosphate or steamed bonemeal will be all that is needed if manure is available.

Getting an Early Start.

Hotbeds.—The hotbed and the flat are important aids toward early vege-

Individual Staked Tomato plants. Fruits of
much superior quality are thus produced.

tables. Many tender plants may be started in a hotbed and then set in a cold frame where they gradually become accustomed to outside weather conditions. By the time frost danger is past, the vegetables have already made a good start and then grow rapidly when set out in the garden. Often two weeks or more may be gained in this way. The method lends itself to the culture of such crops as cabbage, tomatoes, cauliflower, lettuce, peppers and egg plant.

Indoor Plantings.—These plants may be started in the house in shallow, wide boxes. Good rich soil should be used and it should be watered frequently. The box should be set near a window. When the plants are well up they should be thinned to two inches apart for best growth. When such boxes are set outdoors in the daytime for several days before transplanting, the plants become hardened.

Berry boxes may be used to start a hill of beans, melons, cucumbers or sweet corn in the house. Then, when the ground is ready for planting these hills may be transplanted and they have a headstart on others planted from seed in the garden.

Planting.

Time.—After heavy frosts are past early potatoes, spinach, smooth peas, kale and onion sets may be planted. Lettuce, endive, radishes, beets, parsley, turnips and carrots may soon follow. When all frost danger is past and apple trees are in bloom, cucumbers, pumpkins, squash, sweet corn and beans may be planted with safety.

Depth.—The depth of planting, distance between rows, and other information are set forth in the accompanying planting table, based upon recommendations given by the U. S. Department of Agriculture.

Succession of Crops.

By a succession of crops the gardener may have a supply of fresh vegetables over a longer season. Radishes, lettuce, spinach, beets, peas, beans, sweet corn, parsley and turnips may be planted at intervals of about two weeks. The late plantings may not be quite as good as the earliest, but still will be satisfactory.

Early crops of beets, turnips, carrots, corn and cabbage may be out of the way in time for late crops. All those mentioned in the preceding paragraph may be followed by other crops and may follow early crops. Beets, spinach, peas, celery, cabbage, Brussels sprouts, cauliflower, kale, endive and turnips are crops adapted to late planting. In general succeeding crops should differ as widely as possible from the early crop, in order to produce the best vegetables. Root crops should follow fruiting or foliage crops rather than other root crops.

Cultivation.

The soil should be stirred frequently with a horse cultivator or hand tools. A crust should never be allowed to form on the surface of the ground. Hoeing should be practiced to kill weeds and to loosen the soil close up to the plants.

Cultivation after each rain will preserve a mulch. Stirring the soil too soon after a rain makes it compact, or puddled. Shallow cultivation—not more than two inches—is preferable.

If sprinkling is done, it is better to soak the ground thoroughly once a week than to apply some water each day.

GARDENER'S PLANTING TABLE

Vegetable	Seeds or Plants for 100-ft. Row.	Distance between Rows, Feet	Distance between Plants in Row	Depth of Planting, Inches	Time of Planting	
					South	North
Artichoke, globe	½ oz.	3–4	2–3 ft.	1–2	Spring.	Spring.
Artichoke, Jerusalem	2 qts. tubers.	3–4	1–2 ft.	2–3	Spring.	Spring.
Asparagus plants	60–80	3–5	15–20 in.	3–5	Fall or early spring.	Early spring.
Beans, bush	1 pt.	2½–3	1½–3 in.	½–2	Feb.–Apr., Aug.–Sept.	Apr.–July.
Beans, pole	½ pt.	3–4	3–4 ft.	1–2	Late spring.	May–June.
Beets	2 oz.	2–3	2–3 in.	1–2	Feb.–Apr., Aug.–Sept.	Apr.–Aug.
Brussels sprouts	¼ oz.	2½–3	16–24 in.	½	Jan.–July.	May–June.
Cabbage, late	¼ oz.	2½–3½	16–24 in.	½	June–July.	May–June.
Carrot	1 oz.	2½–3	2 in.	½	Mar.–Apr. and Sept.	Apr.–June
Cauliflower	¼ oz.	2½–3	14–18 in.	½	Jan.–Feb. and June.	Apr.–June. (Start in hotbed in Feb. or Mar.)
Celery	¼ oz.	3–4	4–8 in.	⅛	Aug.–Oct.	May–June. (Start in hotbed in Mar. or Apr.)
Chicory	¼ oz.	2½–3	2–3 ft.	½	Mar.–Apr.	May–June.
Collards	¼ oz.	2½–3	14–18 in.	½	May–June.	Late spring.
Corn, sweet	¼ pt.	3–3½	2½–3 ft.	1–2	Feb.–Apr.	May–July.
Cress, water	½ oz.	Broadcast		....	Early spring.	Apr.–Sept.
Cucumber	½ oz.	4–6	4–6 ft.	1–2	Feb.–Mar. and Sept.	Apr.–July.
Eggplant	½ oz.	2½–3	1½–2 ft.	½–1	Feb.–Apr.	Apr.–May. (Start in hotbed in Mar.)
Endive	1 oz.	2½	8–12 in.	½–1	Feb.–Apr.	Apr. and July.
Kale	¼ oz.	2½–3	1½–2 ft.	½	Oct.–Feb.	Aug.–Sept and Mar.–Apr.
Kohl-rabi	¼ oz.	2½–3	4–8 in.	½	Sept.–Mar.	Mar.–May.
Lettuce	½ oz.	2½	4–6 in.	½	Sept.–Mar.	Mar.–Sept.
Muskmelon	½ oz.	6–8	6–8 ft.	1–2	Feb.–Apr.	Apr.–June. (Start in hotbed in Mar.)
Okra	2 oz.	4–5	2–2½ ft.	1–2	Feb.–Apr.	May–June.
Onion, seed	1 oz.	2–3	2–3 in.	½–1	Oct.–Mar.	Apr.–May
Onion, sets	1 qt.	2–3	2–3 in.	1–2	Early spring	Fall and Feb.–May
Parsley	¼ oz.	2–3	3–6 in.	⅛	Sept.–May.	Sept. and early spring.
Parsnip	½ oz.	2½–3	2–3 in.	½–1	....	Apr.–May
Peas	1–2 pts.	3–4	⅘ in.	2–3	Sept.–Apr.	Mar.–June.
Pepper	⅛ oz.	2½–3	15–18 in.	½	Early spring.	May–June. (Start in hotbed in Mar.)
Potato, Irish	5 lb.	2½–3	14–18 in.	4	Jan.–Apr.	Mar.–June.
Potato, sweet	3 lb.	3–5	14 in.	3	Apr.–May.	May–June. (Start in hotbed in Apr.)
Pumpkin	½ oz.	8–12	8–12 ft.	1–2	Apr.–May.	May–July.
Radish	1 oz.	2–3	1–1½ in.	½–1	Sept.–Apr.	Mar.–Sept.
Rhubarb, seed	½ oz.	3	6–8 in.	½–1	....	Early spring.
Rhubarb plants	33	3–5	3 ft.	2–3	....	Fall or early spring
Rutabaga	¼ oz.	2½–3	6–8 in.	½–1	Aug.–Sept.	May–June.
Salsify	1 oz.	2½–3	2–4 in.	½–1	....	Early spring.
Spinach	1 oz.	2½–3	1½–2 in.	1–2	Sept.–Feb.	Sept. or early sp'g.
Squash, late	½ oz.	7–10	7–9 ft.	1–2	Spring.	Apr.–June.
Tomato	⅛ oz.	3–5	3 ft.	½–1	Dec.–Mar.	May–June. (Start in hotbed in Feb. and Mar.)
Turnip	½ oz.	2–3	2 in.	¼–½	Aug.–Oct.	May and July.
Watermelon	1 oz.	8–12	10 ft.	1–2	Mar.–May.	May–June.

INSECT PESTS AND PLANT DISEASES
CHAPTER XXVIII.

THE INSECT DANGER AND GENERAL MEANS OF CONTROL.
THE DANGER.

Loss From Insect Pests.—Hordes of pests threaten our crops each year, and levy a heavy tax on what they do not destroy. Others invade our homes, destroying our food or making life miserable by their poisonous bites or stings. Others carry from place to place man's most dread diseases or annoy his domestic animals. Careful estimates made several years ago indicated that this damage would reach the startling total of one and a quarter billions of dollars annually; a sum that is entirely beyond our comprehension. What it is today when farm crops have doubled or even trebled in value, one would hardly dare say. Certainly they would reach a figure beyond our wildest dreams of greatest wealth. There is, however, another way that the figures become even more startling. Any one, who has observed farm crops at all carefully, is forced to conclude that in general from one-tenth to one-half of the value of the crop is paid each year as a tax to the various kinds of pests which prey upon it. Some crops, it is true, will almost completely escape, whereas other crops will be entirely destroyed. Farmers have become so used to paying this tax, that anything less than complete destruction hardly causes them any concern. This being the case, it is not surprising how little this constant drain is noticed. Yet a similar heavy tax placed by a local, state or national government for necessary expenses for the public good would be followed by a wail both loud and long. But because we have always paid this tax to pests seems to be ample reason in the minds of most of us why we should continue to pay it. It is a pleasure to note, however, that gradually it is dawning that this is an unjust tax and gradually we are learning that such a tax can be avoided. Here, as in many other phases of life, an ounce of prevention is worth a pound of cure; and so in the following pages we will stress as much as possible the desirability of meeting these, our foes, more than half way and giving battle on their own ground.

Foreign Pests.—Many of our worst pests at the present time have come to us from other lands and have found here conditions of climate, of soil or of crops grown to their liking, and have spread, or are spreading gradually until they have occupied, or will occupy, we believe, all the available territory. Such invasions in the past have been made by the cabbage worm and San Jose scale. Such invasions are being made, at the present time, by the Gypsy and brown tail moths on the North East, and by the cotton boll weevil on the South. Two invaders more to be feared than foreign warships and foreign human armies. For, once established, these armies of pests will remain and will continue to levy a heavy tax for many years.

Native Pests.—Not all our pests, however, are of foreign origin. Many of them are native forms which lived formerly on wild plants nearly related to cultivated forms, but which found that the large plantings of cultivated plants offered a much more constant supply of food and a better chance of reproducing their kind. Perhaps the best known case of this kind is that of the potato bug or beetle. It formerly lived on a weed belonging to the potato family, but readily transferred its attention to cultivated potatoes and quickly spread to all parts of the country from its original home in the Rocky Mountains. Thus we are constantly threatened from within and from without by new enemies which may attack us at any time.

GENERAL MEANS OF CONTROL.
Mechanical Means.

In the following pages we will discuss briefly some of the general farm operations from the standpoint of the control of general farm pests. Obviously because of the large number of kinds of crops grown in the United States and also from the fact that we have such a large number of kinds of pests, these operations can be discussed only from the most general standpoint. However, we trust that even with this general treatment the farmer will appreciate the importance of these methods and will so modify his system of farming as to include as many of these methods as possible. Thus he will meet his enemies more than half way, across on their own territory and thus he will find that the battle is more than half won before it is begun.

Destruction of Crop Remnants.—It is doubtful if there is any one farm practice that is of more importance from the standpoint of the control of insect pests than the destruction of the stalks, stubbles or other crop remnants as soon as the crop is harvested. It is usually much easier for the time being to leave these remnants remain in the field. Harvest time is a busy time and certain things must be done. The easiest way is to leave the stubble to take care of itself. Let us look at the danger of this practice from the standpoint of the pests involved. Every stalk or stubble left in the field affords protection for a large number of pests during the winter. We have counted on a single old cabbage stalk in the early spring as many as thirty terrapin bugs and the cocoons of three cabbage worms. Now what does this mean? Just this, that we have left the very best conditions for these fellows to pass the winter. What would happen if these stalks had been removed? These bugs and worms would have been forced to find some other shelter with the chances good that it would never have been found. If it was found, the chances are that it would not have been half as good as the shelter on the old stalks, so a much greater percentage would have perished. What is true of cabbage is true of all other kinds of plants and it is increasingly true of those plants that put out suckers after the crop is harvested. Take tobacco for instance, after the crop is harvested the suckers spring up tender and succulent and furnish an abundance of the best food for tobacco insects until frost. Now if these suckers were kept down, what would happen? The tobacco pests would be forced to leave the tobacco field and find some other food with the chances decidedly against their finding any abundant supply of food. Thus many or most of them would starve to death and of the few that did succeed in living till winter, most of them would go into winter conditions so weak that they would die before another summer.

Thus we might recount case after case of the importance of destroying the remnants of the crop just as soon as the crop is harvested.

Rotation.—This is another very important way of giving notice to all pests that their presence is not desired. If the same crop or similar crops, as far as their pests are concerned, are grown on the same field year after year, each succeeding crop falls heir to all of the pests of the last year. In this way pests become more and more important each year until they cut the crop so badly that it is no longer profitable to grow a crop on that field and the farmer is forced to grow the crop some other place. Formerly we believed that the reason it was not profitable to grow a crop on a field year after year was because the crop exhausted the soil of certain foods. Now we know that all crops take the same kinds of food in slightly different proportions and hence it is suspected that the important thing is not soil exhaustion but rather the increased number of pests which prevent the crop from growing. When it is remembered that fruit trees, for instance, which cannot be rotated, thrive on the same soil year after year, if protected from their pests, and soon die if not thus protected, we see how important this method becomes. Especially when we consider the amount of spraying it takes to keep these fruit trees free from pests.

Since this country embraces a vast extent with much variation in crops grown, it is impossible to lay down any but general rules in regard to rotation. In the case of most pests a three or four year rotation is all that is required, but some require a longer rotation than this to free the soil. This is especially true if the pests that are troublesome are ones that also prey upon common weeds. In general the farmer will have to select his crops from the ones that are profitable in his community. It is well to remember that several of the pests of corn are troublesome to wheat or other grass like plants; that two or three of the worst pests of cotton are also bad pests of corn; and that as a general rule such special crops as beans and peas, clover, cabbage and potatoes are free from general crop pests but have a special set of pests of their own. With these facts in mind the farmer ought to devise for his own locality a system of rotation that would give the best results.

Time of Planting.—As a general rule most crops have a rather wide range of time over which they may be planted and grown fairly successfully. Plantings are usually made early if the season is favorable for the preparation of the land and are usually made late if the season is unfavorable. As a general proposition it may be stated that for most crops, that they should be planted just as early in the spring as it is possible to get the land prepared. Many of our pests are worst when the crop is just coming up and many of them do not emerge from winter quarters until late, so that if crops are planted as early as possible, the plants will be partly grown before the insects attack them and thus they will be stronger and better able to withstand these attacks. A notable exception must be made in the case of cutworms, which are normally much worse early in the spring than at any other time, especially so if the spring is a backward one.

In the fall it is usually best to plant just as late as possible and get a good stand. This is especially true in regard to the Hessian Fly where a delay of a week often means the difference between failure and success. There are a number of local and general factors which make it impossible to state just when any crop should be planted. We know, for instance, that all other things being equal, crops are earlier farther south, and at lower elevations, that they are later farther north, at higher elevations and near large bodies of water. However, if it is remembered that, as a general rule, the earlier the planting in the spring and the later in the fall, the better the results, providing the land is equally well prepared, it will help the farmer to a better understanding of the importance of this factor

Preparation of the Land.—In another section of this work the proper preparation of the land is emphasized. In this connection we wish to emphasize the proper preparation of the land from the standpoint of the control of farm pests. If the land is thoroughly prepared and brought into as high a state of cultivation as possible before the seed is sown, it insures the plant the best possible chance to make a good strong growth. Strong, thrifty plants at the start means that the plant has won half its battles against its enemies. In this connection we must remember that no matter how well the land is prepared, if the plant does not receive the proper fertilization it cannot make a satisfactory growth. So that the study of the proper fertilization of the plant becomes more and more important. This is not the place to discuss in detail just what that fertilization should be as it is discussed in another connection, but the farmer should always remember that proper fertilization is a very important factor in fighting pests of farm crops.

Fall Plowing.—As many of the pests of farm crops spend the fall and winter in the ground in specially constructed retreats, late fall plowing is an important factor in keeping these pests in control, for in this way we break up their winter quarters and place them where winter has a chance to kill them. This fall plowing should not be done too early, for if it is many of the insects will emerge and make new quarters before winter sets in and thus they will escape. As a general rule fall plowing of this kind should not be done until after a heavy frost.

Thorough Cultivation.—As a general rule we cultivate to conserve the moisture and to keep down weeds. and thus encourage the plant to grow. However, we must remember that as we cultivate we also discourage the pests of crops. Especially those that live in or on the surface of the soil, for by frequent disturbances pests are discouraged and driven to some other locality.

Seed Selection.—One would naturally expect that before all care had been taken to rotate crops, to give them proper cultivation and fertilization, the farmer had taken all precaution to see that he had selected the best seeds; but frequently this matter is neglected. As a general rule we need only state that this involves seeds of good germinating qualities. Every farmer should make it a rule to test his seed grain every year to see whether it will germinate properly. If it does, he is sure of good, strong, disease-resisting plants. If it does not germinate properly it should not be used. Some varieties are more disease resistant than others, and these should always be planted providing they are as good in every other way.

Clean Farming.—Many of the things discussed above might be considered as simply good farm practices, and so they are. However, that they are an important factor in the control of all farm pests is easy to demonstrate. All anyone needs to do is to look about in his own neighborhood and see the difference between the farm where fence rows and ditch banks are kept clean, where weeds are kept down and all of the hundred and one odd things are done to make clean farming, to appreciate the difference between such a farm, in regard to the loss due to farm pests, and the farm where these things are not attended to. In this connection it must be pointed out that every plant saved from pest attack is just that much gained. It takes just as much to plant, care for, and harvest a weak sickly plant that only produces a half crop as it does a good plant, free from disease, which produces a full crop. Most of the general things we can do to prevent pest attack are just the things we ought to do for better farming from every other standpoint.

Chemical Means.

Insecticides and Fungicides.—Insecticides are poisons that are used to kill insects. Fungicides are poisons that are used to kill plant pests. Unfortunately we do not have at the present time any one poison that will do for all purposes. Neither is it possible to always use them at the same strength, for while some plants will stand very strong poisons, others will be severely injured by weak poisons. The farmer should therefore study the subject and be sure he is using the proper strength.

Fortunately it is usually possible to use a combination of insecticide and fungicide in the same spray. This should be done whenever possible as by so doing we are killing two birds with one stone. Below we discuss some of the more common insecticides and fungicides and give the strengths at which they should ordinarily be used. Most of these mixtures can be bought already prepared for mixing with water or some other carrier as most of them cannot be made on the farm successfully. Spray mixtures are to be used either as a liquid spray, that is, the poison in water, or as a dust spray, that is, the poison in the form of a dust. The former is used mostly on large trees or where a great deal of poison is to be used. The latter is used on small plants or where not much is to be used.

Paris Green.—This is one of the oldest insecticides and is still used to a great extent. It injures plants more than some other poisons and should, therefore, never be used on tender plants like the peach, plum, cherry and bean. On other plants like cabbage and potatoes it may be used successfully. On cabbage it may be used most successfully as a dust spray. The following are good mixtures:

Dust Spray—
    Air slaked lime or flour..............................1 lb.
    Paris green ....1 oz.

Mix the two together thoroughly until there are no streaks or lumps of darker green. Dust through a thin muslin bag when the plants are wet with dew or just after a shower.

Liquid Spray—

	Large Amt.	Small Amt.
Paris green	1 lb.	1 oz.
Water	50 gal.	4 gal.

Bordeaux Mixture (see page 220) may be used instead of water. The liquid spray is perhaps better for potatoes. Spray plants thoroughly, keeping fresh leaves covered. It is usually better not to spray during the middle of the day.

The following is a poison bait for grasshoppers or cutworms:
    Bran...20 lb.
    Paris green..1 lb.
    Water..3½ gal.
    Cheap molasses...................................2 qt.

Mix the bran and Paris green carefully and then moisten with the water to which has been added the juice and rinds of three lemons. The lemon rinds should be cut up fine. After this add the molasses. Mix this all thoroughly together and scatter broadcast in the fields or gardens wherever cutworms or grasshoppers are bad. If scattered broadcast there is little danger that chickens or other live stock will eat enough to hurt them. The lemons may be omitted, but it is not attractive made up that way.

Arsenate of Lead.—This is a better poison than Paris green in many ways, as it is not liable to injure the plants and it sticks to the leaves much longer. It comes either as a white powder or as a white paste. The powdered arsenate of lead is about twice as strong as the paste.

Powdered Form, Dust Spray—

Strong Mixture.	Large Amt.	Small Amt.
Powdered arsenate of lead	1 lb.	1 oz.
Finely sifted wood ashes	4 lb.	¼ lb.

Satisfactory for use on tobacco, tomatoes, potatoes, and most garden truck excepting beans.

Weak Mixture.	Large Amt.	Small Amt.
Powdered arsenate of lead	1 lb.	1 oz.
Finely sifted wood ashes	8 lb.	½ lb.

Satisfactory for use on beans and other tender garden plants.

Paste Form, Liquid Spray—

Strong Mixture.	Large Amt.	Small Amt.
Paste arsenate of lead	3 lb.	1 oz.
Water or some fungicide	50 gal.	1 gal.

This is satisfactory for spraying apples or most garden vegetables.

Weak Mixture.	Large Amt.	Small Amt.
Paste arsenate of lead	1 lb.	1 oz.
Water or some fungicide	50 gal.	3 gal.

This is satisfactory for peaches or other stone fruits and for beans.

Commercial Lime Sulphur.—This is both an insecticide and a fungicide. It is a clear, reddish brown liquid and requires only to be mixed with the proper amount of water.

Winter Strength.

Commercial lime sulphur	1 gal.
Water	8 gal.

This strength must be used only in the winter after the leaves have shed in the fall and before the buds swell in the spring.

Summer Strength.	Large Amt.	Small Amt.
Commercial lime sulphur	1 gal.	1 pt.
Water	40 gal.	5 gal.

Used on apples and other core fruits, not safe to use on peach or any of the stone fruits.

Self Boiled Lime Sulphur.

Stone lime	24 lb.
Sulphur	24 lb.
Water	150 gal.

Sift the sulphur so that there are no lumps. Put the lumps of lime in a tight barrel and pour over them about four gallons of water. As soon as the lime commences to boil, add the sulphur gradually, stirring the mixture constantly so that the two will be mixed thoroughly together, and adding water slowly so that a thin paste will be formed. As soon as the violent boiling is over, not longer than twenty minutes, depending on the lime, add enough cool water to make the mixture cool, say up to 50 gallons. Then for each gallon of this mixture add two gallons of water, add the right amount of acetate of lead (see page 218) and it is ready for spraying. This is the standard fungicide for peach, plum and cherry, and should be carefully made or it may burn the foliage.

Bordeaux Mixture.—This is an old, powerful fungicide. Its place has been taken largely by commercial lime sulphur or the self boiled lime sulphur, except for potato spraying (see page 255).

	Large Amt.	Small Amt.
Copper sulphate (bluestone)	5 lb.	1 lb.
Stone lime (quick lime or lump lime)	5 lb.	1 lb.
Water	50 gal.	10 gal.

To make Bordeaux Mixture dissolve the copper sulphate in five gallons of water. This is best done by placing it in a coarse bag and hanging it over night in a wooden bucket or tub containing the water.

The lime should be placed in a similar bucket or tub and slaked by using a very small amount of water (hot water is best), just keeping the mixture from burning dry. In this way the mixture will be reduced to a fine paste and it may then be reduced by adding five gallons of water. The bluestone solution and the lime solution are then poured together into a third vessel that will hold fifty gallons and the rest of the water up to fifty gallons added.

It will not do to pour the Bluestone into the lime or the lime into the Bluestone. The two must be poured at the same time into a third vessel. This mixture should be used the same day it is made. As long as the Bluestone and lime solutions are kept separate they will keep indefinitely, so only enough of these two should be mixed for one day's use.

Tobacco Tea.—Place one pound of tobacco stems or tobacco dust into a gallon of water and place on a slow fire. Let it simmer for one hour but do not let it boil. Dilute with three gallons of water and use as a spray.

Liver of Sulphur or Potassium Sulphide.

Liver of sulphur (potassium sulphide)	3 oz.
Water	10 gal.

The liver of sulphur is simply dissolved in the water and then thoroughly stirred. This is an excellent mixture for spraying ornamental plants and for spraying against mildews.

Spray Solution.—An excellent spray solution for plant lice and other soft bodied insects may be made with common soap and water.

Hard laundry soap	1 lb.
or	
Soft home made soap	1 pt.
Water	3 gal.

If hard soap is used it should be shaved fine and boiled with one gallon of water until it is dissolved, then taken from the fire and added to the other two gallons of water. If soft soap is available it should be poured into the water and stirred until thoroughly dissolved.

CHAPTER XXIX.

PESTS OF GENERAL FIELD CROPS.

We consider here a number of pests which attack the general field crops such as Corn, Wheat, Oats and Rye. All of these pests are widespread and are frequently very destructive. They are usually present every year but do not always do enough damage to be noticeable. Outbreaks are apt to occur any year, hence the farmer should become familiar with these pests and be constantly on his guard against them. Here, as in many other cases, "a stitch in time save nine", and an understanding of these pests and their methods of attack will often save a crop from destruction.

Practically all of these pests are well known to the average farmer and practically all of them are pests first of all to grass. Usually we notice them most when grass or sod lands are planted to some other crop.

Key to Principal Pests of General Field Crops.

I. **Attacking the Roots or Sprouting Grains.**
    A. Stout white grubs eating off the roots, killing the plant or stunting it.
                Grubworms (page 222)

    B. Slender yellowish worms which are tough and hard, eating the roots or sprouting grains.
                Wireworms (page 222)

II. **Attacking the Stem.**
    A. Smooth, greasy worms lying beneath the soil during the day, cutting off the stem at the surface of the ground.
                Cutworms (page 223)

    B. Small black bugs often occurring in countless numbers causing the plants to wither and die.
                Chinch bugs (page 224)

III. **Attacking the Leaves.**
    A. Large cutworm-like worms devouring the leaves. Often occurring in devastating hordes destroying crops as they march.
                Army worm (page 227)

    B. Rust colored spots and streaks on the leaves causing the leaves to die.
                Rusts (page 248)

IV. **Attacking the Grains.**
    Causing them to swell to great size and burst open when it is seen that they are filled with a sooty black substance which is either hard or soft.
                Smuts (page 228)

Grub Worms.—The grub worm is one of the most common pests of field and orchard and is familiar to every farmer. Not every farmer knows, however, that these pests are but one stage of the common brown May or June beetle which comes into houses early in the summer and buzzes around the light. These beetles feed on the leaves of various kinds of trees but ordinarily do not become common enough to do much damage.

Grubworms are usually pests of sod lands and ordinarily it is only when these sod lands are planted to some other crop that their injuries are noticed.

One of the chief ways of avoiding injuries by this pest is to avoid planting sod land to corn or other crops that are injured, for the first year or two. In

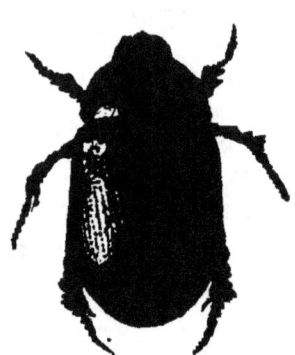

Fig. 1. June bug, the mother of the Grub Worm.

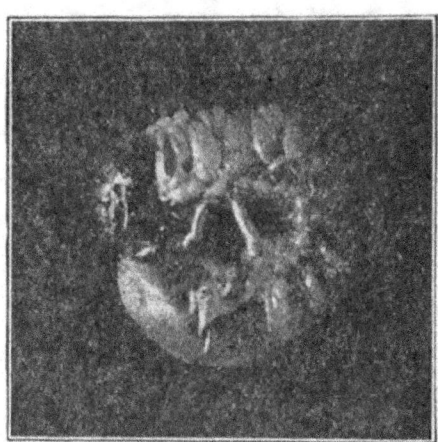

Fig. 2. Grub Worm about half grown, enlarged.

combination with this method it is well to remember that if land is left in sod year after year, the number of grub worms increase, and that if land is left in sod for only a year or two that the number of grub worms is very greatly decreased.

Hogs are very fond of grub worms and if left to root in sod lands to their hearts' content will usually free a piece of sod of these pests. However, it should be remembered that grub worms have in them the so-called stomach worm of the pig and if they are allowed to eat too many grubs there is danger of injury by the stomach worm.

Wire Worms.—Another common pest that hardly needs description to the average farmer is the wire worm. However, not every farmer is familiar with the fact that the full grown wire worm is the common snapping bug, click beetle, or jack snapper. That smooth brown beetle which when placed on its back on a smooth surface will suddenly snap into the air and land on its feet.

These are hard, round, smooth worms, very much like little bits of wire. Like the grub worms they attack grasses, grain and many other crops, feeding on the roots or sprouting grains.

Since this insect lives principally on the roots of grasses, and since injury to grains is always much worse when grains follow sod land, the best method of control is to avoid planting these crops directly after sod and to avoid leaving land lay in sod for very many years at a time.

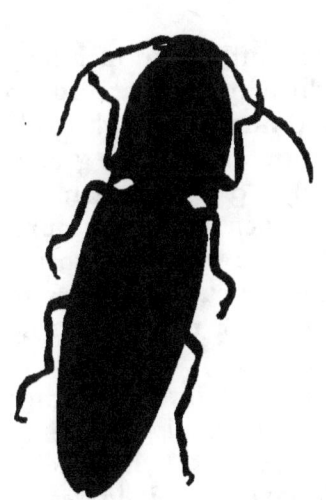

Fig. 3. Click Beetle or Jack Snapper, the mother of the Wire Worm.

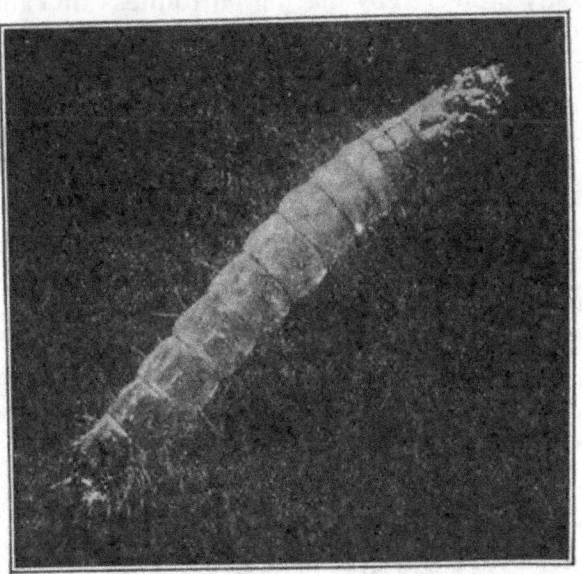

Fig. 4. A Wire Worm, one of the destructive enemies of corn and other grains.

Cut Worms.—The usual work of cut worms which consists of cutting off the young growing plants just at the surface of the ground, is familiar to all farmers. However, the work that they do later in the season when they climb plants and feed on the leaves, is not so familiar and most farmers do not recognize the adults of the cut worms at all. These are grayish moths or millars that come into lights during the summer. The life history of most of the cut worms is fairly simple. The moths lay their eggs in grass land or weedy patches and the eggs hatch out into young cut worms of our more common kinds which become half grown before winter sets in. During the winter they lie buried in the soil. When spring comes they come out of their winter quarters very hungry and attack everything in sight. Frequently they cut down more than they can eat, thus they do very great damage, especially to garden truck.

Most of their work is done early in the morning or at night. During the day they lie coiled up under the surface of the ground and it usually takes only a short time to find one near a plant that has been cut.

Here again is a pest that prefers to work in grass land, hence we should avoid planting such land to crops that are apt to be injured the first year, and especially should we avoid leaving garden spots grow up in weeds and grass in the fall, as this is especially attractive to the cut worm moths.

If one had reason to suspect the presence of cut worms in a field that is to be planted to a crop likely to be damaged by them, the field should be plowed early, thoroughly harrowed and the poison bran mash spoken of on page 218 should be scattered over the field.

In gardens, cabbage, tomatoes, and other plants that are set out may be easily protected by placing bottomless tin cans around the plants or by making little collars out of building paper. These collars are placed in the ground and held in place by the soil.

Chinch Bugs.—These are small black bugs about one fifth of an inch long that are found practically all over the country. Ordinarily they are found every year in grass land and frequently they become common enough to do immense damage to corn and small grains. While each bug can only do a little damage, they frequently occur in such enormous numbers as to do serious damage to wheat, spring oats and corn. It is not unusual for them to collect on such plants in such numbers as to actually make the plant black.

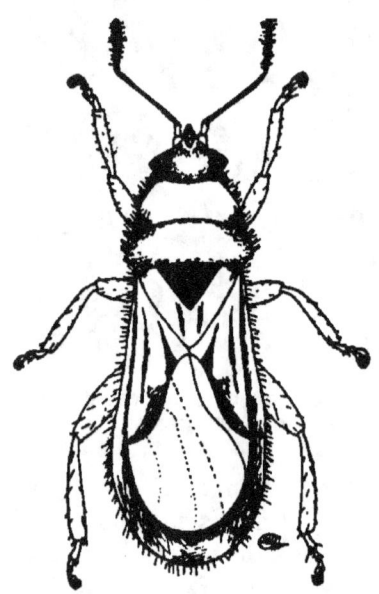

Fig. 11. Full grown Chinch Bug.

This insect lives over the winter in clumps of grass in fields, along ditch banks and fence rows. In the spring it spreads to wheat fields where, if conditions are favorable, it increases to an enormous extent, often completely riddling such fields. About harvest time, when the wheat is getting ripe so that it no longer affords good food, the pests spread to oats or corn. The farmer should always be on the watch for this spread, for he may prevent much damage to his corn or oats by stopping the chinch bug either by making a dust strip around his field, if the weather is dry, or by using a strip of coal tar if the weather is wet. It has long been known that chinch bugs cannot cross a dusty road, and in making a dust strip we make an artificial dusty road. One of the easiest ways of doing this is to plow up a strip eight or ten feet wide around the wheat field, harrow it and drag it with a bush drag, made by tying together the limbs of trees, till the ground is as fine as we can get it.

In wet weather, when it is not possible to make a dust strip, the easiest thing to do is to plow a strip two or three feet wide, throwing it up into a ridge as much as possible and then packing it with a roller. On top of this ridge we pour a narrow strip of coal tar. This strip should be about as wide as the little finger and can be put down easily by putting the tar in a watering pot from which the nozzle has been removed and the spout flattened so that it will pour a small stream. This strip can be put down quickly and can be frequently used to stop the chinch bugs after they have gotten started.

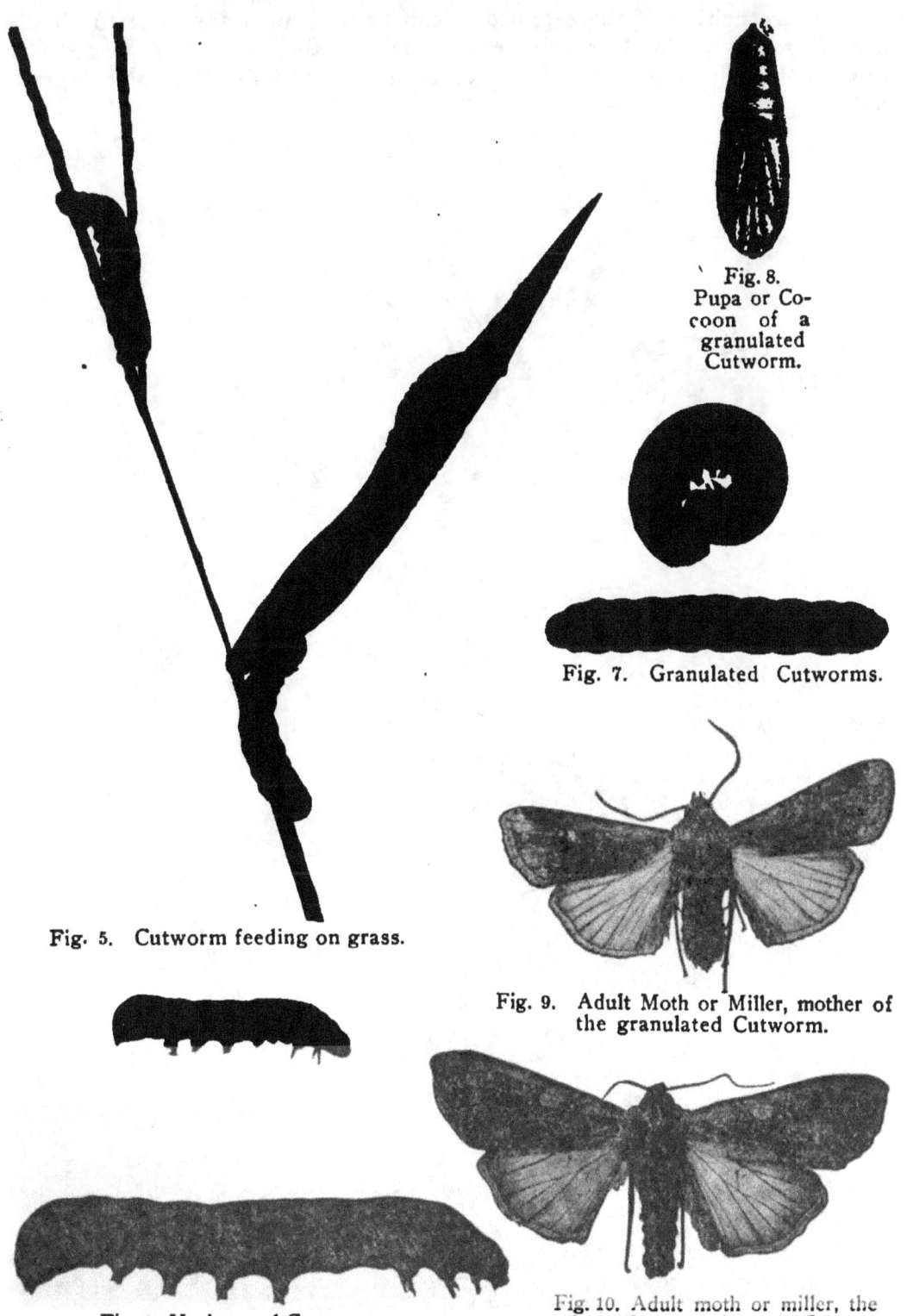

Fig. 8. Pupa or Cocoon of a granulated Cutworm.

Fig. 7. Granulated Cutworms.

Fig. 5. Cutworm feeding on grass.

Fig. 9. Adult Moth or Miller, mother of the granulated Cutworm.

Fig. 6. Variegated Cutworms.

Fig. 10. Adult moth or miller, the mother of the variegated Cutworm.

In case neither of these remedies can be used it is advisable to plow a deep furrow with its steep side next to the field that is to be protected. In this way the chinch bugs will fall into the furrow but will not be able to crawl out.

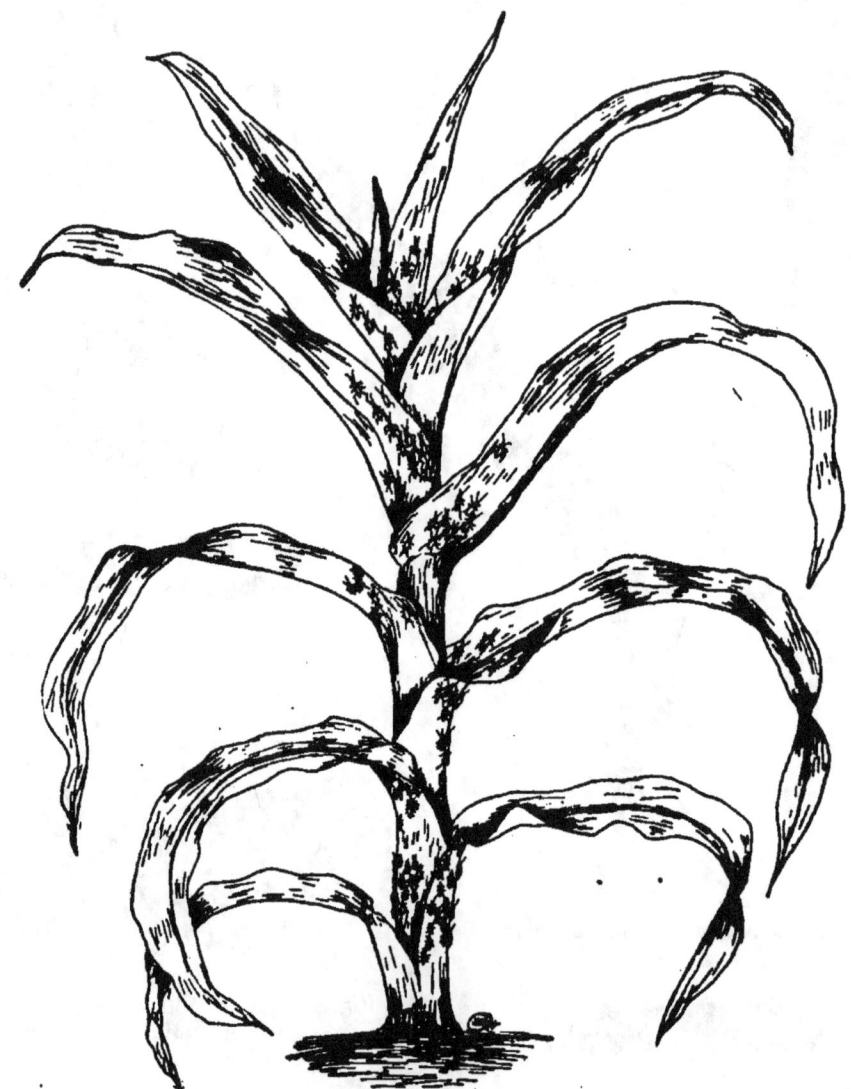

Fig. 12. Corn Stalk attacked by Chinch Bugs.

No matter what method is used, the strip should be watched carefully and if there are any breaks these should be repaired before many bugs have a chance to cross.

In addition to these methods, all grass along ditch banks, fence rows and the borders of woods should be burned each winter. Thus the **number** of chinch bugs passing the winter would be greatly reduced.

The Army Worm.—The army worm is another of those pests like the chinch bug. It is normally present every year in grass land and is apt to overcome its enemies at any time and become a serious pest. We know of no pest that is capable of doing more damage in a short time than is the army worm.

It may be described as a cut worm that has developed the climbing habit and has at the same time acquired the habit of going in immense armies. The army worms look like ordinary cut worms, and the moths are very much like ordinary cut worm moths.

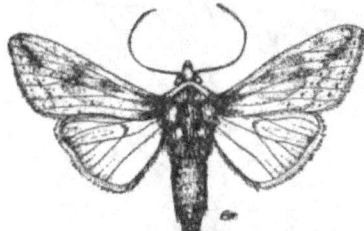

Fig. 13. Adult Moth or Miller, the mother of the Army Worm.

In ordinary years the enemies of this worm keep it in control, but every once in a while the pest for some reason gets the better of its enemies and then does serious damage. Usually by the next year its enemies will have it in control and its ravages will cease for several years, only to break out again.

The only thing that we can do in the case of

Fig. 14. Army Worms destroying Corn Stalk.

this insect is to prevent it from attacking crops not already injured. Usually it starts from grass land such as millet or timothy and spreads to corn.

One of the best remedies to keep it from spreading is to make a steep furrow as recommended for the chinch bug (see page 224). In some cases it is advisable to cut the grass and roll the land as this will crush many of the worms.

Smuts.—These diseases are well known to the farmer as large, black masses on the grain, leaves or stem of the fruit. They do great damage not only by destroying the grain but by stunting the plant and should be prevented as far as possible. On corn it may be controlled to a certain extent

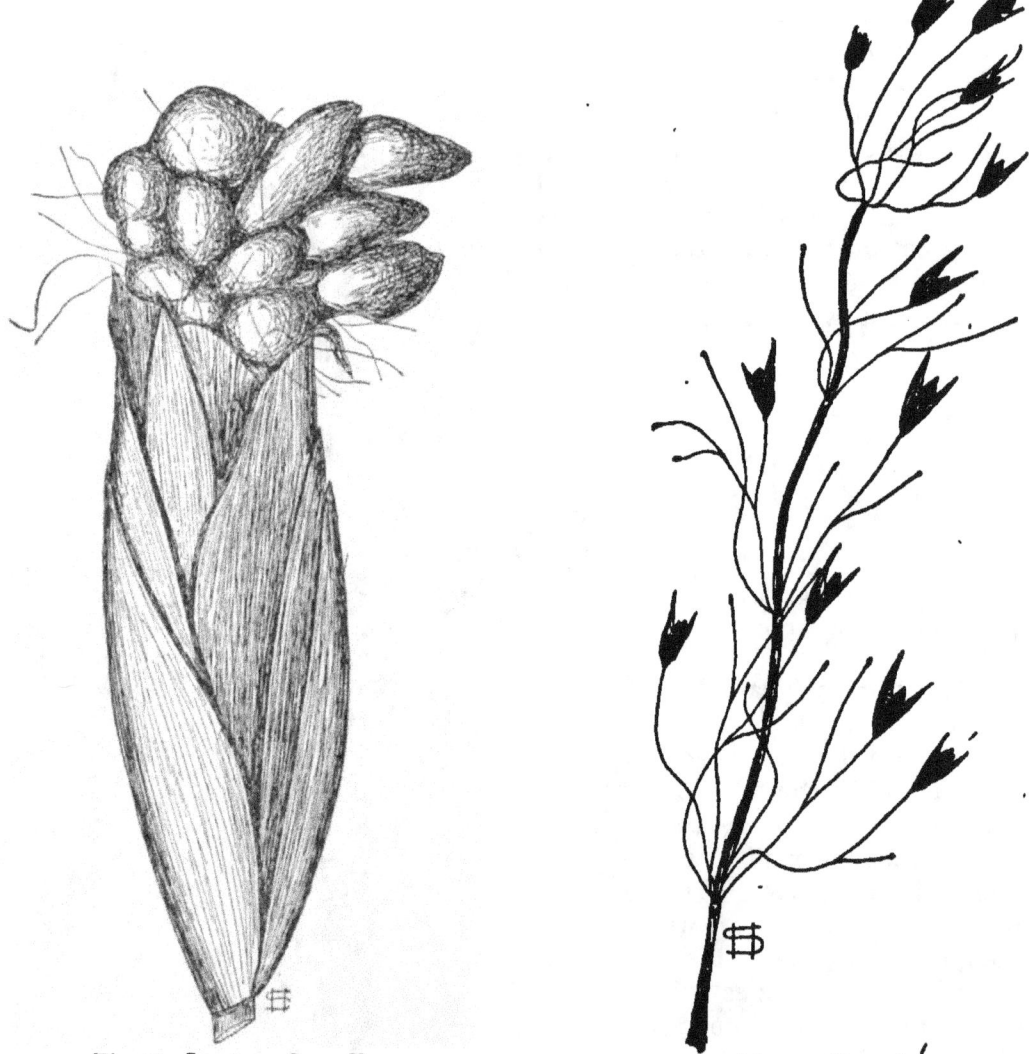

Fig. 15. Smut on Corn Ear.   Fig. 17. Oats Smut.

by cutting out the infested ears and destroying them before the black smuts form.

Seed wheat and oats should be treated with formalin (formaldehyde). One ounce of formalin should be mixed with three gallons of water, (one pound of formalin to 50 gallons of water). The grain should be spread out on the barn floor in a thin layer and one gallon of this mixture sprinkled over each bushel of grain, then the whole thoroughly mixed by shoveling it together. The pile should then be covered with sacks and left over night. In the morning it should be spread out to dry and sown at once. If it is stored it is apt to become infested with the smut again.

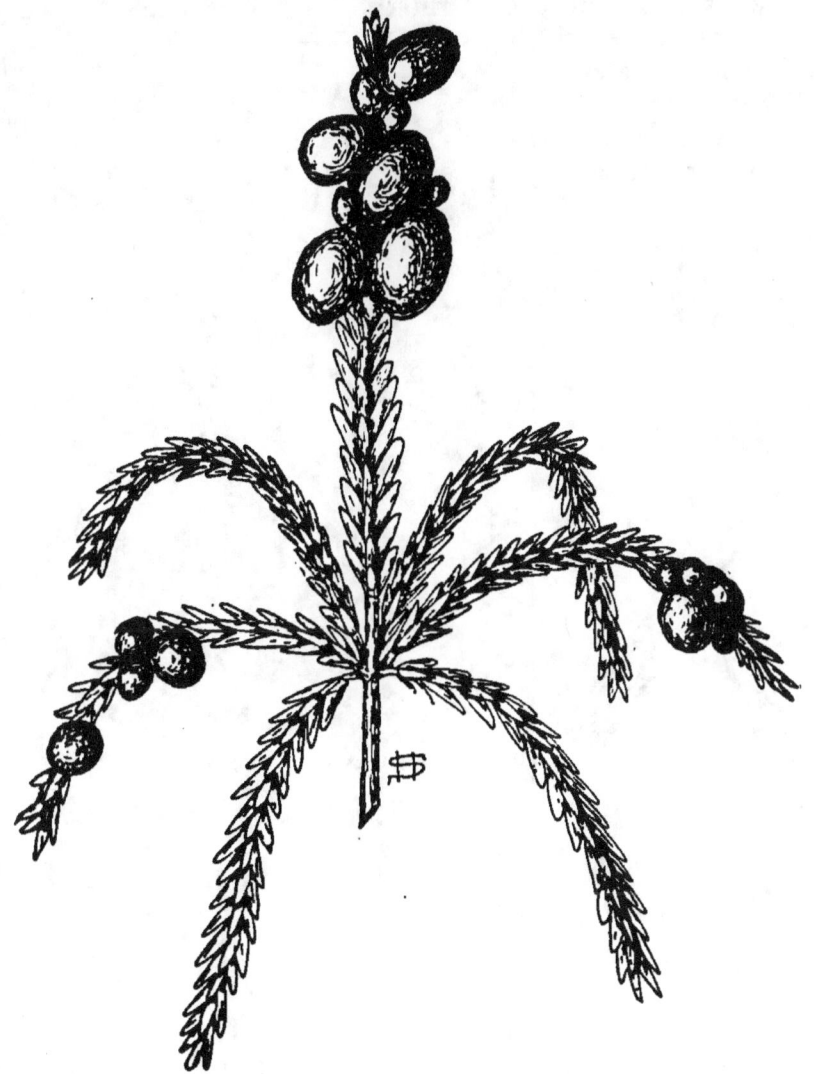

Fig. 16. Smut on Corn Tassel.

Pests of Wheat.

While small grains are troubled by all of the pests mentioned above, wheat especially is troubled by a number of pests that levy an enormous tax each year. A great deal of this damage could be avoided by proper methods of cultivation. A very little change in the ordinary farm practices would reduce this loss from what is almost too heavy to bear to a mere fraction of its present amount.

Hessian Fly.—This is one of the most serious pests of wheat and one of the most destructive insects in this country. There are two main broods each year and each brood does a characteristic kind of injury. The fall brood attacks the wheat soon after it is up. The plants attacked are weakened or killed. If they are not killed outright the central growth is killed and the wheat attempts to recover by sending out a lot of lateral growths. These laterals or tillers as they are called never make a strong growth and

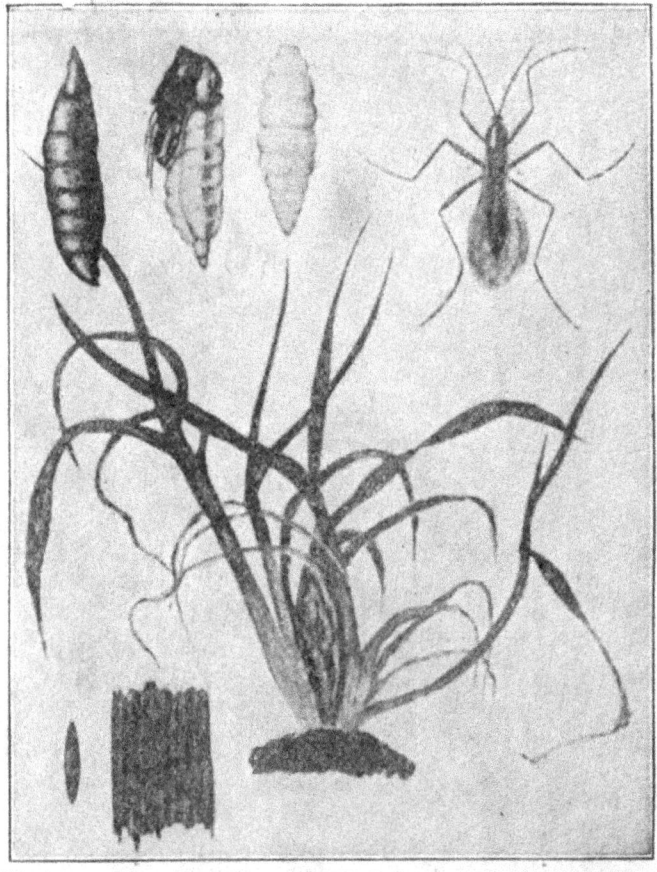

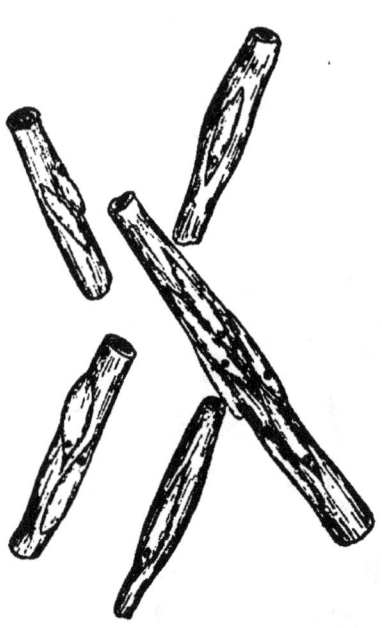

Fig. 18. Bits of straw injured by wheat joint worm found in the threshed wheat.

Fig. 19. Showing above from left to right, Flaxseed, Pupa removed from Flaxseed; Grub and Adult Hessian Fly; below, egg and injured Wheat Plant.

their heads are always small or only partly filled with small grains. During the winter the fly lives at the base of the stalk in what is usually called the flax-seed stage, small seed like pupa that resemble an ordinary flaxseed both in size and color. In the spring the flies come out of these small pupa and lay eggs for the spring brood which attack the stem of the plant, weakening it to such an extent that the wheat falls to the ground. Such plants do not have their heads properly filled out and are usually not harvested.

The pupa remains over summer on the old stubble, the flies coming out in the fall just about the time wheat is coming up. These flies lay eggs for the fall brood and start a new generation.

The control of this pest hinges around a better system of farming. This involves early fall plowing and a thorough preparation of the seed bed with high fertilization. Seeding should then be delayed about one week beyond the usual time for sowing (this time should be devoted to the preparation of the seed bed), for in the meantime the flies have emerged and finding no wheat on which to lay their eggs they will perish. Wheat should be rotated so that each new crop will not fall heir to all the pests of the old crop.

As far as possible the old stubble should be plowed under after harvest or burned off, as in this way most of the pupa that are on the stubble will be destroyed.

The Straw Worm.—Wheat is frequently attacked by straw worms, which cause a very great loss. This loss is felt in two ways, by the small, light grains formed and by the small heads. Hence we have only a very light yield. One of the best ways to judge the presence of the joint worms is to watch the threshed grain. If it contains bits of straw with small knots on them, one can be sure that straw worms have been working on the wheat. The presence of large amounts of shriveled wheat of inferior grade is also indicative of the same thing.

The control of these pests is very simple, consisting of a long rotation for wheat together with moving the wheat field each year as far as possible from the old field. These insects are frail and unable to fly very far and thus we increase the chances that the wheat will not be attacked the following year.

Pests of Corn.

I. **Attacking the Roots.**

A. Small bluish green lice sucking the roots causing the plants to wither. The presence of the lice is often made known by the ants.

Root louse (page 232)

B. Small slender white worms feeding on the small roots and boring into the "bud" below ground.

Root worms (page 233)

II. **Attacking the Stalks.**

A. White worms with black spots burrowing in the center of the stem. Found all winter in the stubble.

Stalk borer (page 233)

B. Hard shelled bugs, with long curved snouts feeding on the stem near the surface of the ground.

Bill bugs (page 234)

III. **Attacking the Ears.**

A. A cutworm-like worm eating the grain usually near the silk end.

Ear worm (page 239)

B. White mold covering the grain.

Dry rot (page 240)

Root Louse.—Whenever large numbers of ants are noticed running up and down the stalk of corn and entering the ground to go down to the roots, one can be pretty sure that root lice are working on the corn, especially if it is noticed that the corn is turning yellow and dying. The only thing necessary to establish the matter beyond doubt is to pull up a stalk carefully and note the small bluish green lice, which are sucking the sap, clinging to the roots.

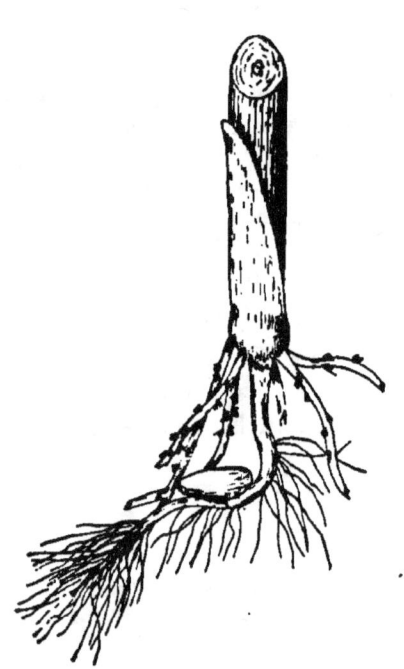

Fig. 20. Root Lice clustered on roots of corn.

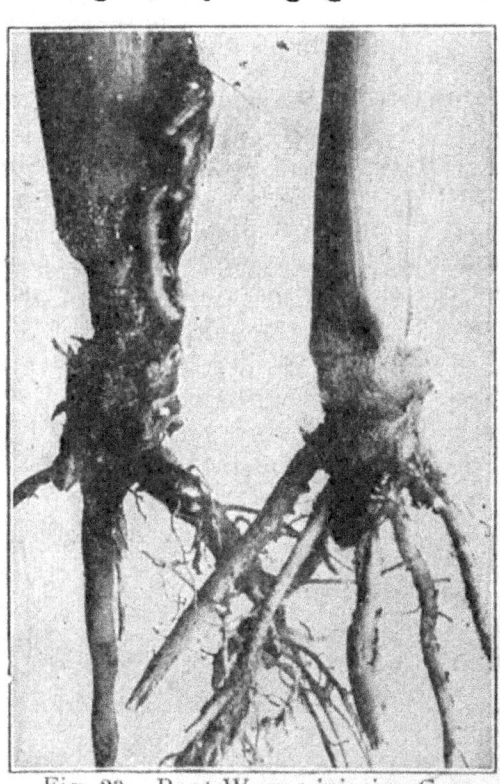

Fig. 23. Root Worms injuring Corn.

The relation between the root lice and the ants reads like a fairy tale. In the fall the root lice lay eggs and the ants take them into their burrows and keep them over winter, taking great care that they get neither too hot nor too cold. In the spring they bring them out on the surface of the ground and let the sun hatch them. The newly hatched lice are then carried away by the ants to the roots of grasses and weeds as there are usually no corn plants at this time of the year. Later when the corn is planted they take the root lice from the weeds to the corn plant. All summer long the ants tend the root lice in this way and in return the lice give them a sweetish liquid called "honey dew".

The successful control of the root louse depends upon a thorough knowledge of its life history. The best remedy that has as yet been devised consists of plowing the fields early in the spring, and then keeping the ground thoroughly stirred up by frequent harrowings. This does two things, it breaks up the ants' nests, killing many of the ants and causing them to loose the eggs of the root lice. It also keeps down the weeds and grass so that the root lice haven't any good place to live until the corn is coming up.

Root Worms.—There are two kinds of root worms attacking corn, the so-called Northern corn root worm and the Southern corn root worm.

So far as is known the Northern corn root worm works only on corn and lays its eggs in the corn field in the fall. These eggs hatch in the spring and the worms feed on the roots, weakening the plant and causing it to die

Fig. 24. Corn Stalk injured by Stalk Borer.

or make unsatisfactory growth. Since this insect feeds only on corn, a simple rotation in which corn is not followed by corn is an entirely satisfactory remedy.

The Southern corn root worm feeds on a variety of plants and its larvae are found in grasses of various kinds as well as corn. Most of its damage is done early in the spring when it attacks the corn just as it is coming through the ground. It eats into the center of the stalk just below the level of the ground. Stalks that have been injured by this pest may be easily recognized, for while the seed leaf is green and remains green for a considerable length of time, the other forming leaves in the bud die.

The control of this pest is not easy because it feeds on such a variety of plants and because the eggs are laid in the spring. A simple rotation as used against the Northern corn root worm will not be satisfactory. Perhaps the most promising remedy is to plant very early in the spring so that the corn will be well started before these pests commence their work.

Stalk Borer.—About the time corn is shoulder high a white worm with black spots can often be found working in the unfolding leaves. A little later this same worm will be found working in the stalk, eating holes into it at several places. Still later it may be found working down usually below the second joint making a tunnel from the second joint down into the tap root. When the corn is cut the stalk borer remains in the stubble all winter. The next spring he changes to a moth which lays eggs and starts a new brood of worms. This worm destroys the leaves to a certain extent, however, its chief damage consists of weakening the stalk so that it is very easily blown over.

All our remedies should be directed towards killing the worms in the stalks after the corn is cut. Perhaps the best way to do this is to disk the land very thoroughly. Thus the stalks are cut up. Many of the worms are killed and most of the rest will be killed by the winter.

Corn Bill Bug.—The corn bill bug is one of the most destructive insect enemies of corn. Many insects are more or less troublesome to corn, but the writer knows of few insects that so completely destroy the corn in the field. He has seen entire fields of corn, in which the corn bill bug has been at work, entirely destroyed. Only a very few stalks are left standing, and these are not apt to survive very long. It is not unusual for the farmer to replant his corn four or five times and then not get a stand. In fields planted at the

Fig. 25. Adult Corn Bill Bugs.

usual time most of the corn is killed by the adults before it has a chance to start. This is especially true if the spring is cold and damp, which is the most favorable time for the bill bugs to work.

The few stalks that escape being destroyed by the adults are usually attacked by the larvae or bugs which eat away the lower part of the stem and tap root, causing the plant to die or to be so badly stunted that it does not make further growth.

The work of this insect is often confused with the work of several other insects which are troublesome to corn. It is therefore necessary to recognize the work of this insect clearly, in order to be sure that the proper remedy is being applied. In the spring when the corn is coming up it is quite commonly attacked by two insects which kill it before it has a chance to make much growth. One of these insects is the Corn Root Worm, often called bud worm, or drill worm, because it "drills" a small round hole into the stalk of corn below ground and causes the "bud" to die.

Plants that are attacked by the corn bill bug early in the season show rows of round holes across the leaves.

This is due to the fact that the bill bug eats into the stalk near the level of the ground where the leaves are rolled up, thus by simply eating into the stalk at one place he cuts a row of holes across the leaf. When young plants are thus attacked, they are so badly stunted that they make no growth and eventually die. These plants are usually killed off at about the same time as the plants which are killed by the root worm, so that unless the farmer is in the field at the proper time, he is apt to think that his poor stand is due to poor seed or damp weather or to any other cause except the right one.

If the plant escapes this first attack, it is likely to be attacked again and again later, and as the leaves unfold they show the characteristic rows of holes across the leaf. Several other insects, especially the stalk borer and the ear worm, work higher up in the bud or unfolding leaves and also eat rows of holes across the leaf. However, it is our experience that these rows of holes are never as regular as the rows eaten by the bill bug, which works down near the surface of the ground.

The farmer should attempt to get as familiar with the work of these various insects as possible, because the remedies which may be used against one are not always effective against the other. One of the best ways to recognize the work of the corn bill bug is to find the insect actually at work on the base of the stalk, sometimes nearly buried beneath the ground. So far as we are aware this is the only insect with a beak about half as long as his body that feeds on the corn stalk near the ground.

A thorough knowledge of the life history and habits of an insect are necessary to an understanding and to a proper application of methods of control. The following observations are based on four years' study of this insect, both in the field and in the laboratory. These observations have been condensed as much as possible and still leave them complete enough so that they may be understood.

The adult corn bill bug (Fig. 25) is a blackish or brownish hard shell bug or beetle a little less than ½ inch long, with a recurved snout or beak about half as long as his body. At the end of this snout are sharp jaws which make it possible for this bug to eat a hole down into the stalk and then eat out the tender growing bud. Thus the insect is able to injure the plant very seriously. These adults go into the ground late in the fall where they stay all winter, coming out the following spring, usually in April, when they commence to feed on the corn. They seem to increase very greatly in numbers, day by day, until the last of May, by which time most of the adults seem to have come out of their winter quarters. Usually by the first of May the adults commence to lay eggs. These eggs are usually placed in cavities which the females have eaten into the stalk of corn, and are usually on a level with or slightly below the level of the ground. In young corn the eggs are usually laid deep in the heart of the plant, but in the older corn they are simply laid in little pockets in the inner sides of the thick outer leaves of the corn plant. Sometimes the adults lay the egg loosely among the roots and the grubs (larva) which hatch from these eggs, feed on the corn roots, causing very severe injury in this way.

The corn bill bug commences to lay eggs in early May and continues more or less regularly through the summer. They are still laying eggs in late corn the later part of September, but they undoubtedly are not as prolific after this date. In a little less than a week after the eggs are laid they hatch into white footless grubs with brownish heads. These grubs eat their way downward and toward the center of the stalk, devouring most of it as they proceed. In this way they work toward the tap root which is sometimes almost completely destroyed.

The bugs are very destructive to corn, perhaps more so than the adults, although the adults are usually held responsible for all of the work of destruction that goes on in the field because they are plainly visible on the outside of the corn stalks, while the grubs are hidden away on the inside of the stalk where they will escape notice unless they are especially searched for. The stalks that escape being destroyed by the adults are usually finished by the grub worms.

After the grubs have fed from four to six weeks they change to the pupa which is yellowish-white in color and a little less than one-half inch long. This stage shows all of the characteristics of the adult, the beak, wings and legs being closely wrapped around the body. The pupa is usually formed in the burrow made by the larva, the burrow being simply packed with chewed up fragments of corn. In this cell the pupa lies for a week or ten days and the adults emerge. The earliest adults to emerge join the old adults in the field and feed with them until they go into winter quarters. The later adults to emerge from the pupa in the fall seem to remain over winter in the pupal cells, not coming out till the following spring.

Thus this insect seems to have but a single generation each year, but each female lays from 200 to 400 eggs, so that the numbers increase very rapidly, especially as this pest has no important enemies.

The adults seem to be most active in the field in the morning from six to about nine, and again in the evening from five to about eight. During the day they seek shelter under sticks, stones and clods in the corn field. At these times the adults are often very common, sometimes being found at the rate of more than 1,800 per acre. This would be an average of about one adult for each five stalks of corn, figuring on the basis of four-foot rows with the stalks one foot apart in the rows, which readily accounts for the rapidity with which bill bugs destroy a field of corn.

The control of the corn bill bug is not an especially easy farm practice. This is due to a number of different factors. In the first place the insect cannot be attacked by any direct remedies, both because of its hardiness and because most of its life is spent in the stalk of corn or buried beneath the ground. The only direct method that might be used to any avail against this insect, is picking off the adults in the early spring and destroying them by throwing them into boiling water or kerosene. This practice would be too expensive to use on the general crop corn, but might be used in small fields of corn which were being grown for roasting ear purposes. If this method were to be used it would be advisable to scatter chips about the field under which the bugs would crawl during the middle of the day and from which they could be collected readily.

In the main, the farmer in the bill bug section of the State will have to depend upon a slight modification of his usual farm practice to control the corn bill bug.

Careful observations made during the past several seasons show quite clearly that the control is not a very easy matter. It has also been clearly demonstrated that no one method can be used with any great success against this insect, but rather that the control is dependent upon a number of factors which, taken together, give us a system of corn bill bug farming, somewhat different from the systems practiced in the corn bill bug section at the present time. These factors are discussed separately in some detail below, however, it must be emphasized again that the farmer cannot put his dependence in any one method alone, but instead he must work out a system of corn farming with these various factors in mind. In this connection it must also be emphasized here that a system which is built upon a basis of the factors here given, is not necessarily a good system for growing corn in all sections of the state and under all conditions, but rather that this is the system the farmer should employ in the corn bill bug sections if he hopes to grow corn with the minimum of loss by corn bill bugs. Neither is it intended to imply that this system will avoid the attacks of other insect enemies of corn, but is simply to be used where the corn bill bug is the most important insect enemy of corn. The writer believes that the farmer who follows this system will also be doing the most important things in the control of many of the more important insect enemies of corn. There are so many factors involved both in the many varieties of insect pests and climatic and other conditions, which have not been studied in detail, that it would be surprising if certain insects were not controlled by following the suggestions as outlined here.

The writer believes that the following factors are the most important factors involved in any system for the control of the corn bill bug, both from the standpoint of ease and cheapness of application: (1) Time of planting, (2) rotation of crops, (3) fertilization, (4) drainage, (5) thorough cultivation, (6) destruction of native food plants. It will be noted at once that all of these factors are what might be called indirect methods of control. From the very nature of the case it would be impossible to apply any direct methods, for all such methods are expensive and the returns from corn are never great enough to justify the expense involved in their application.

The reason that early planting is successful against this insect is undoubtedly because the early corn gets a start and is thus able to keep ahead of the attacks of the bugs. In the fields it always seems that when corn gets to be about waist high that it is no longer much troubled with corn bill bugs, but that it is able to continue growing in spite of their attacks.

Rotation of crops is an important factor in the control of the corn bill bug just as it is in the control of other important insect enemies of farm crops. This is due apparently to two reasons. First, to the fact that if crops are moved from field to field the insects are forced to move also because as a general thing insects which feed on corn do not feed upon other farm crops. In forcing the insects to move in this way, the chances are that they will not be able to find the new location, and as a result will die for the want of feed. Second, when corn is grown continuously on a field for a number of years the food which the corn requires becomes so exhausted that the corn does not make a satisfactory growth, and hence it readily succumbs to the attacks of insects. Whereas, if proper crop rotation is practiced, the corn will get a satisfactory start and will be strong and vigorous enough to withstand the attacks of insects to a certain extent at least.

Fertilization is perhaps next to "time of planting," the most important factor in the control of this insect. This is not of course the proper place to enter into a discussion of the proper fertilizer to use for corn, however, it has been shown that an application of fertilizer will make the plants grow much more rapidly, and hence be less liable to be fatally injured by the corn bill bug than plants which have not been heavily fertilized. In the same way a heavy application of lime to land that is sour will make the plants grow much more rapidly and thus escape the attacks of the bill bug. The importance of lime may be further emphasized by the fact that very much of our corn land in the corn bill bug sections is naturally very acid.

It is a matter of common observation that the corn bill bug is worse in low wet lands than it is in uplands. Naturally this has led to the statement that the most important factor in the control of the corn bill bug is proper drainage of the land, and while this is an important remedy, its importance might easily be over-emphasized. While we must not leave the impression that proper drainage is not an important factor in the control of this insect, yet we must caution the farmer against assuming that drainage is all important. It would not be possible for the corn bill bug to be worse than we have observed it on some well drained fields.

It has been stated that frequent shallow cultivation will keep the corn bill bug in control, and there are many things that would lead one to believe that this might be true. As discussed above, the adults spend considerable time hiding away under clods, etc. Therefore, it would seem but logical that cultivation would have a tendency to break up these retreats, and if the cultivations were frequently made that they would have a tendency to keep the insects disturbed. These factors perhaps have some importance. However, a rather long series of observations has led the writer to believe that the effects of cultivation are not upon the bill bugs directly, but are rather upon the corn, causing it to grow more rapidly, and hence to be much more resistant to the attacks of all its pests. Observations show that corn that is well cultivated will generally make a better growth when attacked by corn bill bugs than corn which is improperly cultivated, but on the other hand the writer has frequently seen corn succumb to the attacks of this insect when so far as he could determine, the corn had had the very best cultural attention. Then again, counts of the number of bill bugs per acre made in widely varying localities failed to show any constant correlation between the state of cultivation and the number of bill bugs present per acre.

The corn bill bug is undoubtedly a native insect, and like many other insect pests, fed formerly on various kinds of weeds. So far as has been determined, these weeds were mostly grass-like plants, especially the grasses which are closely related to chufas. The corn bill bug still continues to feed and breed on these grasses throughout its range in this State. Therefore, corn grown in close proximity to such grasses, or in fields where such grasses have been allowed to grow, is more, apt to be injured by this insect than corn grown in fields that are free from these grasses.

Therefore, anything that can be done to keep down the native food plants of this insect will be important factors in its control. Many of these grasses are swamp-loving forms. Therefore, as the swampy areas of the farm are brought under better and better drainage, these plants will be restricted in area and will therefore furnish less food for the corn bill bug. This will lessen the danger of the insect spreading from its native food plants to corn. In this connection, the farmers in the bill bug sections should pay particular attention to drainage of swampy areas where the native food plants abound. They should also watch closely the edges of ditch banks and similar situations to see that such places do not grow up in these grasses, which furnish these pests with an abundance of food and act as incubators for their increase.

Also it seems hardly necessary to say that, if some other crop be grown on swamp land for the first year or two after it is cleared, it would have a tendency to reduce the damage done by this insect.

Earworm.—The earworm is a troublesome pest over practically the whole country. It is much more troublesome in corn that is grown for table purposes than in field corn, but even in field corn it is very bad practically every year. In general appearances this worm resembles the ordinary cut worm, and it is apt to do very much the same sort of damage as climbing cut worms when the corn is about shoulder high. At this time of the year the earworm attacks the unfolding leaves and the forming tassels and often completely ruins both. A little later the small eggs are laid on the silk and the young worms eat down into the tip of the ear. They continue to eat the forming grains until they are full grown when they leave the ears and burrow a few inches into the ground where they change to pupa and remain over winter.

Since this pest attacks such a wide variety of crops, corn, cotton, tobacco and tomatoes, crop rotation will not be of much value unless the infested crops are avoided. Fall plowing which will disturb this pest in its winter quarters is the best method for its control. If the early worms working on

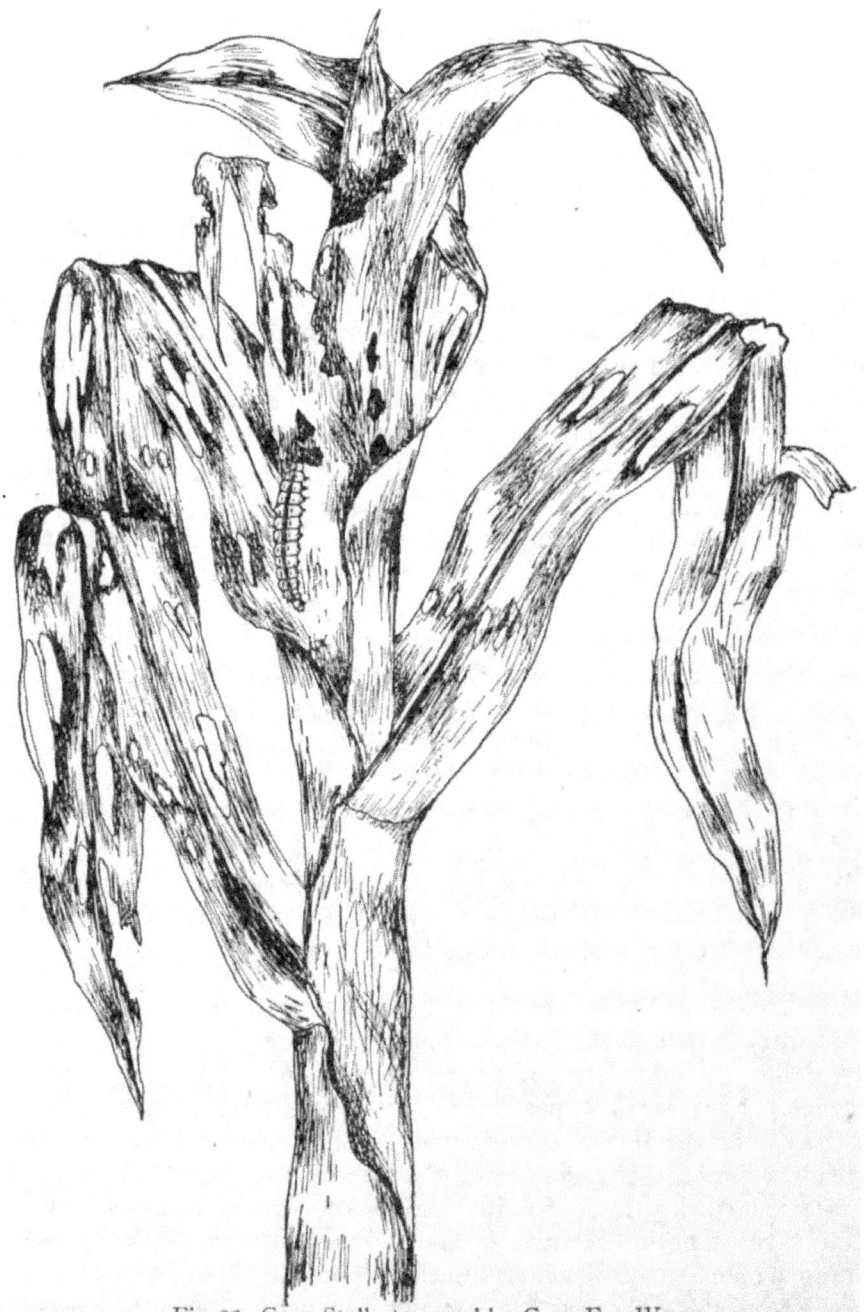

Fig. 26. Corn Stalk damaged by Corn Ear Worm.

the unfolding leaves are bad, they may be killed by mixing paris green, 4 ounces to fine corn meal 1 pound and sprinkling a pinch of this on the "bud" of each plant.

Dry Rot.—This disease is very common, especially where corn follows corn, year after year in the same field. It is readily recognized by the white "mold" which covers the ears. Ears that are "molded" are light in weight, the grains are dark in color, shrunken and not fit for feed.

Very little can be done to control this pest except practice a long rotation so that corn would not be grown in the same field for three or four years.

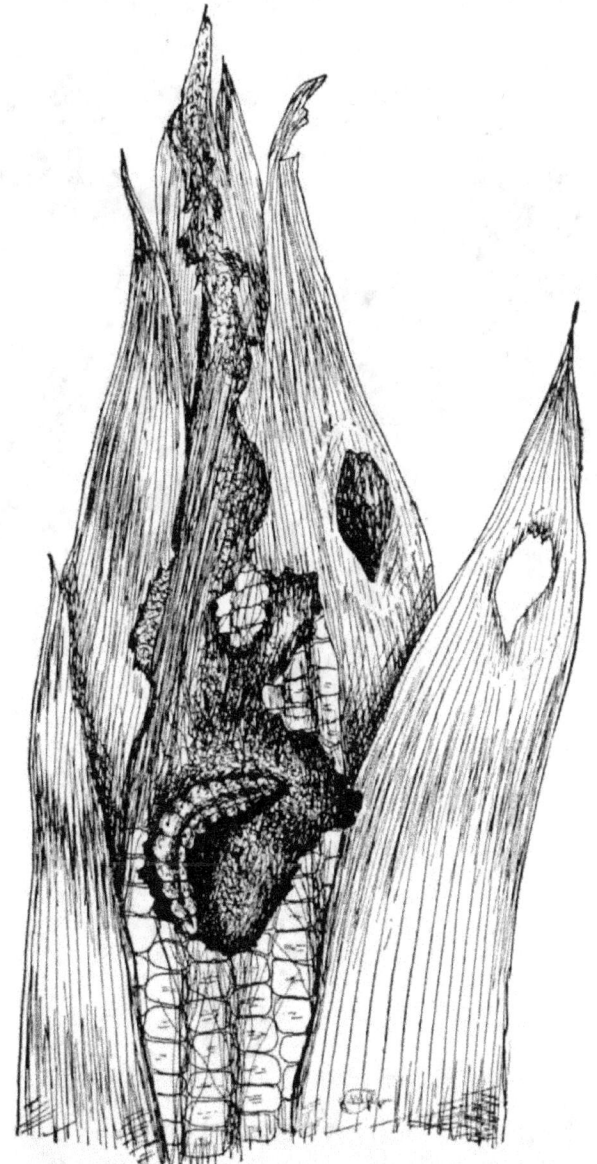

Fig. 27. Corn Ear damaged by Ear Worm.

When noticed on the ears the whole plant should be destroyed so that the disease may not become established in the field.

Clover Pests.

As a general thing it may be said that clover suffers less perhaps from the attacks of pests than most any other crop. Yet occasionally serious outbreaks of pests do occur. Perhaps the most serious damage is to the crop of clover seed. Often the farmer is at a loss to understand why his clover does not yield a greater crop of seed. Apparently it is a good stand and the plants are healthy but the heads are not filled. Usually this is blamed on the weather or some other indefinite cause. Frequently, however, it is the work of pests. (See page 242)

Fig. 28. Clover Leaf Weevil Grubs and Weevils on clover.

Clover Seed Midges.—We discuss here two small insects which live at the expense of the clover seed and are usually the cause of the shortage of the clover seed crop. Both insects are small and unless they are looked for especially they are likely to be overlooked and their work is likely to be blamed upon some other cause.

Fortunately we have a very simple remedy for the control of these pests. If the first crop of clover is harvested early while it is in full bloom practically all of these pests will be destroyed and the second crop or seed crop will be free. This remedy can always be used where clover is grown alone, but where clover is grown with timothy it is better to pasture the hay fields lightly, early in the spring. This will set the crop back so that the clover will bloom later than usual and thus it will escape the attacks of this insect.

Clover Seed Caterpillar.—This is another pest that destroys the seed crop by eating into and destroying the heads of the clover. Fortunately it too can be easily controlled by cutting the hay crop early and by avoiding growing clover on the same piece of land for more than two years. The new fields should be as far as possible from the old fields and they should be pastured lightly in the fall of the year to keep the clover down as much as possible.

Clover Leaf Weevil.—Occasionally clover is injured by stout, yellowish green grubs which cut out circular holes around the edge of the leaves. Usually this pest is not noticed until the field is in clover the second year and ordinarily it does not do much damage until the third year. Hence, clover should never be allowed to stay on the same field for more than two years. This is also a good practice from other points of view.

Clover Hay Worm.—This pest attacks the hay after it has been put in the mow or has been stacked. It is much worse in hay that has been put on top of old hay. Such hay is frequently so badly matted with the webs of insects that it will not be eaten by stock and is a complete loss.

The remedy is a simple one. Clean out all mows well in advance of harvest time, feed up all the hay and sweep up and burn all the trash that is left on the floor. Clean up the old stack before harvest, burn the trash and if possible make the new stacks at some distance from the location of the old stacks.

CHAPTER XXX.

THE PESTS OF SPECIAL FIELD CROPS.

The Pests of Cotton.

Cotton is the principal fiber plant of the world and is very largely grown in Southern United States. It, too, suffers very greatly from the attacks of pests. Interest in this direction has been especially aroused recently by the entrance and spread of the boll weevil which has gradually occupied more and more territory until at the present time it occupies nearly three fourths of the cotton growing area of the United States.

A Guide to the Chief Pests of Cotton.

Attacking the Roots.

Causing the leaves to turn yellow and the plant to die. Ants usually found entering the ground next the plant. Small, bluish green bugs on the slender roots.

Root louse (page 246)

Attacking the Leaves.

Causing the ribs of the leaf to turn dark.

Wilt (page 246)

Eating holes through the leaf.

Cotton worm (page 246)

Causing the leaves to turn yellow or brown and die.

Many small, greenish, louse-like insects especially on the under side of the leaves.

Leaf louse (page 247)

Causing the leaves to turn reddish. Many very small spider-like mites spinning webs on the under side of the leaf.

Red spider (page 248)

Attacking the Bolls.

Large, round holes eaten into the bolls by a worm.

Cotton boll worm (page 248)

Bolls or squares falling to the ground or wilting on the plant with a white grub working at the heart.

Cotton boll weevil (page 245)

Blackish ulcers, large or small, on the bolls, often turning pink.

Anthracnose (page 249)

The Cotton Boll Weevil.—Few insects have attracted as much attention in recent years and few have so threatened a staple crop as this pest. Gaining entrance into Texas from Mexico about 1892, it has spread northward and eastward at a rather steady rate, ranging in the last few years from 20 to

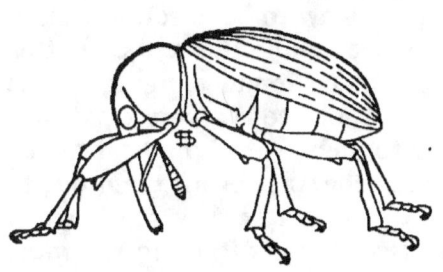

Fig. 33. Adult of the Cotton Boll Weevil.

100 miles, depending upon conditions, until at the present time it covers the principal cotton growing areas of the Southern states except North and South Carolina and will undoubtedly continue until it covers the entire cotton-growing section. (Figure 33). Like the Colorado potato beetle, which at one time threatened the growing of potatoes in this country, this insect is becoming less and less of a terror to cotton farmers as they learn how to meet the situation.

The adult insects live through the winter under various kinds of shelter, frequently some distance from the cotton fields. As soon as the cotton is up in the spring, they find their way to the cotton fields where they feed on the leaves and squares. As soon as the squares are formed on the young plants, the females commence laying eggs and continue to lay eggs for a month or two. They also take care at first not to lay two eggs in the same square or boll so they soon injure a great many. As the grubs develop inside the squares or bolls, the bolls wilt up on the plant or fall to the ground. The control of this pest has led to a revolution in southern agriculture and in this light may be considered as a blessing. The chief weapons used against this pest have been early planting in the spring, together with early harvesting of the crop in the fall and destroying the plants as soon as the crop is harvested. Rotating crops is also a great means of controlling this pest, as it has been proven many times that it is never so bad as when cotton is planted after cotton. Then it has been shown that where bolls or squares drop to the ground and lie in the

Fig. 34. Cotton Boll cut open to show Grub of cotton boll weevil.

sun, the heat will kill the grub. This has led to the introduction of the chain cultivation, which is an ordinary cultivator with heavy chains fastened to it so that the fallen bolls or squares are dragged out from under the shade of the plants into the center of the row in the sun and killed by the heat.

Root Louse, also called "Blue Bug".—When the plants suddenly commence to turn yellow in the field and die in irregular patches, the wise farmer realizes that the root louse is working on his cotton. To be sure that this disease is not confused with any other, he should watch the plants carefully and if many small brown field ants are found entering the ground next to the plants, he can be doubly sure that it is root louse. Now to be absolutely sure, he should pull or spade up some of the plants and examine the roots. If this is done carefully, small bluish gray plant lice will be found clinging to the roots. The relationship between the ants and the root lice is a very close one, as the ants carry the eggs of the lice into their tunnels where the heat of the earth keeps them warm all winter. In the spring the ants bring them out to the surface where the heat of the sun hatches them. After they are hatched the ants carry them away to the roots of various weeds and then to the roots of corn or cotton. In return for this tender care the ants get a sweetish liquid called honey dew from the root louse.

The best method of control consists of plowing the field early in the spring and harrowing it frequently before planting, as this breaks up the nests of the ants and destroys many of the root louse eggs. To control it on cotton, many farmers plant a row of corn in the cotton fields very early so that the corn can make a good growth before the cotton is up. This causes the ants to carry all the root lice to the corn roots. The corn is then plowed up before the cotton is up, care being taken to turn the roots up to the sun as much as possible, which will destroy many of the root lice. Obviously cotton should not be planted after corn as the danger from this pest is very much greater under such conditions.

Wilt.—This is sometimes a serious disease of cotton as it plugs up the ducts of the plant so that water and food cannot circulate properly. This causes the leaf to turn yellow or brown and die. The veins through the leaf are black in color. This disease lives over the winter in the soil. Therefore we should never plant cotton on land that has grown diseased plants the year before. Much good could be done if the diseased plants were pulled up and destroyed by burning as fast as they were noticed in the field, for in this way disease would be kept from spreading and from gaining a foothold in the soil.

Cotton Worm, also called "Army Worm."—Before the cotton boll weevil became so destructive, thus forcing farmers to plant early, much of the late cotton was seriously injured by a blackish worm which ate the leaves off the cotton plants. Now, fortunately, more cotton is grown early and thus largely escapes the attacks of this insect. If cotton is threatened by the attacks of this worm it may be dusted over with a mixture consisting of one pound of Paris green mixed with four pounds of air-slaked lime. The ingredients should be thoroughly mixed so that the whole mixture has a light green color without any streaks or spots of darker green. It should be applied early in the morning while the dew is still on or just after a shower while the leaves are still wet. The poison mixture is placed in thin bags and tied on the ends of poles eight or ten feet long. A man carries these through the field on horse back, shaking the pole as he goes, thus causing the poison to sift down on the plants. If he laps his rows as he goes across the field, enough of the poison will stick to the leaves to protect them very well.

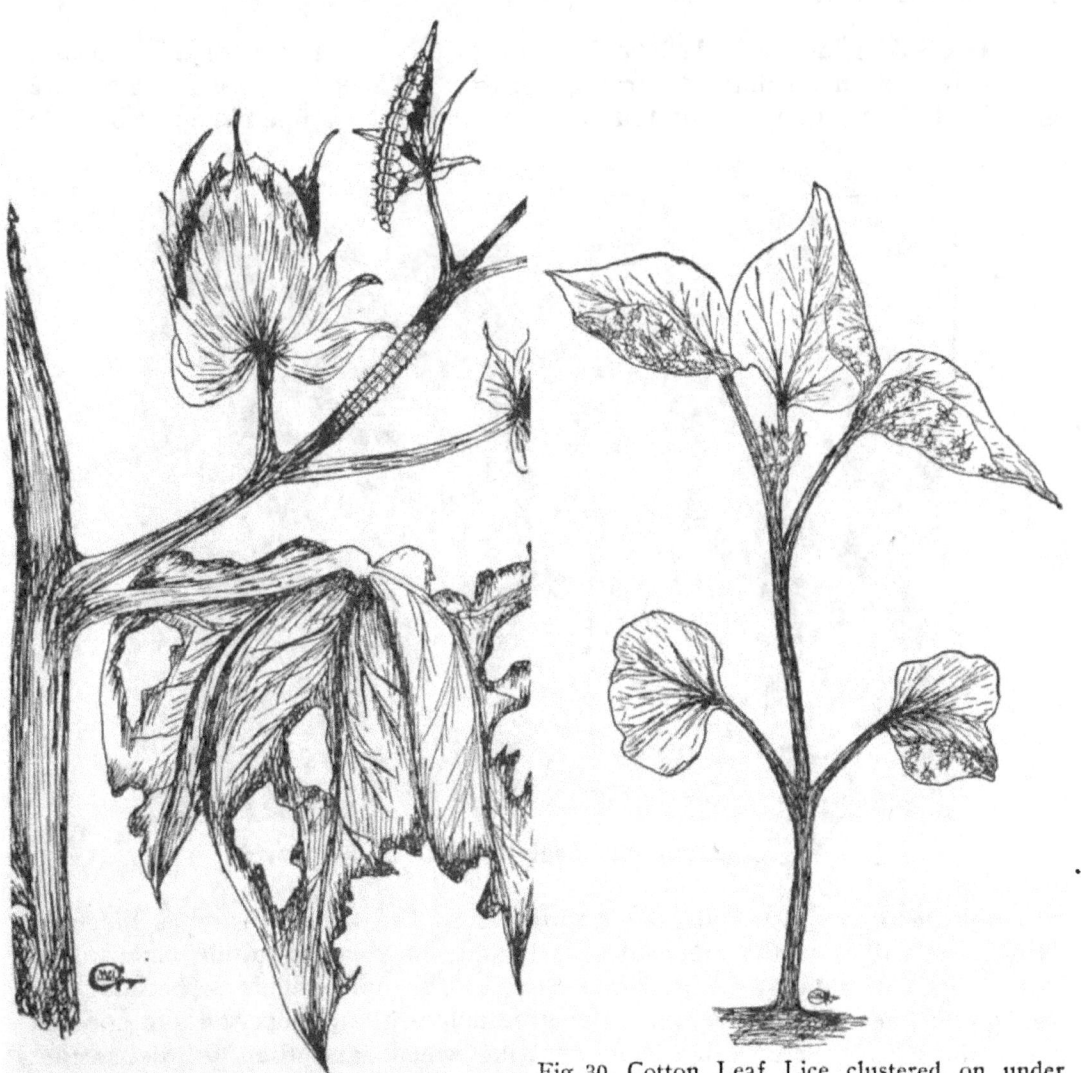

Fig. 29. Cotton Worms damaging cotton.

Fig. 30. Cotton Leaf Lice clustered on under sides of cotton leaves.

Leaf Louse.—Very closely related to the root louse is a greenish louse that lives principally on the under side of the leaves of the cotton plant, sucking the sap and causing the leaves to turn yellow and die. This insect is worse in cold, damp weather and usually is checked by dry, hot weather. Hence it is apt to be bad in the early spring when the plants are small. Delaying plantings will usually escape the attacks of this insect. These lice are always preyed upon by hordes of lady bugs if the weather is favorable and farmers are apt to confuse causes and blame these lady bugs for the injury to the cotton, whereas, as a matter of fact, they are only doing good by destroying the lice. Since this same louse is a very bad pest of melons, it is not good practice to grow melons in cotton fields as the lice are apt to spread from the melons to the cotton.

Red Spider, also called "Rust."—Late in the season the leaves of cotton plants will often be noticed turning a rusty red color. If such leaves are examined closely they will be found to be covered by a fine silken web made

Fig. 31. Cotton field badly damaged by Red Spider.

by very small, reddish, spider-like animal. If the farmer watches his field closely he will note that such attacks usually start along ditch banks or in the vicinity of rank-growing weeds such as the polk. This is because the red spider lives over the winter on weeds such as the golden rod and spreads from these to polk or other rank-growing weeds and then to the cotton. Obviously anything that is done to keep down the weeds along ditch banks, fence rows and the edges of fields will help to control the pest. Perhaps the best way is by burning over the winter. Fields should be watched and as soon as the pest is noted on a few plants these should be pulled up with others close by, piled up on the spot with dry weeds or straw and burned.

Cotton Boll Worm.—Frequently cotton bolls are eaten into by a worm about 1¼ inches long, and the whole inside of the boll destroyed. This is a notorious pest, attacking corn, tobacco and tomatoes besides some other crops. This pest passes the winter as a pupa a few inches below the surface of the ground, hence the deep fall plowing will kill most of them, and this should always be practiced in a field that has been much injured by this insect.

Anthracnose.—Hand in hand with the work of the boll worms goes the work of this disease which causes the bolls to rot on the plants. The disease usually starts as small, blackish ulcers which spread rapidly until the whole

Fig. 32. Cotton Boll Worm eating into cotton boll.

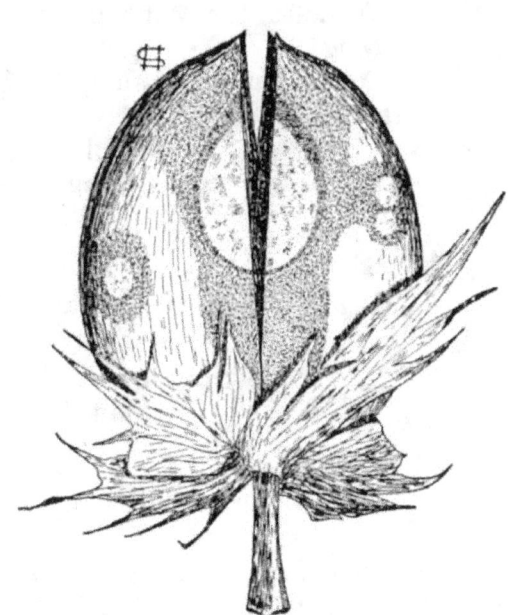

Fig. 35. Cotton Boll injured by Anthracnose.

boll is affected. Later these ulcers turn pinkish in color. This disease is carried from year to year in the seed, so that only seed that has been grown in a field that is free from the disease and that has been ginned separately from other cotton should be sown. No other treatment known is of any use against this disease. Fortunately such seed may be readily secured from certain dealers who take special care in this matter.

THE PESTS OF IRISH POTATOES.

Irish potatoes form one of the staple vegetables for this country. Not only are they grown in nearly every garden, but in many states they are grown on a commercial basis. Few crops suffer as greatly from the attacks of pests as this one and few crops can be so completely protected against the ravages of these pests as the potato.

Guide to the Pests of Potatoes.

Attacking the Leaves.

Large holes eaten by insects through the leaves.
Soft, slug like, reddish grubs present on the plant as well as hard, stripped beetles or bugs.

Colorado potato beetle (page 250) ,

No grubs, large grayish or blackish beetles or bugs with long legs.

Blister beetles (page 252)

Small holes, usually circular in shape.

Flea beetles (page 252)

No holes, the leaf wilting or turning yellow, brown or black and dying.

Leaves, especially along the border, and sometimes the stem, turning black.

Blight (page 253)

Leaves and stem wilting.

Wilt (page 254)

Attacking the Stem.

Stem turning black, small potatoes formed above ground.

Stem rot (page 254)

Stem hollowed out, usually a whitish grub can be found.

Stalk borer (page 254)

Attacking the Tubers.

Skin of tuber rough and scabby.

Scab (page 255)

Round holes, about the size of the little finger, eaten in or through the potato.

Mole crickets (page 255)

The Colorado Potato Beetle, also called Potato Beetle and Potato Bug.— This is one of the most widely known pests of crops in this country. Its history is interesting as it forms a striking example of the gradual spread of an insect until it has covered the entire country. Unlike some of our other bad pests, this insect is a native of this country and formerly lived on a kind of wild potato in the Rocky Mountains. But as soon as potatoes were placed extensively in that region (about 1855), it turned its attention from the wild potatoes to the cultivated crops, as there were many more plants to feed on and the chances of multiplying were thus increased. It soon found almost a continuous string of potato fields from the mountains eastward and took up a steady march in this direction. In nine years it had crossed the Mississippi River and in ten years more it had reached the Atlantic Ocean. Since that time it has spread to all parts of the country and has become one of the most familiar insects. Every farmer, man or boy, knows the potato bug.

Most farmers recognize two kinds of potato bugs. These are called "hard-shelled and soft-shelled bugs" or "grubs." However, not every farmer is familiar with the relations between these two forms. Not everyone knows that "hard-shelled bugs" lay eggs which hatch into "soft-shelled bugs," which later go into the ground and change into "hard-shelled bugs", but such is the case. The "hard-shelled bugs" or adults live over the winter, usually in the ground, coming out in the spring as soon as the potatoes commence to grow.

Fig. 36. Potato Plant showing eggs, grubs and adults of
the Potato Bug

They feed on the tender leaves and often do much damage to the plants, especially if the spring is backward, so that the plants cannot grow rapidly while the bugs keep steadily at work. These adults, however, do not usually do enough damage to attract the attention of the farmer. Neither do the irregular patches of orange red eggs which are laid on the leaves usually attract much attention, perhaps because they are mostly placed on the under side of the leaves. These eggs hatch into small, reddish grubs which also usually escape notice because they are small. However, the female bug lays so many eggs and these eggs hatch into grubs which eat so fast and destroy so many of the leaves that they usually attract the attention of the farmer about the time they are half grown. From this time on until they are full grown they eat an amount of food that is past belief, often completely stunting the plants, that is, eating off all the leaves and leaving nothing but the stems. When the grubs are full grown they go into the ground where they change into an entirely different form, the pupa. This remains in the ground for a week or two and then changes into the adult, which crawls out of the ground into the plant where the females lay more eggs and start a new crop. There are from one to three crops each year, depending upon the climate. But the different crops are pretty badly mixed up, because the females lay a few eggs each day over several weeks, so that one usually finds eggs and all sizes of grubs, pupa and adults in the field at the same time.

These pests have many enemies. Certain kinds of birds, such as the crow and turkey, eat them with relish. There are also a large number of different kinds of insects that prey upon them, however all of these friends of the farmer cannot keep this pest in control without help from the farmer himself. They do a great deal of good by getting the few that are left after the farmer has done his share.

The control of the potato bug is a fairly easy matter, although it is a constant fight from the time the potatoes are planted until the plants are ripe. At no time during the growing season can the farmer feel free to slacken his vigilance, for if he does, almost as by magic vast numbers of bugs will appear and, unless he is prompt in applying some remedy, they will quickly do much damage. On small potatoes in gardens the bugs may be collected by hand by brushing them into a can filled with kerosene (coal oil) and water, or by knocking to the ground and crushing. Especially should the eggs (figure 36) be looked for and destroyed. However, this is not the most profitable way and in large fields it is almost out of the question. Under these conditions spraying is the best thing to do. (Page 255)

Blister Beetles, also called "old fashioned Potato Bugs."—These insects feed on a large number of different kinds of vegetables. They come in vast swarms at times and are bad on potatoes. They are rather slender, blackish, or grayish beetles about half an inch long, with long, slender legs. They have a very complex life history which it is not necessary for us to go into in detail. It is interesting to note that at one stage the grubs feed upon the eggs of grasshoppers. Thus we have the interesting case of an insect being beneficial in one stage and injurious in another.

The same remedies can be used against these fellows as are used against the Colorado potato beetle. The ordinary sprayings keep them in control.

Flea Beetles, also called "Flea Bugs."—These are so small and so active that they are seldom seen by any one not familiar with their habits, so that frequently the potato grower is at a loss to understand just how so many little round holes got into the leaves of his potatoes. But if he will crawl up to a potato plant carefully and give it a sudden jar he will more than likely be surprised to see many small insects leaping in all directions. If some of these insects were captured they would be found to be small, blackish beetles with strong hind legs. There are really a number of different kinds working on potatoes, but they all do the same kind of damage, that is, they eat small round holes through the leaves, and as their numbers increase they often completely riddle the leaves. The grubs of these beetles are all found in the ground where they live on the potato tubers, eating tunnels through them in various directions. These tunnels look like some one had pierced the potato with a small, hot wire. Usually this damage is not serious, but sometimes it is very bad.

This is one of the most difficult pests of Irish potatoes to control, and it often does more damage than all the other pests put together. On small potatoes a large number of the adults may be collected by means of the sticky box. To make this box, a light store box about two feet square is taken and about half of the front and back are sawed out. The whole interior of the box is covered with sticky fly paper which is tacked in. To this box two

Fig. 38. Leaf of potato injured by Flea Bugs.

Fig. 39. Leaf of Potato showing beginnings of Blight. Leaves turn light green, yellow and then brown and die.

handles are fastened so that two men can carry it along the rows, one man on each side of a row. The plants readily pass through the box where the parts have been sawed out and this will cause the beetles to leap from the plants, many of them being caught on the sticky sides of the box. After the papers have become full of bugs they must be removed and fresh papers added. In this way many beetles will be captured and if the work is done early it will help the young plants very much. Complete spraying as outlined on page 255 will also help in controlling this pest, but it will not be as successful as against the other pests.

Blight.—This is one of the commonest potato diseases. It usually starts on the tip of the leaf, causing it to die and curl up in dry weather, or to decay in wet weather and give off a strong odor. The disease is carried in the seed and attacks the young as they come through the ground, causing them to be stunted and turning them brown. From these unhealthy shoots the disease quickly spreads to others if the weather is warm and wet. Thus the whole field may be quickly injured, the disease spreading from the leaves to the stem and even to the tubers, which dry rot or wet rot.

Since the disease is spread through the seed, only good seed should be planted. Diseased seed may easily be recognized by the brownish color of the inside of the potatoes. Such diseased potatoes should be discarded when the potatoes are cut for planting. Spraying, as outlined on page 255, should always be used, as this spraying not only controls the bugs but the plant diseases as well.

Wilt.—This is also usually noticed first by the curling leaves. The disease as a rule spreads quickly, causing the plants to wilt down. This cuts the crop short because the tubers are usually not mature, and such potatoes, as a rule, dry rot very badly in storage. This disease is also carried in the seed and all such diseased potatoes should be thrown away.

Besides careful selection of seed, care should be taken not to grow potatoes in the same field two years in succession.

Stem Rot, also called "Little Potato."—Frequently potato plants will be found with small potatoes formed above the ground. This is quite generally believed to be the seed of the potato by some growers, while others consider it an abnormal growth but not due to a disease. These small potatoes above ground would be of no importance if they did not tell us of the presence of a disease. Plants with these small potatoes above ground usually produce small potatoes under ground which are apt to be used for seed and thus the disease is spread. The disease is most destructive to young plant just about the time they are coming through the ground. Such plants are sickly, do not make a good growth and show large or small black spots on the stem near the surface of the ground. As these spots increase in size, they kill the stem or cause it to become very much twisted, so that it does not make good growth.

Only clean seed should be used or the seed should be treated by dipping it in formaldehyde solution, 1 pint of formaldehyde to 30 gallons of water, for two hours. Potatoes should not be grown in the same field year after year as this disease also lives over in the soil.

The Stalk Borer is sometimes a troublesome pest of potatoes as he works in the stems causing them to be greatly weakened or to die. The damage is done by a yellowish white grub which is about one half inch long when full grown. This grub hollows out the main stem or branches of the potato, horse nettle, jimpson seed and nearly related plants, late in the summer. The adult lives over the winter in these stalks. This gives us the best means of control for this pest. For if the stalks are raked together and burned as soon as the potatoes are harvested, practically all the insects will be destroyed. Since the insect breeds in certain common weeds, every effort should be made to keep these down, especially in and near the potato field.

Scab.—This is perhaps the best-known disease of potatoes, both because it is very prevalent and because it attacks the tubers in a conspicuous way. Every housewife is familiar with scabby potatoes, and the market price of such potatoes is always very much less than that of potatoes free from the disease. Scab, like other diseases of the potato, is carried both by the seed and in the soil. Unfortunately it can live in the soil for a long time, even more than five years, so that the common practice of growing potatoes in the same field for several years is a bad one. However, a four-year rotation in which plenty of cover crops, such as the clovers or rye, are turned under, will help to free the soil. Especially if no stable manure, wood ashes or lime is added, as these things tend to increase the damage from scab. On the other hand, commercial fertilizers, weak in potash and land plaster, tend to decrease the amount of scab. Even if the soil is free from scab, scabby potatoes will result if scabby seed is sown. To avoid this, the seed potatoes may be dipped

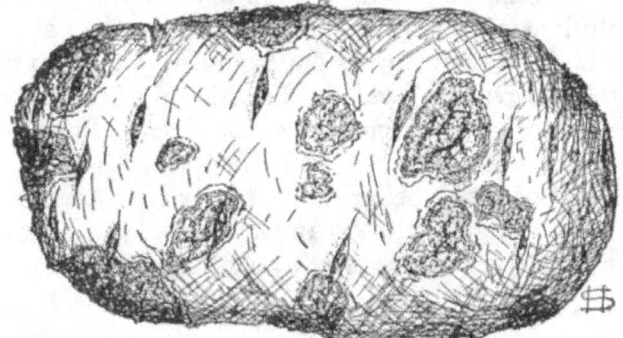

Fig. 40. Potato Scab.

in formaldehyde before they are cut as recommended under Stem Rot (page 254). It is better to do this just as the potatoes are being planted, then there is no danger of their becoming reinfested with the disease before they are planted.

Mole Crickets.—Frequently potatoes grown in low, mucky soils will have large round holes, almost as big as the little finger, drilled through them in various ways. Such potatoes are, of course, worthless for market purposes. Mole crickets fortunately are confined to low, wet fields so that two courses are open, either drain the fields thoroughly or avoid planting potatoes in such situations.

Potato Spraying.
It is doubtful if there is any farm operation that will pay as big returns for the time and money involved as potato spraying. Work carried on in Vermont and New York show that the cost of spraying an acre of potatoes will average about $5.00, and that the average increase in yield in an acre of sprayed potatoes over unsprayed potatoes ranges from eighteen bushels per acre to more than 200 bushels per acre, depending upon the weather and other local conditions. In North Carolina the yield of early potatoes has been in-

creased 33 bushels per acre by proper spraying. Potatoes, like many other crops, should be sprayed in anticipation of the pests that attack them rather than wait until they have made their attacks. It is best to spray the potatoes first just as soon as they are up far enough so that the eye can follow the rows across the field. Then let the potato beetle indicate when it is best to spray again. In other words, spray the plants when they are a few inches high and then repeat the sprayings as often as the bugs become numerous. This will usually mean from three to six sprayings each season. If it is preferred, the sprayings after the first may be given at regular intervals, every two weeks, until the plants are full grown.

The best material for spraying potatoes is Bordeaux mixture and arsenate of lead, or some other like poison. The Bordeaux mixture is made by dissolving 1 pound of copper sulphate (Bluestone) in 5 gallons of water in one vessel and slaking 1 pound of quick lime or stone lime, in as little water as possible, in another vessel, then add five gallons of water to the slaked lime. After the two solutions are made in separate vessels they are poured together into a third vessel. If the pouring is done carefully the mixture will be light blue in color without streaks of white. To this mixture is added 5 pounds of arsenate of lead or arsenate of lime or two pounds of Paris green, which has been mixed with just enough water to make a thin paste. The arsenate of lead (or the arsenate of lime) being much better than the Paris green. This mixture should be thoroughly stirred and then sprayed on the potatoes, taking care to cover the leaves as thoroughly as possible (see page 218).

Tobacco Pests.

Tobacco is a specialized crop and tobacco farmers are generally more familiar with the pests of their crop than almost any other group of farmers. The following guide includes only the most important pests of tobacco. Besides these there are a host of others which may be locally important.

I.   **Attacking the Roots** causing them to turn black.
Root rot (page 257)

II. **Attacking the Leaves.**
A.   Causing the leaves to be lop-sided or of uneven texture.
Frenching (page 257)
B.   Large, irregular holes eaten through the leaf. Large, greenish worms usually present.
Tobacco worms (page 257)
C.   Small, irregular holes eaten through the leaf.
Flea beetle (page 259)

III. **Attacking the Stem.**
Cutting it off near the ground soon after the plants are set.
Cutworm (page 257)

IV. **Attacking the unfolding leaves of the bud of the plant, also the seed pods.**
Bud worms (page 259)

Root Rot.—This disease is frequently bad in the seed beds, causing the plants to die in large numbers while very small. From the beds, it is carried to the fields, causing the plants to wilt or die or stunting them so badly that they do not produce a satisfactory crop.

In controlling this disease care should be taken to have a new seed bed each year and burning off the bed carefully before planting the seed.

Frenching.—This disease is another that is spread from the bed to the field. In the field it is spread from one plant to another by the usual farm operations in the tobacco field. Diseased plants have leaves that are irregular in shape and very uneven in texture, being thick and green in one spot and thin and yellowish in other spots.

In controlling this disease it is well to remember that it is readily spread from one plant to another. If the diseased plants are few, they should be destroyed. If many, the farmer should avoid handling or working with them and then working with healthy plants.

Cutworms.—These are frequently bad in tobacco fields just after the plants are set, cutting off the stems near the surface of the ground. They work principally at night and one worm may destroy several plants in this way. During the day they lay curled up under sticks or clods or in the ground near the plants. They may be recognized by their smooth, greasy appearance. The adults are the common grayish moths or millers with wings an inch and a half across, that are commonly seen around

Fig. 41. Tobacco Leaf to show injury caused by Frenching.

lights at night. Very often they lay their eggs in grass lands and tobacco is always most seriously damaged when it is planted in fields that have been in sod. Therefore, avoid planting tobacco in sod lands when cutworms are thought to be present. The land should be plowed early and thoroughly harrowed some weeks in advance of planting. Then about a week before planting use one of the poison baits (page 218). These poison baits may be used after the plants are set and will help destroy the worms, however, they are not as effective as when used before the plants are set.

Tobacco Worms, also called "Horn Worms," "Hornblowers," "Tobacco Flies."—These worms are very familiar objects to all tobacco growers, and "worming" tobacco is one of the most expensive operations in connection with raising a crop of tobacco.

Most tobacco farmers recognize the large, strong-flying, narrow-winged "candle flies" as the insects which lay the small, light green, ball-shaped eggs on the under side of the leaves. They know that these eggs hatch out "horn worms" which later go into the ground and change to "brown pitchers," which later issue as "candle flies."

It is in the "pitcher" stage that this insect passes the winter just a few inches below the ground, so that if the tobacco fields are plowed during the fall, most of the pitchers will be killed. The most common way of controlling tobacco worms is to pick them off the plants by hand. This is not only a very laborious way but a very expensive way to keep tobacco worms in check, although it is usually done in connection with cultivating, sucking, or tapping the tobacco. The worms are not usually noticed until they are very large and have done much damage, and even when done very carefully a large number of worms are missed. Realizing this fact, many tobacco farmers have commenced to dust their plants with Paris green or powdered arsenate of lead. They use from one to two pounds per acre of Paris green and three to five pounds of the arsenate of lead, usually making three or more applications each year depending somewhat on the numbers of worms present. The arsenate of lead is a little more expensive than the Paris green, but it is much better for a number of reasons. It does not burn the leaves of the plant and it sticks to them longer as it is not easily washed away by rains. Both of these poisons should be put on the plants by means of a blow gun with an extra large fan

Fig. 43. Young Tobacco plant badly damaged by Flea Bug.

and a long nozzle. The farmer should work with the wind not against it so that the poison is blown away from him. If arsenate of lead is used it should be the kind made for tobacco work and should be thoroughly mixed with four parts of dry, finely sifted wood ashes.

Flea Beetles, also called "Flea Bug."—These insects are very small and are hardly ever seen, although the work they do is familiar to all tobacco farmers. Flea beetles are very small and have strong hind legs which give them great power for leaping, hence the name "flea beetle." They live over the winter under trash and leaves around the tobacco field, coming out very early in the spring and attacking the young plants as soon as they come through the ground. They follow the plants from the seed bed to the field where they feed and multiply until frost drives them to shelter for the winter. Usually they work more on the first leaves, however, in bad outbreaks they spread throughout the whole plant, injuring every leaf. The holes are usually round and about the size of a pin head at first, but as the beetles continue to work these holes become much larger and very irregular.

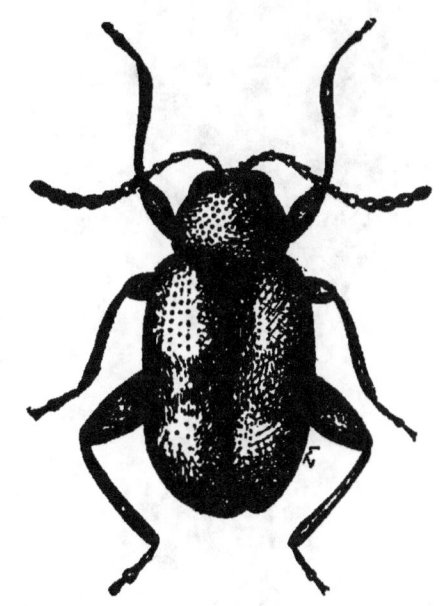

Fig. 44. Adult Tobacco Flea Bug.

Since the flea beetles come out of winter quarters and seem to gather on the seed beds first, our efforts towards controlling them should be directed towards keeping them in control on the seed beds. This is more profitable than attempting to control them in the fields as the plants are, at this time, confined to a small area and set close together, so that they may be sprayed or dusted quickly and without wasting the solution. Furthermore, it is evident that the destruction of beetles at this season not only means stronger and healthier plants but it also means that there will be fewer beetles go to the fields and increase on the plants there.

For this purpose the plant beds should be dusted with arsenate of lead at the rate of one pound to one pound of finely sifted wood ashes (see page 218).

Bud Worms.—These injure tobacco in two ways, by eating the unfolding leaves or bud of the plant and by eating into the seed pods. When these insects work in the bud of the tobacco, they often do a great deal of damage. as small holes eaten into the unfolded leaf means very large holes when the leaf is fully expanded. It has been noted that when plants are properly dusted for the tobacco worms, there is very little damage by these insects. If the tobacco is not dusted and the worms are not abundant, they may be controlled by hand picking, or if they are abundant they may be controlled by sprinkling the plants with poisoned corn meal. This should be prepared as follows: To a quart of finely ground corn meal add half a teaspoonful of Paris green and mix until there are no streaks or spots of darker green, but the whole is

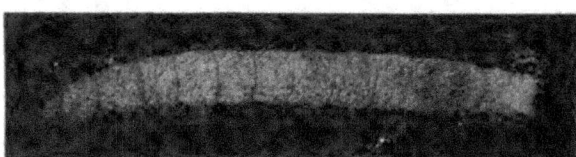

Fig. 45. Grub of Tobacco Flea Bug.

Fig. 46. Seed pods of tobacco injured by Bud Worm.

Fig. 47. Moth or Miller of Bud Worm.

thoroughly mixed. A sprinkler can easily be made by punching many small holes in the bottom of a small tin box so that when the box is filled with the poison and shaken over the plants the buds will be thoroughly sprinkled. This should be applied just after a rain if possible and it may be necessary to repeat it frequently.

CHAPTER XXXI.

THE PESTS OF VEGETABLES.

In this chapter we have discussed the common pests of the more common vegetables. The general pests of vegetables are placed first as they are the ones most commonly met with. These are followed by a discussion of the pests of various vegetables. As far as possible these are arranged in groups according to the plants that they attack; thus, the pests of melon-like plants: watermelons, cucumbers, squash, gourds, and canteloupes are arranged together.

Cutworms.—One of the most serious pests of garden crops is the cutworm, familiar to everyone that has ever attempted to have a garden. Many folks, however, do not realize that the smooth, greasy cutworms are but one stage of the common moths, millers or candle flies that fly into houses especially early in the summer and again in the fall and fly around the lights, but such is the case. These cutworms breed every year in large numbers in grass lands or weedy places, and it is only when we plow up such places and set out plants at regular distances that we begin to notice the damage caused by these insects. Most farmers do not realize the damage that cutworms do to corn, wheat, etc., because the number of plants are so large in these cases; however, where we have only a few cabbage plants or beans or melons, if some of these are cut, the damage becomes very noticeable.

The characteristic way in which this insect works is to cut the stem of the plant near the surface of the ground, and then devour the leaves later if it is left undisturbed. Cutworms work mostly at night and lie curled up in the ground during the day.

The seasonal history of the cutworms is important because it shows why they are so destructive in the early spring. In the case of some of our most important cutworms, the eggs are laid in the fall and hatch out so that the cutworms get about half grown before winter sets in. They remain in a half-grown condition all winter and are very hungry when spring comes, so that they eat a great deal before they become full grown, especially of the very early vegetables.

From what has been said above, it will be clear that it is a bad policy to let the gardens grow up in weeds and grass in the fall, for these simply serve as good feeding ground for the cutworms until spring, when they destroy the vegetables that we plant where the weeds were growing.

In the garden individual plants may be protected by putting around each plant a little collar of stiff paper such as writing paper, extending down to the roots and up to the leaves. These can easily be cut the right size and placed around the plants while they are being set. The part that extends below the

261

surface will hold the collar in place and before the plant has gotten big enough to be crowded by it, the collar will have been destroyed by the weather.

Poison bait may be used to destroy cutworms (page 218). They are best put out before the plants are set, but are partially successful if put out when it is noticed that the plants are being injured.

Fig. 48. Cabbage Plant Lice on stem of plant.

Flea Beetles, also called Flea Bug.—Flea beetles are small beetles or bugs that feed on various kinds of garden plants belonging to the potato family such as the potato (page 252), pepper, egg plant and tomato, although certain flea beetles injure other kinds of vegetables. They do their damage by eating small holes through the leaves which stop the growth of the plant. The fact that they are so small, together with the fact that they are so very active, makes these insects very hard to see, however, their work is not hard to see and may be easily recognized. For the most part the holes are small, about the size of a pin head, but later when the plants are riddled the holes may run together, thus causing larger holes. For the most part these holes go through the leaf. The effect is the same, however, the plant ceases to grow or grows very slowly.

Flea beetles are among the hardest insects to fight because they eat such small holes that it is hard to cover the leaf thoroughly with the spray to kill them, and as new growth is being formed continually the beetles attack this in preference to the old growth, so that one would have to be spraying continuously. However, it is only by spraying or dusting with arsenate of lead or arsenate of lime that we can hope to keep these pests subdued. For this purpose the mixture should be double strength (see page 218) as most of the plants that are injured by flea beetle will stand heavy applications of spray mixtures without being injured.

Plant Lice.—Practically all vegetables are attacked by small, greenish or blackish louse-like insects which suck the sap, thus weakening and stunting the plants. Sometimes these plant lice get numerous enough to kill the plant outright, but usually their enemies (see page 147) hold them in check so that the plants are not killed but do not make satisfactory growth. These plant lice are usually attended by ants which come to lap up the sweetish honey dew which they produce, and since the plant lice are small and do not move actively, the presence of the ants is one of the best indications of the presence of the plant lice, which can generally be discovered if the ants are watched closely.

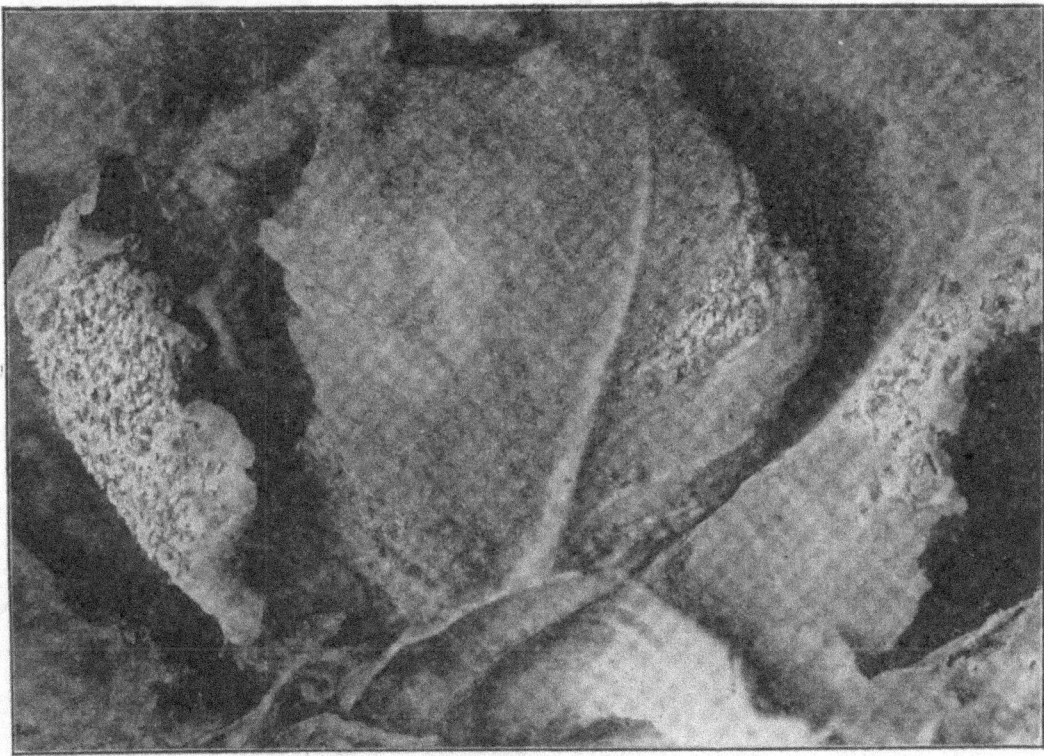

Fig. 49. Cabbage Plant Lice on cabbage.

In spite of the fact that the plant lice have so many enemies, they often become destructive, especially if the weather is cold and damp, for they can multiply in severe weather that stops the work of their enemies completely. This being the case, the gardener usually has to fight these pests every year. One of the simplest remedies is to drench the plants with soap solution (page 220). Since the soap solution only kills the plant lice that it actually touches, it is necessary to apply it very carefully and in bad cases it may be necessary to make two or three applications before all the lice are killed. It is well to remember that plant lice breed with great rapidity and that if a few are left on a plant it will not be long before the plant is badly infested.

Asparagus Rust.—This is a very destructive disease of asparagus, appearing as reddish or blackish blister on the stem. It causes the plants to turn yellow very early, thus not very much food can be stored for the next year and the crop for the succeeding year is greatly decreased. If the proper treatment is not given, the disease gets worse and worse, and eventually causes the total destruction of the plant.

The diseased tops should be cut and burned in the fall.

Bean and Pea Weevils, also called "Bugs."—The seeds of beans, peas and cowpeas are frequently attacked by small weevils or bugs which live inside the seeds for the greater part of their history. The adults emerge from the seeds by eating small holes through the seed and come out to mate and lay eggs. They all have about the same life cycle except the form that commonly breeds in peas. In the case of those that breed in beans and cowpeas, there

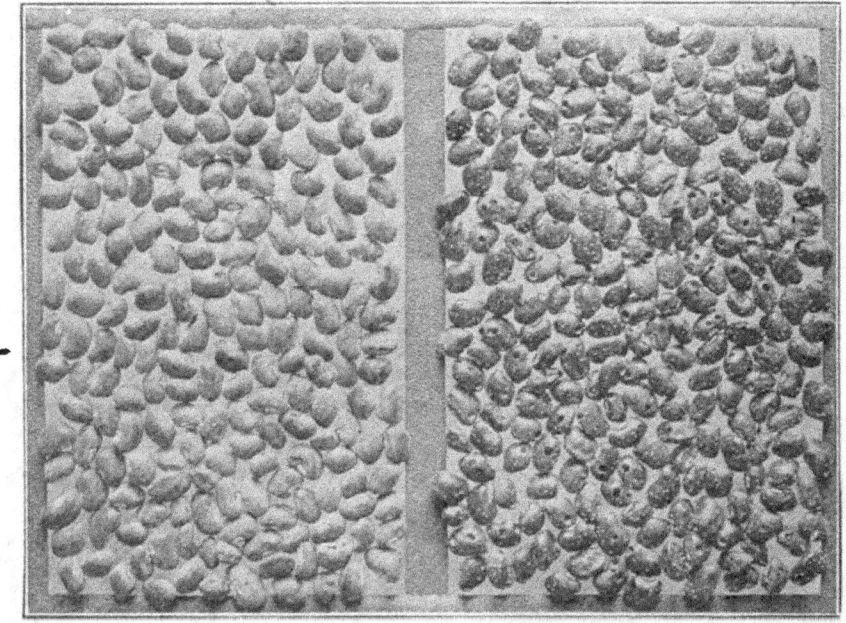

Fig. 52. Buggy Cowpeas on right, uninjured on left.

are several generations each year. They continue to breed in the same lot of seed until the seeds are completely riddled. In the case of the pea weevil there is only a single brood each year. The adults escape from the seed peas and fly away to the garden, where the female lays a single egg in each green pea in the pod.

When these peas are gathered for seed, the grub is already present in the pea and continues its development until the next spring. The reason that this is important is this, peas may be stored from one year to the next and if placed in tight sacks the adults will be shut in, and since they cannot breed in seed peas they will be killed. This same remedy cannot be used against the bean and cowpea weevils because they can continue to breed in the seed beans and cowpeas until the seed is completely ruined for seed or eating purposes. Beans may be treated by throwing them in water at a temperature of from 130° to 140° F. if the beans are to be used for seed, or hotter water will do no harm if they are to be used for food. For those who do not have a thermometer that registers 140° F. it may be stated that this temperature is a little hotter than one can bear the hand in comfortably. After the beans have been in the water for several minutes, they are removed and dried thoroughly, then stored in tight sacks in an airy room so that they will not mold. Dry heat may be used in the same way, but it is not safe to use on beans for seed owing to the danger of overheating.

Seed peas may be treated in the same way, thus avoiding the trouble of holding them over for two years.

If one has a number of beans or cowpeas to store, the simplest way is to mix them with an equal amount of air-slaked lime and place in sacks, or, better still, mix them with one-half or one-fourth as much lime and store in bins with a layer of air-slaked lime on the top of the peas. This operation is very simple, all that is required is to measure or weigh out the same amount of cowpeas and air-slacked lime and then shovel them together and store either in the bins or in sacks.

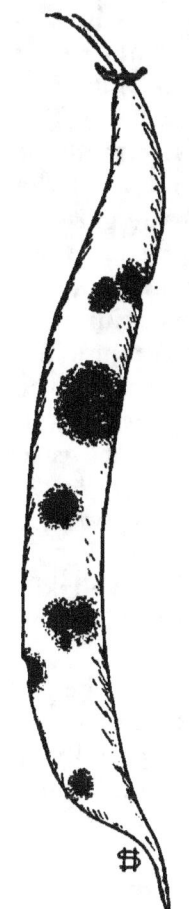

Fig. 54. Bean showing Pod Spot.

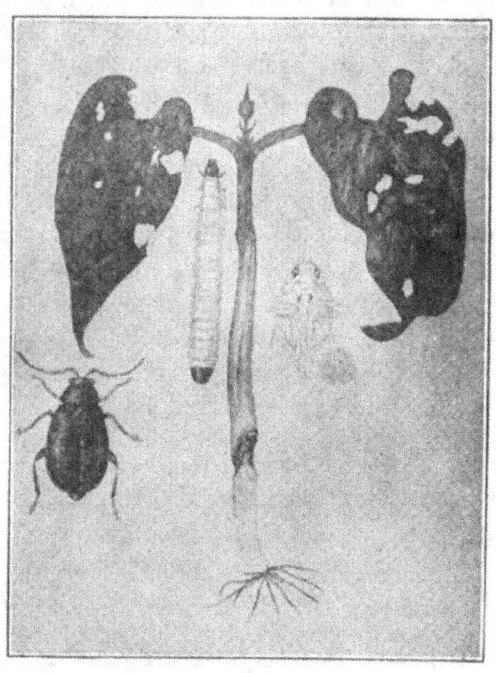

Fig. 53. Bean plant showing leaves injured by Bean Leaf Beetle and stem injured by Grub. Below Bean Leaf Beetle, Grub and Pupa.

Pea Pod Spot.—This disease attacks various parts of the bean but is most noticeable upon the pods where it causes dark spots sometimes nearly half an inch in diameter. Only clean seed should be sown if possible and the diseased vines should be removed from the field and burned.

Bean Leaf Beetle.—The most evident mark of the attacks of this insect is the small irregular holes eaten through the leaves by the adult beetle, but while the adults are feeding on the leaves the grubs are feeding on the young stem below ground, often cutting off your plants much as cutworms do. On the older plants the grubs feed on the roots and nodules, thus injuring the plants very severely.

Since the adults are small and very active, the only remedy that is left open is to poison the leaves. And since the bean is very easily injured by poisons, it is necessary to use the half-strength formula for spraying (page 218) or to dust the plant very lightly with the powdered arsenate of lead used at half strength (page 219).

Cabbage Worms.—There are two main kinds of cabbage worms in this country, both of which are very destructive to all cabbage-like plants. One has the measuring worm habit of crawling and the other does not. The first we may call the cabbage looper, the other the common cabbage worm. The cabbage looper belongs to the cutworm family and the adult is a moth or miller. The common cabbage worm belongs to the butterfly family and the adult is the common white butterfly that one sees everywhere in the gardens and fields and along road sides in the summer. With these differences in mind the two insects are about the same. Both spend the winter in the cocoon stage, usually on the old cabbage stalks, and both worms injure the plant in the same way, that is, by chewing large irregular holes through the leaves and by leaving their dirty castings everywhere on the plants. On the young plants the injury caused by the worms is usually very severe and checks the growth of the plants. On the older plants, especially in the case of cabbage and cauliflower, the worms work mostly on the old outer leaves and the damage they do would not amount to anything if it were not for the fact that they leave their dirty castings everywhere on the plant.

A great many things will kill cabbage worms. Fine road dust will kill them fairly well, especially when they are young. Water may be used in the same way, but the best remedy is to dust the plants with arsenate of lead (page 218) or Paris green and lime (page 218). Dusting is much better than spraying because the leaves of the cabbage plant are smooth and waxy and spray mixture simply rolls off, whereas the dust will stick, especially if it is applied early in the morning while the dew is on or just after a rain while the leaves are still wet. For a discussion of the dangers of using poison on plants see page 277.

Since the cocoons of the cabbage worms are found principally on the old stocks in the field, these stalks should be gathered up in the fall and burned, as they cannot possibly do any good and since they harbor several other pests.

Cabbage Blackrot.—This disease may be recognized by the blackish areas on the leaves which travel towards the stem of the leaf and thence to the stalk, until the whole plant is diseased and gives off a strong, offensive odor. This is a soil disease and care should be taken not to grow cabbage in the same field year after year. Avoid using barn yard manures, for these are apt to carry infection. Do not use tools in diseased fields and then in free fields without first treating them with formalin. Soak seed in formalin solution as recommended for oats smut (see page 228).

IMPORTED CABBAGE BUTTERFLY
The four successive stages of its life

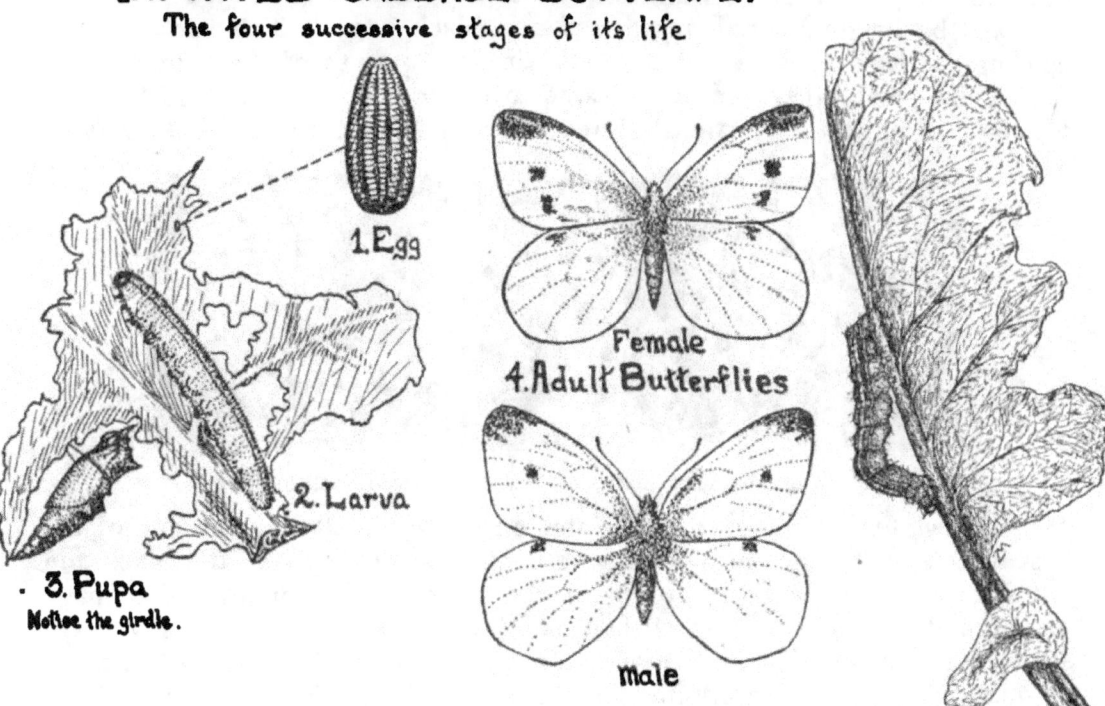

1. Egg

2. Larva

3. Pupa
Notice the girdle.

Female
4. Adult Butterflies

male

All about natural size except "1, Egg", which is much enlarged
Fig. 55. Life history of common Cabbage Worm.

Fig. 58. Cabbage Looper on cabbage leaf.

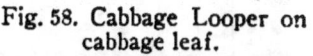

Fig. 59. Adult Moth or Miller of Cabbage Looper.

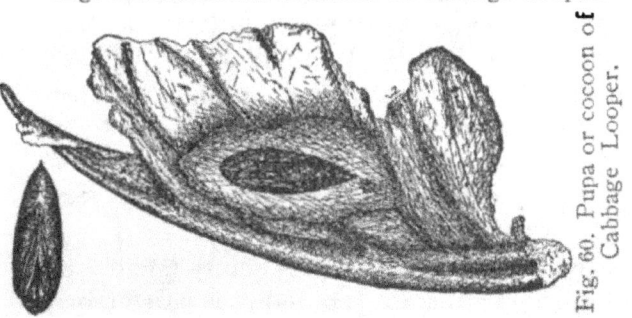

Fig. 60. Pupa or cocoon of Cabbage Looper.

Fig. 56. Adults of imported Cabbage Worm. Male above, female below.

Terrapin Bugs, also called Collard Bugs, Cabbage Bugs and Calico Back.
All over the South and Southwest, cabbage collards and related plants are
attacked by bright-colored orange-red and black bugs, which suck the sap
causing the leaves to die and thus stunting the growth of the plants. This
insect, like some other bad pests, came to us from Mexico, and, fortunately
for those living in the Northern States, it does not seem to be able to survive

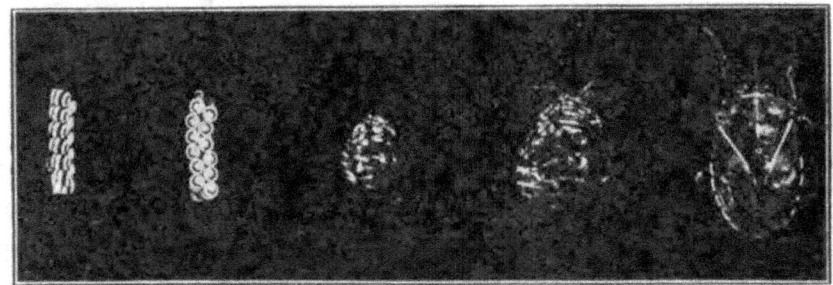

Fig. 61. Eggs, nymphs, adults of Terrapin Bug on cabbage leaf.

the cold winters. It passes the winter on the old stalks and is one of the
first insects out in the spring. It soon commences to lay its little black and
white barrel-shaped eggs in groups on the leaves of the collard and cabbage
plants, and soon after the small bugs, no larger than the head of a pin, are
swarming over the plants. These small bugs grow rapidly and suck a great
amount of sap, thus severely injuring the plant. It is this stage of the insect

Fig. 62. Eggs of Terrapin Bug on cabbage
leaf.

that should be watched, for it is
comparatively easy to kill the young
stages but very difficult or almost
impossible to kill the old insects.
As this insect is a piercing one we
must use one of the oil sprays for it.
Paris green or any other poison will
have no effect whatever.

Here again we have an insect
that passes the winter on the old
stalks in the field and nothing will
help quite as much in fighting it as
burning these old stalks after the
crop is harvested. One reason this
pest is so bad on collards is because
it is the common practice to leave a few plants stand in the garden all winter.
Nothing could be better from the standpoint of this pest. Therefore destroy
the old stalks just as soon as the crop is harvested. Or leave a few plants
standing in the field and then spray them with pure kerosene (coal oil) very
early before these bugs commence to spread.

Club Root.—This may be easily recognized by the enlargements of
the main root and the side roots. Such plants do not make a satisfactory

growth, have a sickly look and never produce a head. This is another soil disease, hence cabbage should not be grown on the same field year after year. If the soil becomes infected some other crops should be grown for several years before cabbage is grown again. All the young plants should be examined at setting time and all diseased plants should be destroyed. It has been found that a heavy application of lime to the cabbage field will greatly aid in reducing this disease.

Fig. 63. Cabbage plant injured by Club Root.

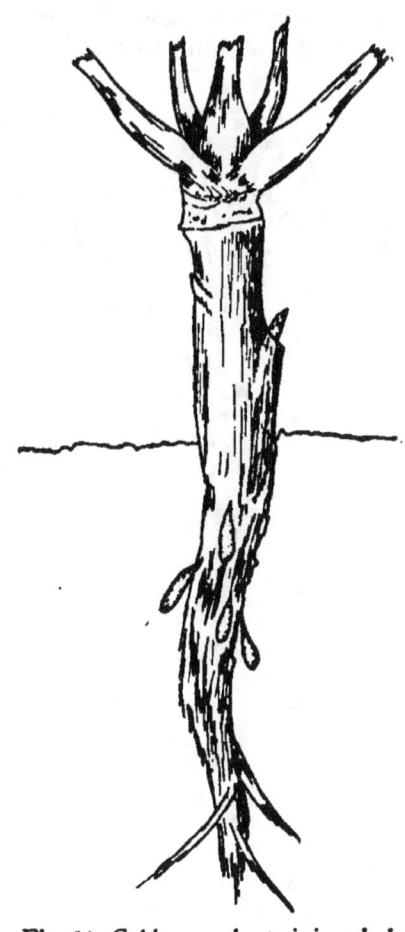

Fig. 64. Cabbage plant injured by Cabbage Maggot.

Cabbage Maggot.—Frequently it is noticed that after cabbage is set out it turns yellow, stops growing and dies. If such cabbage plants are pulled up and examined it is found that the roots are badly misshapened or destroyed. Usually one can find slender, white maggots like the maggots of the house fly, but smaller, clinging to the roots. This is the work of the cabbage maggot and frequently as high as 90% of the plants in a garden or field will be destroyed in this way. The eggs of the maggot are laid by a fly which looks very much like a small house fly, just at the time the cabbage plants are being set. This fly takes care to place the eggs near the stem of the plant on the ground. These eggs hatch and the young maggots commence to burrow into the stem and roots. Later the maggots change to little brown capsules in the ground. It is in this capsule stage that the insect passes the winter, coming out the following spring as a full-grown fly about the time cabbage is ordinarily set.

There are several ways of controlling this pest that can be used successfully on the average farm. Usually it will take a combination of all these methods to keep the pest in control. All the stalks and refuse from the field or cabbage patch should be gathered in the fall and burned. The field should then be plowed to a depth of seven or eight inches. In the spring planting should be delayed as late as possible, in the hope that many of the flies will die before the plants are set and hence not so many eggs will be laid. If planting is delayed, the plant beds should be well covered to keep the flies from laying in them.

Many farmers hill the earth up around the plants carefully when they are set and then in about five days pull this earth down level. In this way most of the eggs will be pulled away from the plant before they hatch and the young maggots will not be able to reach the plants before they die of starvation.

Perhaps the best way to prevent the fly from laying eggs is to place small pieces of tarred felt paper (ordinary tarred paper will not be satisfactory) around the plants as they are set. The odor of the tar is distasteful

Fig. 65. Cabbage Plant protected by tarred paper.

to the flies and they will not lay their eggs on such plants. These bits of tarred felt can be cut from a sheet of felt with a sharp knife. The simplest way is to cut the felt into pieces about three inches square. Then each square has a slit cut from the middle of one side, about two inches long, and across this slit another slit is cut about one inch long. The plant is then inserted into the long slit and the slit makes flaps which fit snugly around the stem of the plant. The plant with its bit of tarred felt is set with the felt resting smoothly on the ground. While this involves a little extra time in setting, in regions where the cabbage maggot is bad it is worth the cost in preventing the attacks of this pest.

Celery Leaf Worms.—There are several worms that eat the leaves of celery, but as most of them are large and make conspicuous webs in the leaves it is a simple matter to destroy them by hand.

Celery Leaf Spot.—Yellowish spots appear frequently on the leaves of celery. These spots usually are about ⅛ to ¼ inch in diameter. This disease is very destructive at times, stunting the plants quite severely and spoiling the sale of the plants.

Careful watch should be kept of the young plants and all diseased ones should be pulled from the seed bed and destroyed. Only healthy plants should be transplanted and if the disease has been bad in previous years the plants should be sprayed every ten days or two weeks with Bordeaux mixture. (See page 220).

Lettuce Drop.—This is a serious disease of the lettuce. The first indications of the disease are that the outer leaves wilt and fall flat on the ground. A little later the whole plant wilts down just as if it had been scalded with hot water. This disease is especially troublesome as it lives from year to year in the bed. The best method of control consists in removing the plants as soon as they show any signs of the disease and destroying them. If this practice is persisted in the disease can be controlled.

Fig. 66. Striped Cucumber Beetle.

Cucumber Beetles.—Just at the time melons, squashes and cucumbers are coming through the ground, and before they have a chance to vine or run, they are attacked by two pests known as the striped and spotted cucumber beetles. Both of these pests work by eating holes through the leaves and frequently they destroy the plants before they get started.

One of the best methods of preventing the attacks of these pests is to cover the hills of plants with a cone made of fine wire fly screening. This should have 18-20 mesh to the inch and if they are all made over the same pattern they can be nested together after they are used and saved to be used again and again for several years. These should only be left on the plants until they start to run or vine for after that time the plants will usually be able to grow in spite of the attacks of these pests. Several things may be dusted on the vines to protect them against these pests.

Melon growers use tobacco dust or strong smelling commercial fertilizers successfully in this way and as both of these substances aid the plant in growing rapidly, they are a further help in that way. It is usually asserted that poisons have no effect on these pests, but that is a mistake. It is usually troublesome to keep the leaves all covered with poison as the plants grow very rapidly. But by frequent applications of powdered arsenate of lead or arsenate of lime the plants may be thoroughly protected. The applications should only be made often enough to cover the new leaves and should be made in the form of a light dust.

Melon or Pickle Worms.—Boring into the buds and fruits of melons, squashes, and cucumbers are a couple of pests that we call melon or pickle worms. Their attacks are worse on squashes, cucumbers and canteloupe. Frequently they destroy more than half of the crop in this way.

Perhaps the most important thing to do for their control is to rake up all the trash and old vines just as soon the the crop is harvested, for in this way all of the late worms will be destroyed before they go into winter quarters.

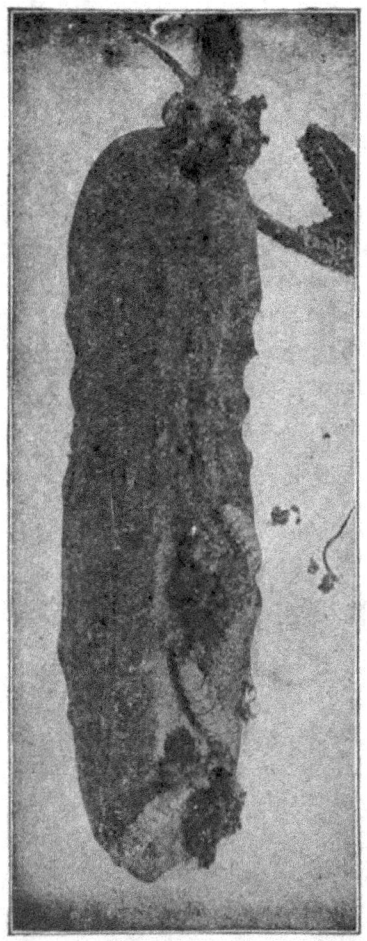

Fig. 68. Cucumber injured by Pickle Worm.

Fig. 71. Pickle Worm.

Melon Leaf Blight.—This blight starts on the leaves as small brown spots which increase very greatly in size, causing the leaf to curl up and dry, the fruit ripening before the proper time.

Rotation should always be practiced with melons as this will aid greatly in keeping the disease in control. Spraying with Bordeaux mixture (page 220), first, when the vines begin to run and then repeating the spraying every week or ten days, being careful to cover both the upper and under surface of the leaves thoroughly, is entirely effective but rather difficult **to** apply.

Melon Wilt.—This disease may be recognized by the fact that the leaves commence to wilt and then die and soon the whole vine is dead. It is one of the soil diseases, hence melons should not be planted in the same field two years in succession. If the disease appears in a field, no attempt should be

Fig. 69. Canteloupes injured by Pickle Worm.

Fig. 70. Adult Moth of Melon Worm.

made to grow melons there for several years. There is great danger of the germs being carried through the manure, therefore it is better not to use stable manure on melons.

Squash Vine Borer.—Frequently it will be noted that about the time the squash vines are coming out into bloom, the whole plant withers and dies. If such plants are examined carefully, it will be noted that there is a large

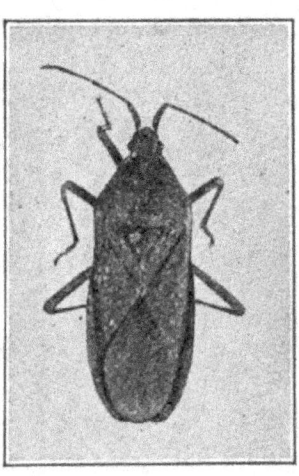

Fig. 75. Adult of common Squash Bug.

Fig. 72. Adult Squash Vine Borer.

Fig. 74. Adult Horned Squash Bug.

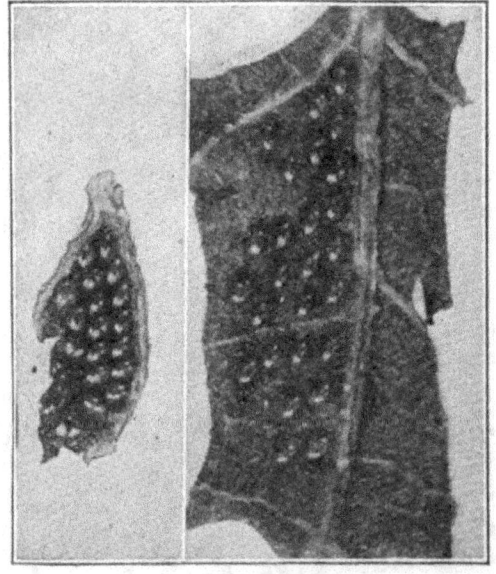

Fig. 76. Eggs on squash leaf.

amount of sawdust at the base of the vine and that the vine is badly mis-shapened and decayed. If the vine is split open usually one or more white worms about ⅞ of an inch long will be found working in the vine.

As this insect works in the vine there is no poison that can be used against it. However, all the old vines should be raked up and burned as soon as the crop is harvested and the ground thoroughly harrowed, as in this way many of the insects will be destroyed.

Squash Bug.—Just after the cucumber beetles get through with their attacks on melons, cucumber, and squash plants, they are attacked by the squash bug. The first indications of the attacks of this insect will be a few leaves wilted here and there on the plant. A little later it may be noted that the whole plant is withering. If a careful search is made, especially on the under sides of the leaves early in the morning, a dark brown stink bug will usually be found. A little later the brown eggs will be found in groups on the under or upper sides of the leaves and still later the young red and green bugs will be found.

The eggs are easily found and should be crushed on the leaves. The adults should be killed wherever they are found on the plants. They can be collected to best advantage by placing chips under the vines, the squash bugs will collect under these at night and can be killed easily in the early morning.

Sweet Potato Bin Rot.—This disease is very destructive to sweet potatoes in storage. When a potato is attacked by the disease it turns soft usually at one spot, but the disease progresses very rapidly until the whole potato is reduced to a rotten mass. It usually starts in potatoes that have been injured at digging time, therefore these should be carefully sorted out and not stored with the main crop, but used at once. All care should be taken in handling the sweet potato to see that it is not bruised or otherwise injured. When only a few sweet potatoes are grown, as on the average farm, it pays to wrap each potato separately in newspaper and store in a dry closet or room where the temperature does not ordinarily fall below 40° or rise above 70°.

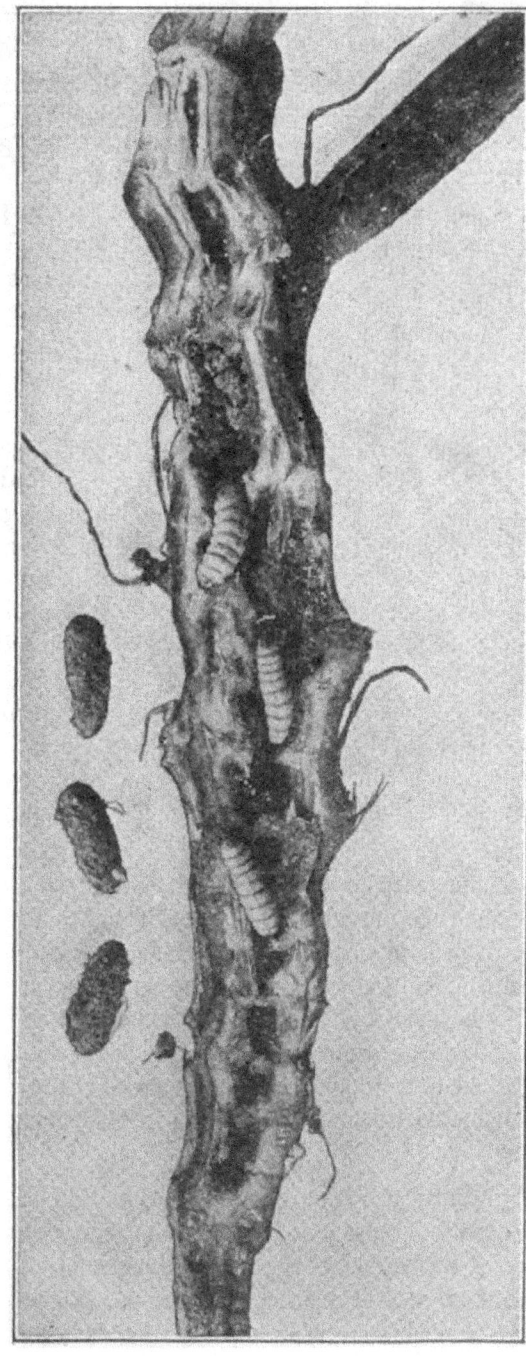

Fig. 73. Grubs in squash vine.

Tomato Wilt.—In this disease the leaf turns dark green, then brown and then black, the plant wilts and dies. It is thought that the disease is spread by insects, hence these should be controlled as far as possible. Since the disease lives over the winter in the soil, care should be taken not to plant the same part of the garden in tomatoes year after year.

Tomato Worms, also called "Tobacco Worms" and "Horn Worms."—This is the same pest that is so destructive to tobacco (page 257). They are found frequently on tomatoes, ragging the leaves very badly. These worms get to be very large in size which, together with the conspicuous horn near the hind end of the body and the peculiar way they have of throwing the head from side to side, leads many people to believe that they are poisonous or even deadly. But nothing could be farther from the truth as these worms are perfectly harmless and can be handled without any danger whatever. After these worms are full grown they crawl into the ground where they change to pupa which are often called pitchers because of their peculiar shapes. From these pitchers the full grown moth or miller will emerge the following summer. This moth or miller flies away to lay her eggs on the leaves of various kinds of plants especially jimson weed, tomato and tobacco. From these eggs the little worms hatch.

Fig. 78. Tomato Fruit Worm injuring tomato.

To protect tomato plants against this pest it is best to dust the plants with a mixture of dry arsenate of lead; one ounce to a quarter of a pound of dry sifted wood ashes. These two are mixed together until the whole mass is light gray in color without streaks. The mixture is placed in a thin muslin bag and shaken over the plants, care being taken not to breathe too much of the poison as it is very irritating to the nose and throat. This is much better than waiting until the worms get large enough to be seen and then picking them off by hand, for by the time they get large enough to be readily noticed they have done an enormous amount of damage, whereas the poison will kill them while they are young, before they have a chance to do any real damage.

Tomato Fruit Worm, also called "Corn Boll Worm" and "Tobacco Bud Worm".—This pest feeds upon a long list of cultivated plants and is frequently found boring into the green fruits of tomatoes. One worm may bore into a number of fruits as they have a very great capacity for doing damage. The fact that this same insect also feeds on certain field crops such as corn, cotton, and tobacco, makes the danger of its attacking tomatoes just that much greater. This insect belongs to the same family as the cutworms and has much the same life story as those pests. The eggs seem to be laid by preference on the silk of corn, on the young grains of which this insect commonly feeds. After the worms are full grown they go into the ground a few inches, where they change to brown pupa or cocoons, in which stage they remain over winter, coming out the following summer about the time corn is in the silk to lay eggs again. Thus the weakest point in this cycle is during the fall and winter when they are in the ground, in the form of naked pupa. If the ground is plowed during the fall many of these pupa will be disturbed and killed. In order that this remedy may be of some benefit to the tomato crop in the garden, the fields that are in corn, cotton or tobacco should be fall plowed, as otherwise the insects will simply spread from these fields to the garden. Besides this remedy the only other method that is of any benefit, is the destruction of the worms as they are found in the garden. This becomes more important when we remember that one worm may bore into several tomatoes.

Danger from Using Poisons on Vegetables.—In the minds of some people there is expressed a doubt as to the advisability of using poisons on vegetables. Of course this should apply only to those plants where the leaves are used for food. Any serious question of course should not arise where it is the roots or the peeled fruit that is used for food as the poison is not absorbed by the plant. Perhaps the best reply that can be made to this objection is to state that poisons have been used for years on such plants as lettuce and cabbage where the leaves are used for food and there has never been an authentic case of poisoning from this source. If poisons are used as directed, that is, when the plants are young, the pests will be controlled better and only a small amount of poison will have to be used. Even on large cabbage plants it is never necessary to use the poison so very heavily. In heavy applications I have made, I used only enough poison to make about three injurious doses to man per plant. That is figuring all of the poison that has been used. Any one who has ever attempted to dust or spray plants knows that he wastes a great deal, say one-fourth of all he uses. Careful analyses show that of the amount used more than 90% is washed off by the rain or blown away by the wind in less than three weeks. But if none were removed in this way, it must be remembered that all vegetable leaves that are used for food are washed before being cooked, so that not even a trace could possibly remain.

CHAPTER XXXII.

PESTS OF ORNAMENTAL PLANTS AND SHADE TREES.

There are a large number of different kinds of ornamental plants and shade trees, and it is not our purpose here to discuss each one of these but rather to deal with the large groups and to discuss some of the remedies for their pests.

Herbaceous Group.

This group includes all of our common flowers except a few flowering shrubs. As used here it is intended to include such common flowers as the violets, carnations, asters, hollyhocks, verbena, phlox, begonia, peony and many other kinds.

Guide to the Chief Pests of Ornamental Herbaceous Plants.
I. **Attacking the roots.**
    Eel worms (page 278).
II. **Attacking the stems.**
    A. Cutting off the stem near the ground.
        Cutworms (page 279).
    B. Small greenish or blackish bugs sucking the sap. These bugs often attended by ants.
        Plant lice (page 279).
III. **Attacking the leaves.**
    A. Large ragged holes eaten through the leaves.
        Leaf worms (page 279).
    B. Dead spots in the leaves usually surrounded by a yellow border.
        Leaf spot (page 279).
    C. Rust colored spots especially on the under side of the leaves.
        Rust (page 279).
    D. Leaves covered by a white powder.
        Mildew (page 279).

Eel Worms.—Frequently plants are attacked by eel worms which cause swellings of various sizes on the roots. The eel worms are so small that they can only be seen with the aid of a microscope so that the grower is often at a loss to understand just what is causing the disease. Diseased plants are stunted and never make satisfactory growth.

The only method of treatment consists of destroying the infested plants and planting new seeds or cuttings in fresh soil. If the plants attacked are potted plants, the pots should be scalded and the new soil should be baked in an oven.

Cutworms.—Cutworms are often very destructive to ornamental plants. They work by cutting off the plants near the surface of the ground. Usually a little search will locate the worm in the soil near the cut plant where it may easily be destroyed. These worms should always be searched for as they will continue their attacks often destroying many plants.

Care should also be taken not to plant ornamentals in land that has long laid in sod without first spading up and plowing the ground and then scattering poison bait. (See page 218).

Plant Lice.—Plant lice are too well known to need much description. They are variously colored bugs, some green, some purplish, with or without wings, that attack the stems and leaves of various kinds of ornamental plants. Their presence is often revealed by ants crawling up and down the plants in search of the sweetish honey dew that these insects give off. These pests often do serious damage by sucking the sap from plants. They may easily be controlled by spraying with soap solution (see page 220).

Leaf Worms.—The leaves of ornamental plants are often badly eaten by various kinds of worms. Some of these are large and occur in several numbers and may easily be destroyed by hand picking, others are small and occur in large numbers. In such cases the plants should be sprayed with weak arsenate of lead (see page 219) or dusted with arsenate of lead powder (see page 218).

Leaf Spot Rust and Mildew.—The leaves of ornamental plants are frequently rendered unsightly and not infrequently the leaves are killed or the plant seriously checked or killed by one or all of these diseases. They should be watched for carefully and whenever noticed the affected leaves or plants should be removed or destroyed. If these diseases have proven bad in the past, the plants should be sprayed several times early in the season, say every two weeks for a couple of months, with either liver of sulphur (see page 220), concentrated lime sulphur (see page 219), copper sulphate (see page 220) or Bordeaux mixture (see page 220).

Groups of Shrubs and Vines.

This group includes a number of well known ornamental plants such as the rose, privet, snow ball, spires, lilacs, ivy, clematis and many other kinds. They are troubled by many diseases similar to those found in the herbaceous group (see page 278) especially plant lice, leaf worms, leaf spot, rust and mildews. These diseases should receive the same treatment in both cases. Besides these diseases, the plants of the shrub and vine group are troubled by leaf beetles, which eat the leaves. These beetles, with the exception of the so-called rose beetle which attacks the leaves and flowers of many kinds of plants, may easily be controlled by spraying or dusting with arsenate of lead (see page 218).

These pests come early in the summer and usually disappear in a few weeks. During the time they are common, plants should be protected by being covered with cheese cloth bags or the beetles should be brushed from the plants into shallow pans of kerosene each day.

Many of the plants of this group have their limbs and branches encrusted with scale which suck the sap and cause the plants to die. Such plants should be sprayed in the winter with concentrated lime sulphur. (See page 219).

Tree Group.

Trees of various kinds are troubled by a variety of pests, chief among which are **scale insects** attacking the branches and which should be sprayed with concentrated lime sulphur (see page 219) in winter; **Leaf worms** which

Fig. 79. Elm Leaf Beetles injuring elm leaves.

attack the leaves often building webs or nests in the trees, should be cut out and burned or sprayed with arsenate of lead (see page 218).

There are also a large number of borers which eat into the trunk and branches, often doing serious injury. However, such borers cannot be reached by any known methods.

Trees are also subject to the attack of rots which cause the decay of the branches or trunk. Such rots usually start where a branch has broken off or been sawed off, or where the tree has been injured by insects or by rabbits, or has been carelessly chopped into by some thoughtless person. Such causes should be prevented whenever possible, however, if the tree is injured accidentally, the wound should be covered with good paint and then allowed to heal over as soon as possible.

Fig. 80. Tent Caterpillars in tree.

Fig. 81. Walnut Caterpillars — such groups of worms should be cut out and burned.

CHAPTER XXXIII.

PESTS OF SMALL FRUITS.

Small fruits, berries or grapes are grown to a limited extent on practically every farm. Needless to say they suffer from the attacks of a great many pests. Most of these attacks can be prevented easily if the proper remedies are applied at the right time.

Gooseberry Mildew.—This disease gives the gooseberry leaves, stem and fruit, the appearance of having been dusted with flour. It is a very destructive disease and threatens gooseberry growing in this country. It may be controlled by spraying the bushes just at the time the leaves are appearing in the spring with liver of sulphur (see page 220).

This application should be made every two weeks for at least three months. It is very essential that spraying be commenced early, as it is very much more effective when the plants are sprayed just as the bushes are commencing to show green leaves.

Currant Girdler.—Currant stems frequently wilt from the attacks of another pest, the currant girdler, which girdles the stems near the tips. Usually such girdled tips hang on to the stem, but sometimes they fall to the ground. The eggs in this case are not placed in the tip beyond the girdle, but in the main stem about one inch below the point where it has been girdled. The grubs tunnel down in the stem to a distance of usually not more than six inches.

The control of this pest is fairly simple as it consists of trimming out the girdled stems about eight or nine inches below the girdle and burning them.

Leaf Bug.—This is a very small bug that usually escapes detection. Its work, however, is conspicuous. It attacks the leaves of gooseberries and currants and sucks the sap so that the leaves turn brown, curl up and die. Frequently the whole stem of the plant stops growing and dies. In severe cases it is said that the plants look as if they had been scorched by fire. Clean cultivation is said to aid in keeping this pest in control as it feeds on other plants and weeds. The collecting of the stunted tips, in which the eggs are usually found, and burning them is also effective.

Currant Worms.—There are a number of worms that feed on the leaves of currants and gooseberries wherever they are grown. They sometimes get abundant enough to actually strip the plants. These worms can be collected by hand but it is much better to spray the plants while the worms are small, using one ounce of arsenate of lead to every two gallons of water. If this treatment is applied early it will effectively check these pests and no further treatment will be necessary.

Tree Crickets.—Most farmers have never seen the whitish tree crickets, which are common everywhere in weedy patches, in thickets of raspberries and blackberries, and which make our summer evenings melodious with their

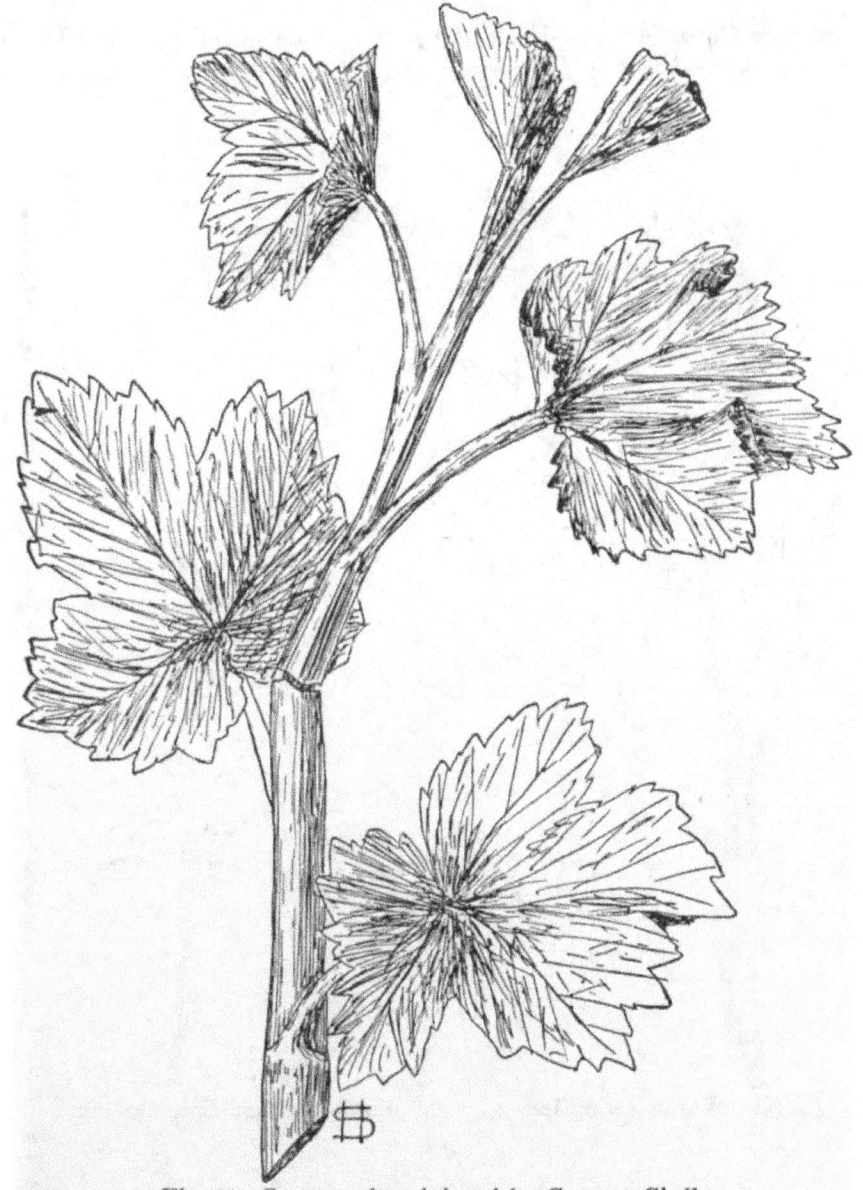

Fig. 82. Currant plant injured by Currant Girdler.

songs. The injury that they do has been observed commonly on raspberries and blackberries. This injury consists of thrusting their eggs into the canes, thus the plants are weakened and frequently they break off at this point. When they become abundant, as they are apt to be in neglected berry patches, steps should be taken for their control. This may be accomplished by burning all of the infested canes at pruning time. It is much better to keep the berry patch in good cultivation, the vines well pruned and all weeds down. Such berry patches are much less attractive to the tree cricket than weedy neglected patches.

Raspberry Cane Borer.—Frequently it will be noted that the tips of raspberry and blackberry canes are withering. If such canes are examined it will

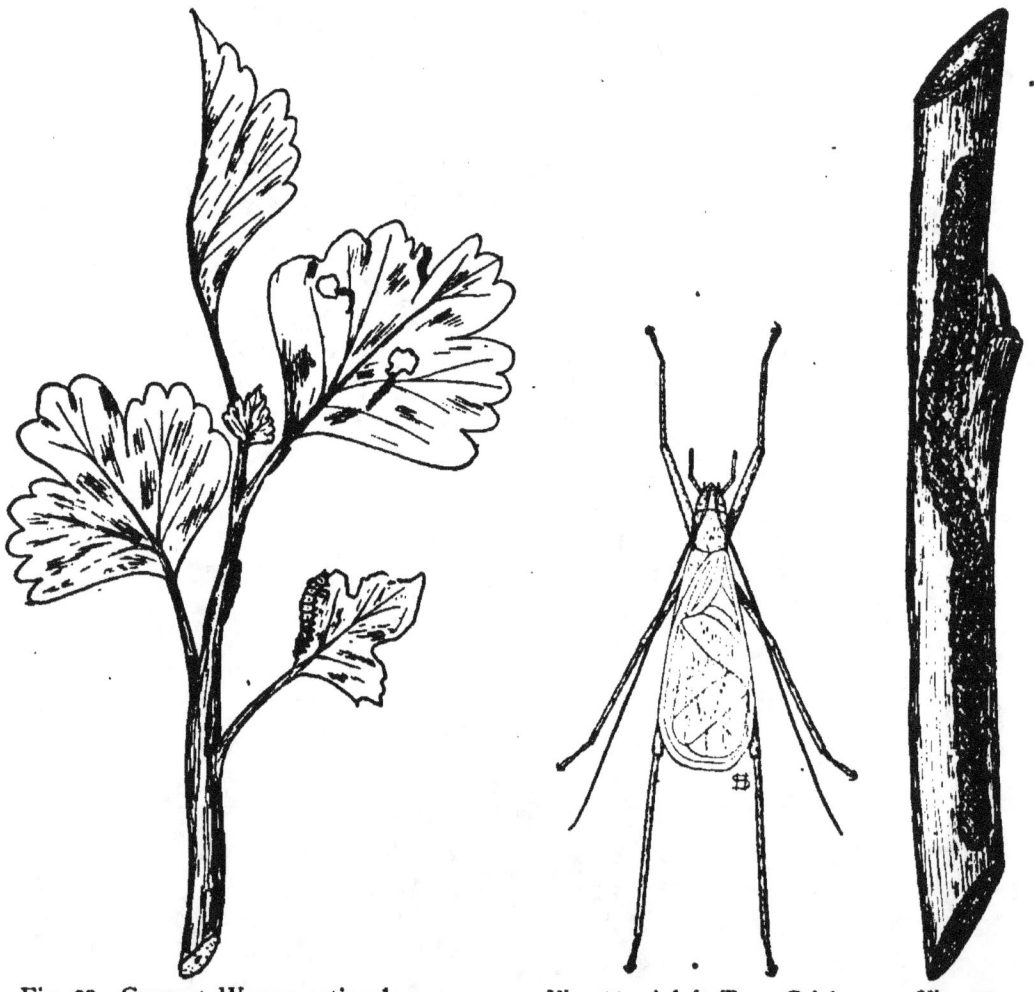

Fig. 83. Currant Worms eating leaves.   Fig. 84. Adult Tree Cricket.   Fig. 85. Blackberry stem injured by Tree Cricket.

be noticed that they have been partly or completely girdled. This is the work of the cane borer which lays its eggs in such girdled canes, inside of which the grub works its way down toward the base of the cane, which it usuall reaches by the second fall after the eggs have hatched. The burrows of the cane borer can usually be discovered by the large amount of sawdust which is forced out through holes in the cane. This work is very destructive to the bearing canes, which are usually killed.

The wilted tips of the girdled canes should be watched for and pruned out and burned. Also the bearing canes showing the work of this pest should be pruned out and burned.

Blackberry Rust.—This disease is common on blackberries and raspberries. It appears in the spring as rust colored spots under the sides of the leaves. Since this rust lives on the plant year after year and is easily scattered to new plants, the infested plants should be dug up and burned.

Fig. 86 Blackberry Rust. Just beginning on left, badly injured plant in center, healthy plant on right.

Stem Wilt.—There are two or three diseases which attack the stems of raspberries and cause them to wilt.

These diseases are always worse in patches that have been left in berries for a long time, hence the berries should not be planted on the same patch year after year. The old canes that are pruned out should be carefully gathered and burned.

Grape Berry Worm.—These small greenish or purplish worms feed among the blossoms and young fruit of the grape, early in the season. Later in the season, they bore their way into the green grapes and feed on the pulp and seed. Often a single worm will destroy several grapes in this way. Every attempt should be made to destroy the young worms that are feeding on the blossoms and young grapes. This is best done by spraying the grapes at the time the fruit is setting, using arsenate of lead at the rate of one pound to fifteen gallons of water. If this work is done early and thor-

oughly, it will prevent the later worms from attacking the fruit to a very
great extent. Although in severe attacks it may be necessary to spray the
grapes again in about ten days and then when the grapes are about half
grown.

Leaf Worms.—There are a number of worms which attack the leaves of
grapes. They sometimes do great damage, however, as most of them are
large or make conspicuous holes in the leaves; they are usually easily detected and can be destroyed by hand. If this practice is followed each year,
the number will soon be greatly reduced.

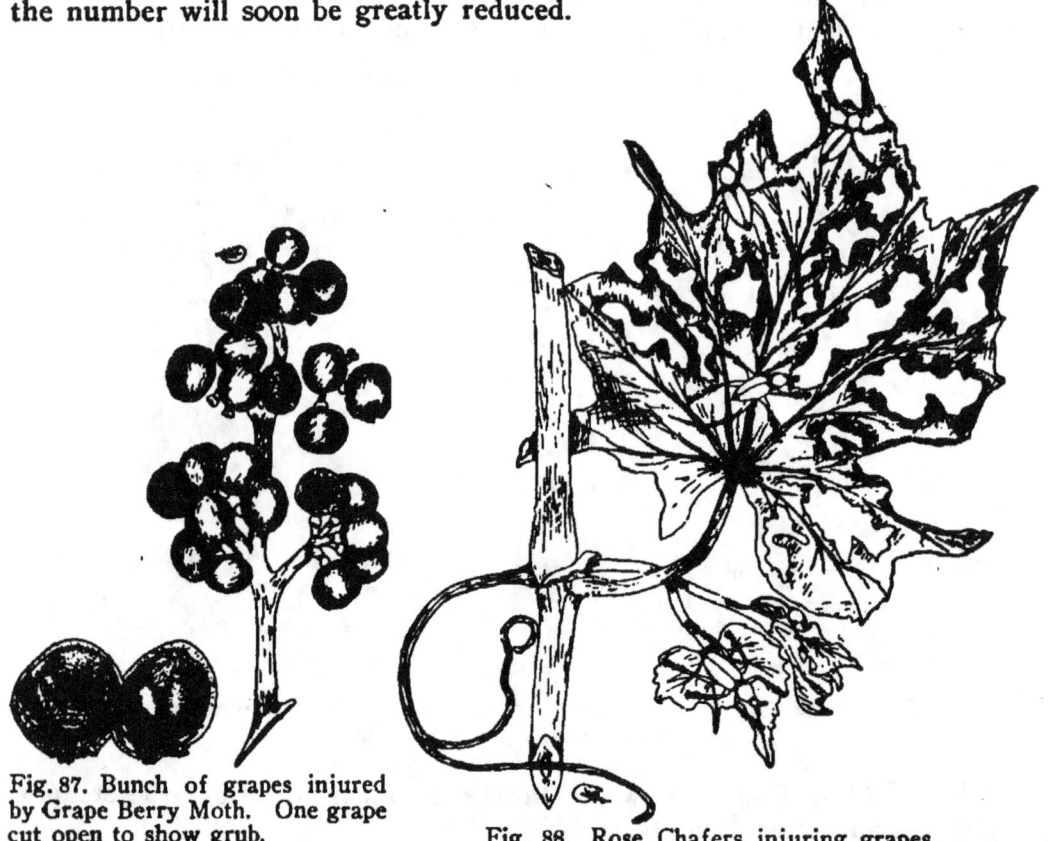

Fig. 87. Bunch of grapes injured
by Grape Berry Moth. One grape
cut open to show grub.

Fig. 88. Rose Chafers injuring grapes.

Rose Chafer.—Frequently this grayish-brown bug, with long spradly
legs, will descend upon the grape in vast hordes and eat the blooms, buds
and leaves. It also attacks other fruits and flowers. They come in such
swarms that it is very difficult to do anything with them. These pests breed
in grass lands where the grubs feed upon the roots of grasses. Something
perhaps might be done to protect fruits and flowers by plowing up sod lands
late in summer. Covering choice vines with cheese-cloth bags, or enclosing
each bunch of grapes in ordinary paper bags, is troublesome but effective.
Spraying the vines with arsenate of lead one pound, glucose two and one-
half pounds, to water ten gallons, gives a great measure of protection. It is
necessary to add the glucose as the arsenate of lead alone is not effective.

Black Rot.—This is the' most destructive disease of the grape, attacking the stem and leaves, but it is chiefly on the fruit that we notice its work. It attacks the green grapes first as small black or brown spots, which grow very rapidly until the whole berry is soon covered. These berries hang on

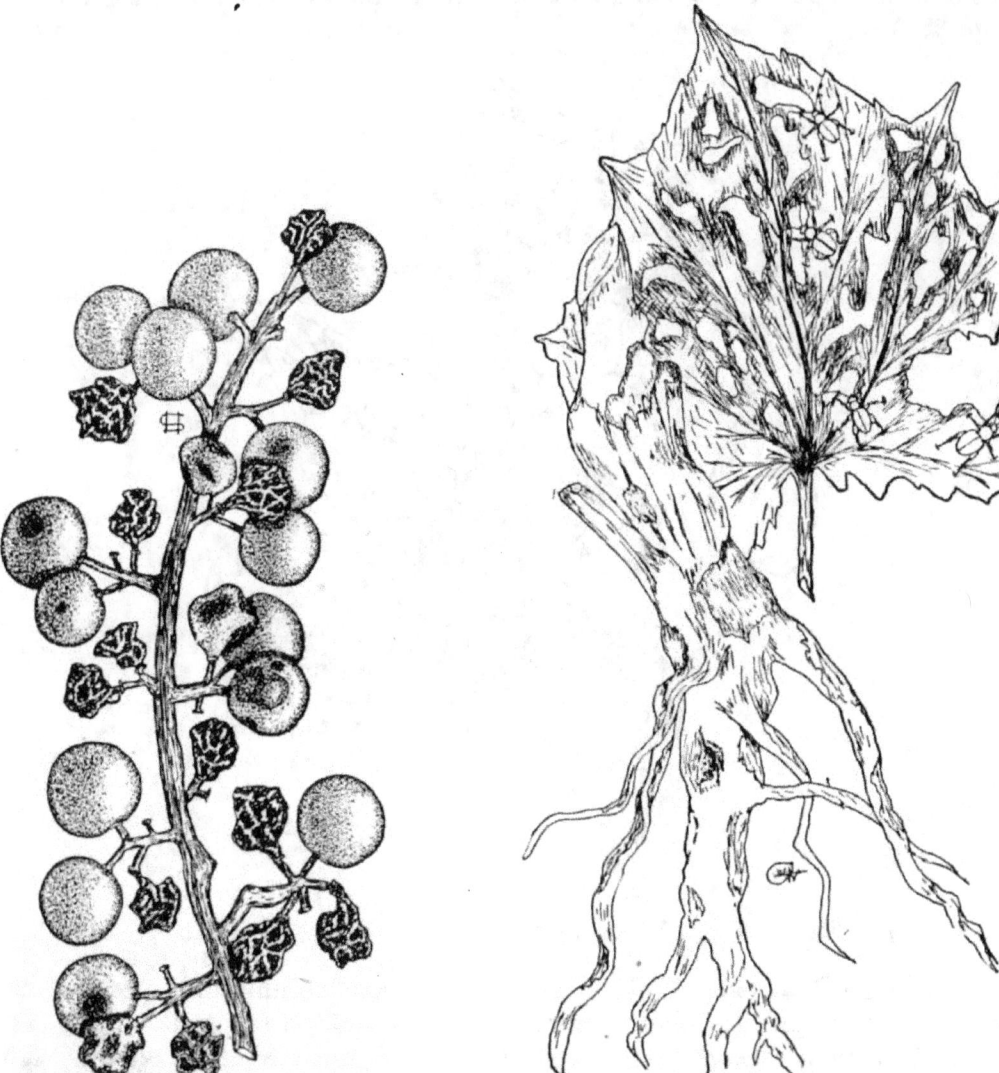

Fig. 89. Bunch of grapes showing injury by Black Rot.

Fig. 90. Roots and leaf of grape injured by Grape Root Worm.

the vines and carry the disease from one year to the next, hence they should be picked off and destroyed by burning. All the dead leaves and pruned vines should be raked up and burned and the vineyard should be deeply plowed in the spring to prevent the spread of this disease. Spraying with Bordeaux mixture (see page 220) is effective for this disease if commenced early in the summer, spraying thoroughly five or six times, taking great care to reach the bunches of grapes. This spraying will not be effective if it is delayed until the rot commences to show on the green grapes.

Grape Mildew.—This disease gives the grape the appearance of having been dusted with flour. It attacks the stems, leaves, flowers and fruits, and can be controlled by the same sprayings as recommended for black rot (see page 287).

Fig. 91. Grape leaves injured by Grape Vine Flea Beetle.

Grape Root Worm.—This is one of the most troublesome pests of the grape in both the full grown or beetle stage as well as the grub stage. The beetles live on the leaves and frequently reduce them to mere shreds. The grubs live on the roots and often greatly injure the plants by cutting off the young roots and by severely injuring the main stem. When the roots are attacked the plant is given a very severe set back, or may be killed. Such plants usually remain unthrifty for a year or two and then die.

Close watch should be kept for the beetles and as soon as they are found feeding on the leaves, the vines should be sprayed with arsenate of lead one pound to fifteen gallons of water. All the leaves should be covered as well as possible so that the beetles will not have a chance to feed on any other unprotected leaves. Our only hope is to kill the beetles before they have a chance to lay eggs, for there is no successful way of killing the grubs working on the roots and it is in this way this pest does most of its damage.

Grape Vine Flea Beetle.—Early in the season the buds of the grape are frequently eaten into by a small bug called the grape vine flea beetle because the insect can leap very much like the flea. In neglected vineyards where the grass is allowed to grow up around the vines and along the fence, or in vine-

Fig. 92. Grape leaves injured by Leaf Hoppers.

yards next to woods, this damage is usually much worse than it is in well cared for vineyards. Later on the grubs of this flea beetle feed on the leaves or on the clusters of young grapes and often do quite serious damage.

The best method of control for this pest is to keep the vineyard clean and to spray the vines thoroughly with arsenate of lead one pound to ten gallons of water about the last of June, when the grubs are noticed feeding on the leaves. This will protect the crop for the next year. In severe cases the flea bugs may be jarred into pans containing a little kerosene. This is tedious but is about all that is left if the vines have not been sprayed the year before.

Grape Leaf Hopper.—This small pest of the grape is very widely distributed in the United States but is very variable in numbers. Some years it attacks the grapes in vast swarms and does a great deal of damage, in other years it seems to be nearly absent.

This bug, which is only about one-eighth of an inch long when full grown, is very active and may escape notice, however, as it occurs in vast numbers and sucks a great deal of sap; it usually makes its presence felt. The leaves when severely attacked turn brown and fall off. If this occurs to any very great extent the vine is weakened and does not mature its crop perfectly.

A great deal may be accomplished in controlling this pest by burning the grass along fence rows and about the vineyard where these pests spend the winter.

Most of the early eggs are laid on the lower suckers usually some time in June; if these suckers are left and cut out in early July and burned, many of these pests will be destroyed before the eggs hatch.

Fig. 93. Strawberry plant, to show berries injured by Strawberry Weevil.

Strawberry Weevil.—In certain sections this is the most destructive pest of strawberries. It works on the flower buds of the plant, laying its eggs in them and then cutting the stem of the bud so that it falls to the ground. If very plentiful this cuts short the crop to a considerable extent and frequently makes it unprofitable to grow strawberries. Strawberry growers meet the situation by planting varieties with a great deal of bloom so that even if part of them are cut there will still be enough buds left to make a profitable crop. Plants that are thoroughly dusted with a mixture of arsenate of lead one part, to sulphur five parts, seem to be pretty well protected against this pest.

Strawberry Leaf Roller.—This worm folds the two sides of the leaf together and fastens them with a web. In this shelter the worm lives a very protected life. Our only hope of controlling this pest is to spray the plants with arsenate of lead, one ounce to one gallon of water, before the worms commence to roll the leaves.

Strawberry Root Louse.—This is a very destructive pest of strawberries in some localities. The presence of the louse can usually be detected by the large numbers of ants about the plants and by the unhealthy color of the plants. Frequently plants die in irregular patches.

The best remedy for this pest is to plant nothing but clean plants in new fields. Do not plant strawberries in the same field year after year but

Fig. 94. Root Lice on strawberry roots.

transfer them to a new field each season. Destroy the old fields as soon as the crop is harvested so that it will not be a breeding place for root lice to spread to new fields.

Strawberry Leaf Spot or Rust.—This disease is common on the leaves of strawberries and frequently does serious damage. It may be recognized by the white spots on the leaf, each spot being surrounded by a purplish ring. If this disease proves serious, care should be taken to plant only plants free from the disease. Rotate the crop and burn over the old beds as soon as the crop is harvested.

CHAPTER XXXIV.

THE PESTS OF ORCHARD FRUITS.

The so-called tree fruits like the apple, pear, quince, peach, plum and cherry are usually badly attacked by a number of pests. Certain of these pests are apt to occur on any of these fruits and these are discussed below under the heading "General Orchard Pests." Then there are two main classes of fruits, those with hard seeds or stone fruits and those with cores. To the latter group belong the apple, pear and quince and to the former group belong the peach, plum and cherry. Each of these groups besides being attacked by the general orchard pests, are attacked by their own special pests. Therefore in looking up the pests of any orchard tree, we should look first among the "General Orchard Pests," and if it is not found there, then look among the stone fruits or the core fruits as the case may be.

Most of these pests can be controlled by a system of spraying which differs for each group, that is to say, the same spraying is applied to the apple, pear and quince, but it takes a slightly different spraying for the peach, plum and cherry. These sprayings have been reduced to the simplest possible terms in order to produce fruits free from disease and it is only by spraying that we can produce disease-free fruit, at the present time.

General Orchard Pests.

The following pests are, for the most part, distributed generally throughout the country and occur more or less commonly on all kinds of orchard fruits, although they may be much more destructive to one kind of fruit than another.

Guide to General Orchard Pests.

I. **Attacking trunk, limbs and branches.**
   A. Giving the trunk, limbs and branches an appearance as if ashes had been poured on them.
      San Jose Scale (page 293).
   B. Eating out the soft sap wood and boring small round holes through the bark, giving the tree the appearance of being riddled by shot.
      Shot hole borers (page 294).

II. **Attacking the leaves.**
   A. Worms webbing the leaves together.
      1. Webbed nests containing worms in the winter.
         Brown tail moth (page 295).
      2. No worms in the nests in winter.
         Fall web worm (page 295).

B. Leaves not webbed together.
   1. Small web in crotch of tree.
         Tent caterpillar (page 296)
   2. No web in crotch of tree.
         Gipsey moth (page 295).

San Jose Scale.—This is a pest that was introduced into California from China, and from California it has gradually spread until it occupies practically all of the United States, except the extreme northern border and the higher parts of the Rocky Mountains. It is gradually extending its range

Fig. 95. San Jose Scale on twig.

into these regions. It may be recognized on leaf, branch, or fruit, by the nearly circular outline and the light ashy color of the scale covering, which is about as big around as the lead in a pencil. Under this scale a yellowish insect lives. The insect is little more than a sack with a long slender beak, which extends down into the bark and through which the insect sucks the sap. This usually weakens the tree very greatly and unless the increase in the insect is checked, the tree usually dies in from two to five years. Peach trees are more easily killed than apple trees, and the writer has seen hundreds of peach trees planted, that were killed before they ever bore a crop of fruit. The rapidity with which this insect increases and covers a tree, is almost beyond belief. Since the insects are so small they easily escape notice when the trees are planted, and unless very close watch is kept, they will take the tree before it is completely grown. Before buying trees, insist that your nurseryman show that his stock has been inspected by the proper State official (State Entomologist or Nursery Inspector) and declared free from all dangerous pests. This cannot be taken as an absolute guarantee but is a good indication. Then spray the tree once each winter with lime

Fig. 96. Twig showing injury by Shot Hole Borer.

Fig. 97. Tree showing gum exuding, the work of the Shot Hole Borer.

sulphur from the time the trees are planted until they cease bearing and are chopped out. Bear in mind that the scale is a very small insect and that the lime sulphur only kills the scale that it actually touches.

Shot Hole Borers, also called "Bark Beetles."—Under this head we discuss a couple of so-called bark beetles which work as small white grubs in the tender green sap wood just under the bark. When these grubs get full grown they change to small hard shelled beetles. These bore small round holes through the bark in order to escape, then they bore back through the bark again and construct a tunnel in the sap wood where the female lays eggs for another brood. These small holes eaten through the bark give the tree the appearance of having been struck by a load of small shot. More or less sap flows from these holes, thus the tree is weakened and as it becomes weaker, these insects as well as others increase their attacks until the tree is killed.

Trees that are kept in good healthy condition are not apt to be attacked by these pests. This means that the trees must be properly sprayed, pruned and cultivated. The trees should be watched closely and if the presence of the shot hole borers is noticed, the limbs or branches containing them should be pruned out and burned at once. If a whole tree is badly infested it should be chopped down and burned. The limbs and branches cut out in general pruning should be burned at once otherwise this insect is apt to continue to breed in the cut branches and later spread to healthy trees.

Fig. 101. Fall Web Worm.

Brown Tail and Gipsy Moths.—We discuss here two insects that are natives of Europe and which were first introduced into Massachusetts. They have spread gradually until they now cover most of the New England States. They are discussed briefly here because they give every indication of spreading throughout the Northern United States at least and because they are so very destructive to all kinds of trees including the fruit trees. The brown tail moth lives over winter as half grown worms or caterpillars in web-like nests on the trees. Fruit trees should be watched and if such nests are noticed containing worms in the winter they should be cut out and burned. The Gipsy moth passes the winter in the egg stage. Female moths lay several hundred eggs on the trunks of trees, on fences or buildings. These egg masses are about one inch long and about half as wide; they are light yellow in color and should be looked for and painted with creosote.

In addition to burning the winter nests of the brown tail moth and painting the egg masses of the gipsy moth, fruit trees in the areas infested should be sprayed with arsenate of lead at the rate of five pounds to fifty gallons of water, the first week of May and again the first week of August.

Fall Web Worm.—This insect appears late in the summer or early in the fall in the form of hairy caterpillars which build webs over the leaves of fruit trees and some forest trees. They extend their web as they feed on the leaves until they cover good sized branches. The ordinary late sprayings of fruit trees will usually keep this pest in control. The nests containing the worms may be burned out of the trees by means of a torch made by wrapping old rags tightly around a pole long enough to reach the highest part of the trees, and saturating the rags with coal oil (kerosene). By moving this rapidly there will be no danger of injuring the branches and the worms will be burned to death.

Tent Caterpillar.—Tent caterpillars are rather hairy worms that appear in the spring before the leaves are well out on the tree and build a small web in a crotch. When the caterpillars are young they return to this nest at

Fig. 99. Adult of Tent Caterpillar.

Fig. 100. Eggs of Tent Caterpillar.

night, going out to feed on the young leaves during the day. A spraying made just as the leaves are coming out on the trees, consisting of two pounds of arsenate of lead to fifty gallons of water, will control this insect. Fortunately, however, this pest is usually kept in control by its natural enemies, so that the few nests that do occur in orchards may be readily burned out with kerosene torch as recommended for the fall web worm (page 295). In this case it will be necessary to burn the nests late at night or very early in the morning, as the worms stay in the nest only at night.

The Spraying of Orchard Fruits.

Unfortunately there are so many pests of the orchard fruits at the present time that no one can hope to raise these fruits successfully without spraying them. It is also unfortunate that it takes a different kind of spraying for the two main classes of fruits, the stone fruits and the core fruits. We have attempted to reduce this spraying to as simple proportions as possible in the following outline and yet have a successful spraying practice.

If orchards are carefully sprayed as here outlined they will be comparatively free from the ordinary pests and in this way a saving of a great many dollars may be effected.

Pests Attacking the Core Fruits.
(Apple, Pear and Quince.)

I.   **Attacking the roots, causing small knots.**
> Woolly Louse (page 300).

II. **Attacking the trunk or main branches.**
> Borers (page 300)

III. **Attacking the smaller limbs.**
> A. A scale-like insect encrusting the branches.
>> Oyster Shell Scale and Scurfy Scale (page 300)
>
> B. Causing rough cankers on the bark.
>> Canker (page 302)

IV. **Attacking the leaves.**
> A. Rust red or orange spots on the leaf.
>> Rust (page 303)
>
> B. The leaves and usually the young green branches withering and turning black as if they had been burned.
>> Fire Blight (page 304).
>
> C. Worms or caterpillars eating the edges of the leaves.
>> Caterpillars (page 307)
>
> D. Small worms feeding between the upper and under surface, making trumpet shaped thin places on the leaf.
>> Leaf Miner (page 307)
>
> E. Small louse-like insects on green branches and leaves, causing the leaves to curl and drop early.
>> Plant Lice (page 309)

V.   **Attacking the fruit.**
> A. Pinkish white worms eating around the core.
>> Codling Moth (page 298)
>
> B. Causing brownish or blackish rotten spots.
>> Rot (page 311)
>
> C. Causing the skin to be scabby and often cracked.
>> Scab (page 310)
>
> D. Causing small irregular scabs on the skin of the fruit.
>> Blotch (page 309)

Codling Moth, also called "Apple Worm."—This is the common cause of wormy apples, a condition known throughout the country wherever apples are grown. It is the chief pest of the apple, not only resulting in great loss

Fig. 115. Adult Codling Moth.

Fig. 117. Cocoons of Codling Moth from underbark.

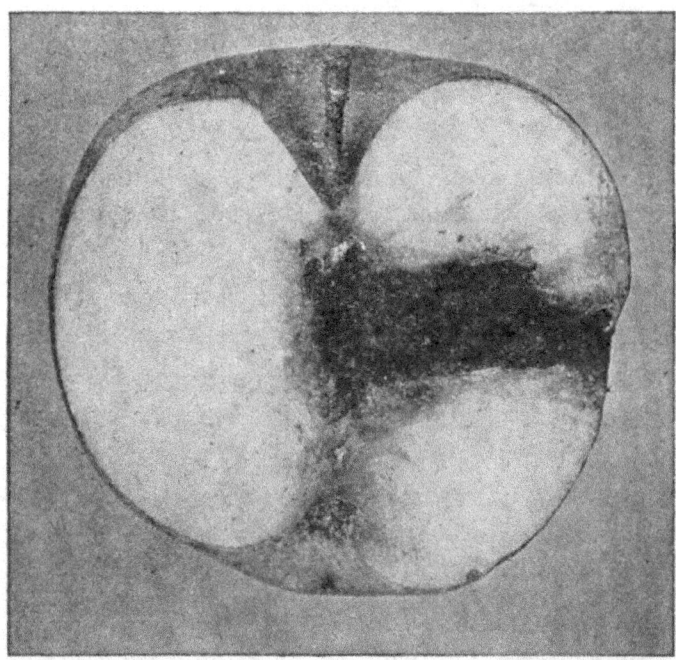

Fig. 116. Apples injured by Apple Worm or Codling Moth.

by causing the early dropping of the fruit, but also causing it to be small, stunted, misshapened and of small value on the market or for home use. The farmer is often at a loss to know how these worms get into the apple without leaving any indications of the place where they entered. This is explained by the fact that the worm enters from the blossom end when it is quite small and feeds around the core until it gets full grown. When it is full grown it eats its way out and leaves a small round hole through the side of the apple. After it leaves the apple, it crawls under a loose piece of bark

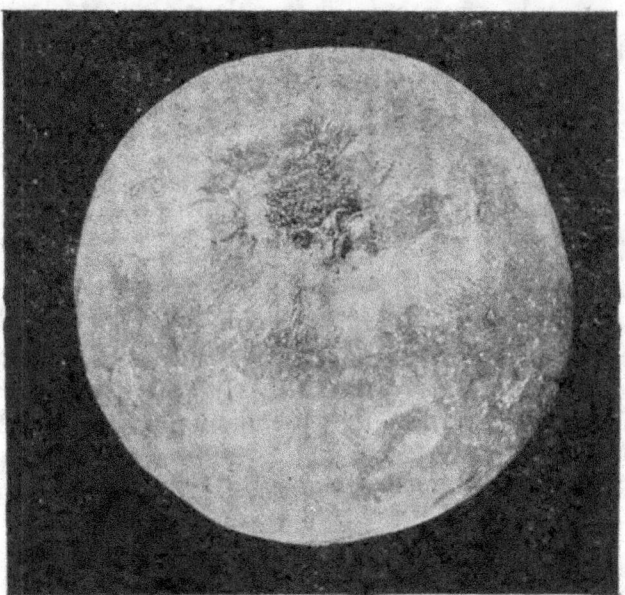

Fig. 118. Blossom end injury.

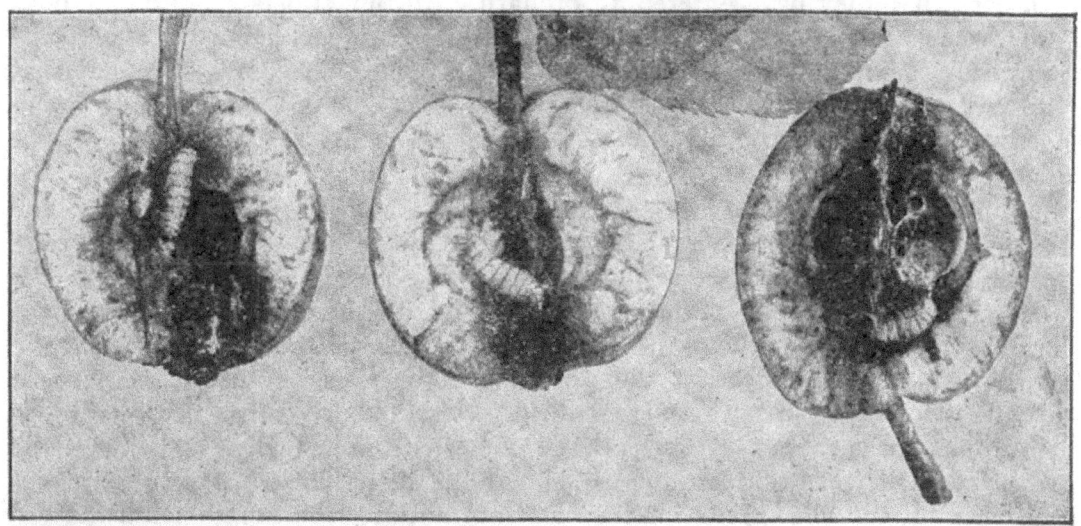

Fig. 119. Codling Moth Worms in apple.

on the trunk of the tree where the next spring it changes to a delicate little moth about the time the apple trees are in full bloom. This moth flies away to lay tiny little eggs, a bunch of which looks like a drop of milk on the leaves. From these leaves the little worms hatch which bore their way into the apples. These worms nearly always enter at the blossom end of the apple and spend a day or two in the blossom cup feeding. It is for this reason that it is important to spray **promptly** just after the blossoms have fallen from the trees as outlined under "Apple Spraying" (page 312). This little blossom cup only stays open a few days then it closes so tightly that we can no longer get the poison into the cup, but the worm can still eat its way in. Then, too, if we spray much later than this, most of the worms will have found their way into the apples and our spraying will do no good.

While spraying is the chief remedy to be used against this pest, it will help very much if hogs are allowed to roam the orchard and pick up the wormy apples as they fall. Or these wormy apples or wind falls may be picked up and destroyed.

Woolly Louse.—Often it will be noted that trees in the orchard are stunted and lack vigor without any apparent cause, save perhaps a few small louse-like insects covered with a white powdery substance in the crevices of the bark or on the water sprouts. If, however, such trees are dug up the roots are found to be badly decayed and covered with small round nodules. This of course indicated the real cause of the lack of thrift in the tree. These swellings on the roots have been caused by louse-like insects similar to those found above ground. There is this difference in the two forms, the ones above ground seldom do any damage while those below ground often do very great damage.

There is very little that can be done in a practical way to control these insects after they have entered an orchard. All our efforts therefore, should be directed to seeing that no infested trees are planted. Examine every tree carefully before it is set out and discard every one that shows any abnormal growths on trunks, branches or roots. An apple tree is set for a good many years, therefore, take great care to see that you start with a healthy young tree.

Borers.—Under this head we consider several pests which bore into the trunk and larger branches and are often very destructive to young trees, completely girdling the tree and killing it. The presence of these borers can often be discovered by noticing little piles of fine sawdust around the base of the tree on the ground. Often large or small spots on the bark will die and turn brown, indicating the presence of borers underneath. The best remedy to use against these pests is to keep the trees healthy by proper pruning and spraying, as trees that are healthy are much less liable to be injured than trees that are neglected. If the little piles of sawdust are noticed, or if the dead spots on the bark are noticed, the borer should be cut out with a knife, taking care to make no more cuts than necessary and to make all cuts lengthwise of the limb or trunk as otherwise we are apt to do more damage than the borers. Often the borer can be followed in its tunnel by means of a piece of soft wire so that it will not be necessary to do so much cutting.

Oyster-shell and Scurfy Scale.—These two pests are considered together here as they produce the same effect on the apple, causing the bark to become

Fig. 103.
Round headed Apple Tree Borer.

Fig. 104.
Flat headed Apple Tree Borer.

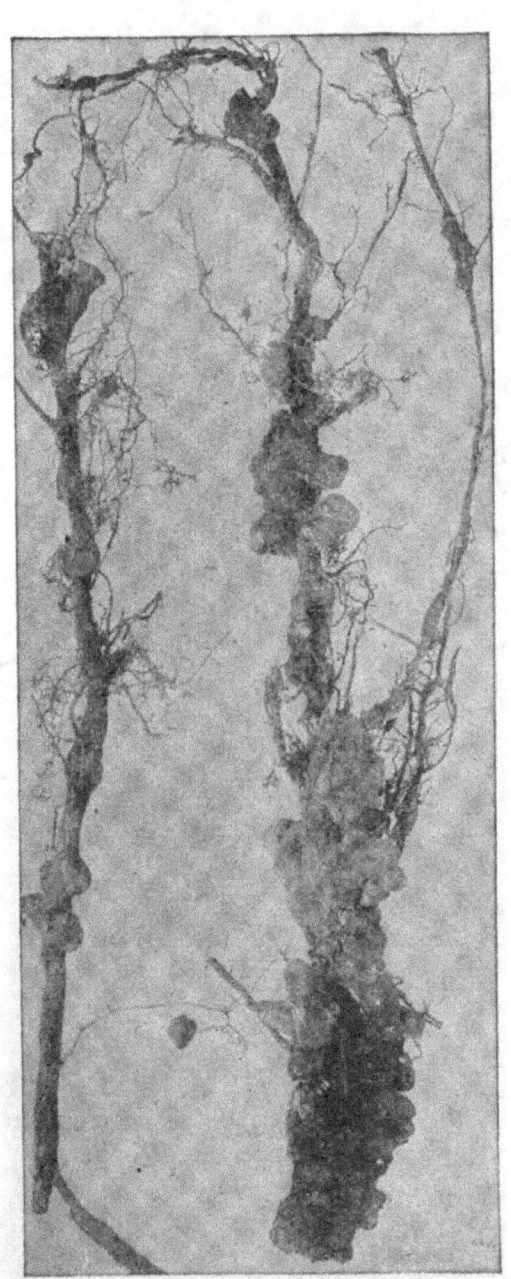

Fig. 102. Roots of apple tree injured by
Woolly Louse.

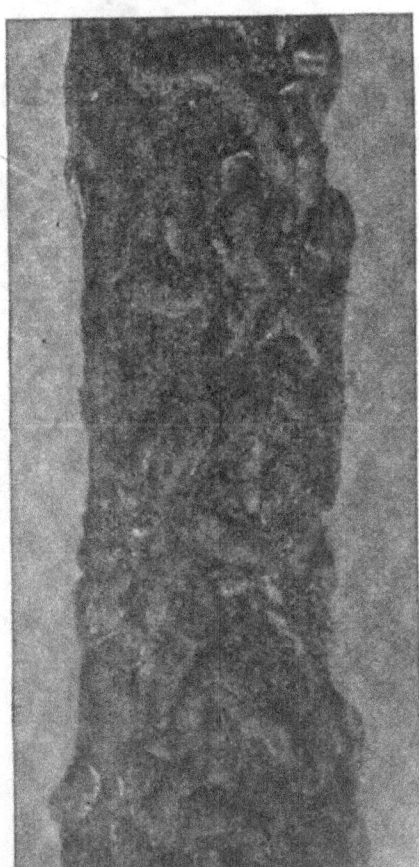

Fig. 105. Oyster Shell Scale on twig.

rough. They both weaken the tree just as the San Jose scale does by suck-ing the sap. The Oyster-shell scale is dark gray in color and is shaped some-thing like the shell of the oyster. The scurfy scale is white in color and not so long but broader than the oyster-shell scale.

Fig. 106. Canker on apple twig.

Although they are harder to kill than the San Jose scale, both of these insects are ordinarily kept in control by the usual spraying for the San Jose scale. If the usual method is not sufficient, the number of sprayings should be increased or the strength of the mixture increased. It has been found that the soluble oils are more satisfactory for controlling these pests than is lime sulphur.

Cankers.—Canker is not a single disease but several diseases that are not in themselves always serious. They may, however, be the starting points of some of the serious rots of the fruit, or of some of the diseases of the leaves. Therefore, as the trees are being pruned, the farmer should watch carefully for these cankers on the limbs, and should cut them out and burn them. Some of the cankers are really serious and soon destroy young trees. Most of these cankers start where the tree has been injured by the plow, hame of the harness or by the careless use of the ladder or the saw during pruning time. Needless to say these causes should be avoided as far as possible.

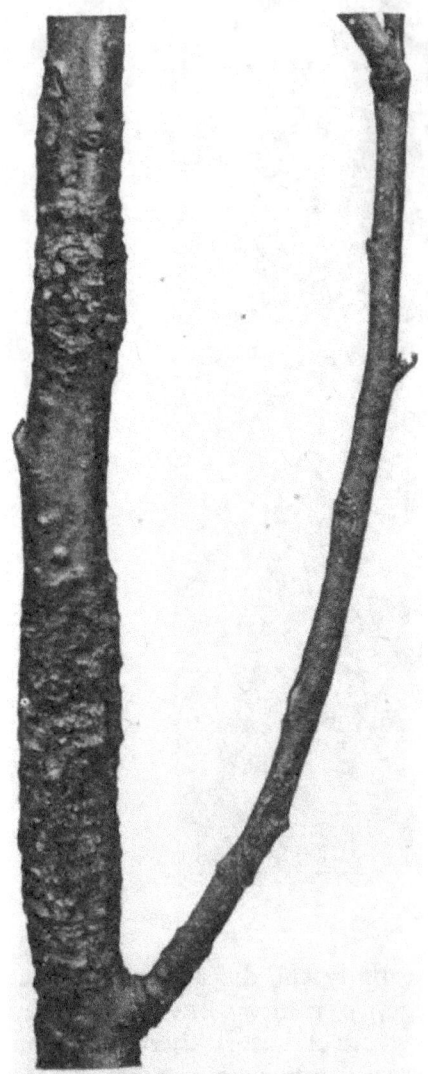

Fig. 108. Canker on apple tree twig.

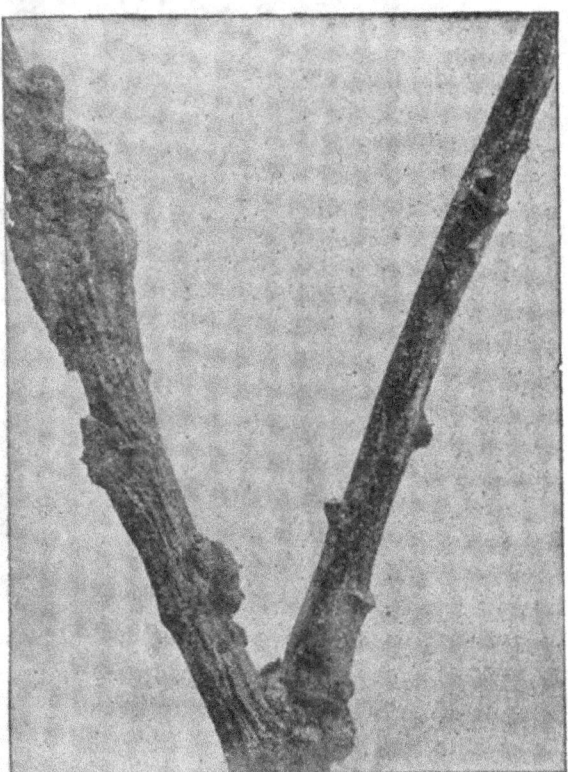

Fig. 110. Apple Canker.

Rust.—The rust of the leaves of the apple has a peculiar history in that it spends the winter on the red cedar or juniper tree in the form of the so-called cedar apple, then in the summer the spores (seeds) of the disease are carried back to the leaves of the apply by the wind. This is one of the really serious diseases of the apple, especially so in some sections where the cedar is much grown. In certain states laws have been passed making the red cedar a public nuisance, which may be destroyed by the proper authorities. We have here the solution of this trouble, that is, the cutting down and burning of the red cedar trees in the vicinity of the orchard. This usually applies

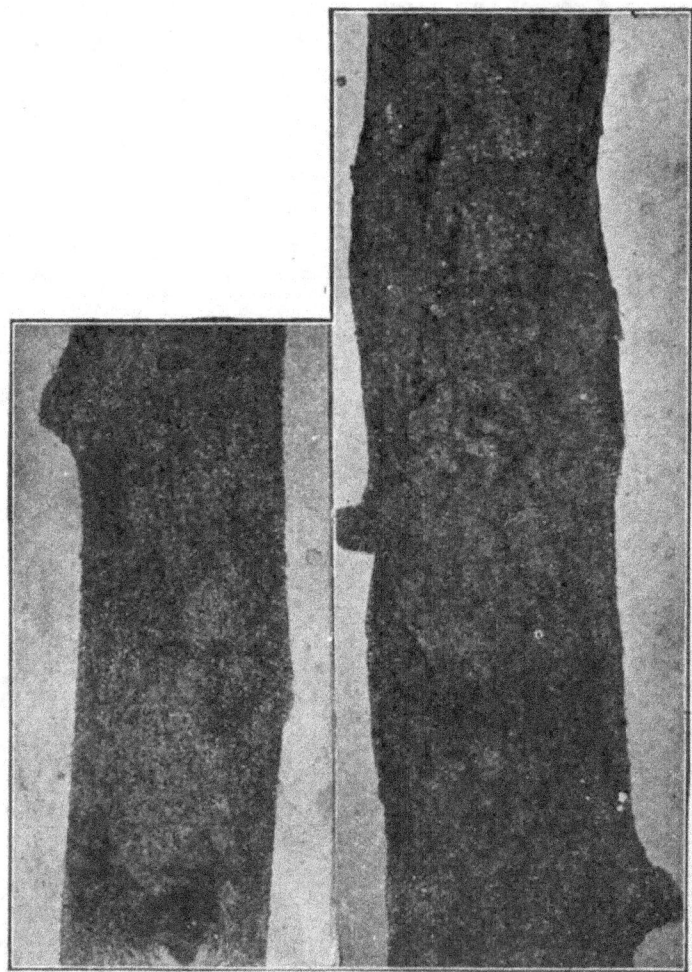

Fig. 109. Pear Canker.

only to those trees near the orchard, but occasionally the disease may be carried long distances by the wind. The spraying program outlined for the apple on page 312 will usually keep this pest in control but if there are many cedar trees near the orchard, even thorough spraying will not be entirely effective.

Fire Blight.—This disease is almost too well known to need description. It is worse on the pear, however, it attacks other fruits as well, especially the apple. The characteristic way in which the twigs die back is familiar to most orchardists. Its effects vary with different varieties of fruit. With some it is progressive, continuing until the whole tree is dead. In other cases it will kill only a few branches and will not become serious. Still other varieties seem to have a very high resistance and are little or not at all troubled by the disease. In general the disease seems to be spread from one blossom to another by any of the insects that feed on nectar (honey),

Fig. 111. Apple leaf infested with Rust.

Apple.
Fig. 112. Apple Rust on cedar, so called Cedar

thus the disease is often spread over wide areas and usually not a single susceptible tree escapes. There are several different things that seem to tend to increase the disease. Thus too heavy pruning in the winter seems to cause the trees to grow too rapidly in the spring, thus making them more likely to fall a victim of this disease. In some orchards that were heavily fertilized with a fertilizer that promoted too rapid growth, it was noted that the disease was worse than usual. Water sprouts are especially likely to have this disease and they should be kept down. Plant lice (page 309) seem to spread the disease from one part of the tree to another, or from one tree to another. The orchards should therefore be kept free from these as much as possible.

Fig. 113. Fire Blight. Injured and uninjured twig.

It is a constant fight to keep this disease in check. The farmer should go over his orchard frequently during the summer and cut out the diseased limbs, taking care to get well down below the part that is diseased   If only the diseased parts are cut out the chances are very likely that the disease has really started below the place where it shows, hence, in a very short time this part will also be dead and will have to be cut out. If the limb had been cut back far enough the chances are that we would have gotten below the diseased part and it would have been checked. All the tools, pruning knife, shears or saw, should be dipped in a disinfectant solution after each limb is cut, so that there will be no danger of leaving some of the germs of the disease on the cut. This solution is made by dissolving a one-grain tablet of corrosive sublimate (Bichloride of Mercury) in one pint of water. The sublimate tablets may be secured at any drug store. This is one of the most deadly poisons known and must be handled with great care. Be especially careful to keep it away from children.

Caterpillars.—Apple leaves are eaten by a large number of different kinds of caterpillars or worms, all of which have different names and different histories, however, as they are all subject to the same measures to keep them in check, they are all considered together here. Ordinarily the usual spraying as outlined for the apple (page 312) will keep all of these caterpillars in check. Once in a while when the conditions are just right, they may increase in spite of the usual sprayings. In that case a spraying of two pounds of arsenate of lead paste in fifty gallons of water, one ounce or one teaspoon

Fig. 114. Plant Lice curling apple leaves.

level full in a gallon and a half of water, will stop most of them. If it does not, spray again using twice as much poison.

Leaf Miners.—Sometimes, especially in orchards that are not well cared for, the apple leaves will fall very early in the fall. If these leaves are examined they will show curious little trumpet shaped areas where the inner part of the leaf has been eaten away between the upper and the under surfaces. This is the work of the leaf miner, a very small insect which cannot do much damage except where it comes in large numbers, which is some-times the case.

Since this insect passes the winter in the fallen leaves, it is usually a rather easy matter to kill them by plowing these leaves under in the fall or in the spring.

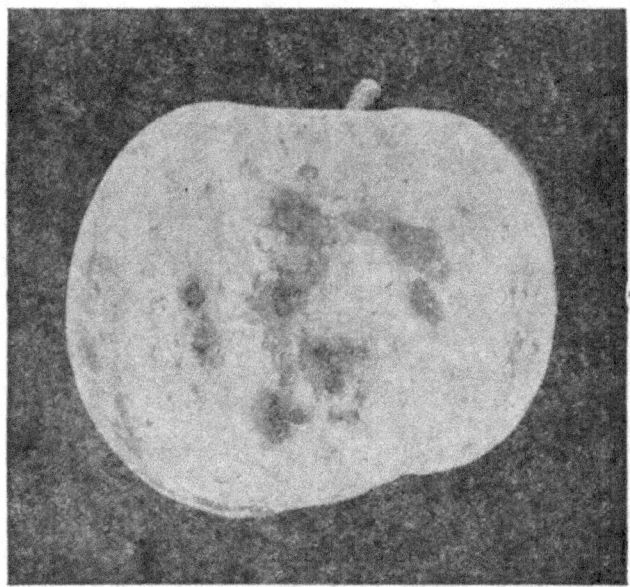

Fig. 125. Scab on apple.

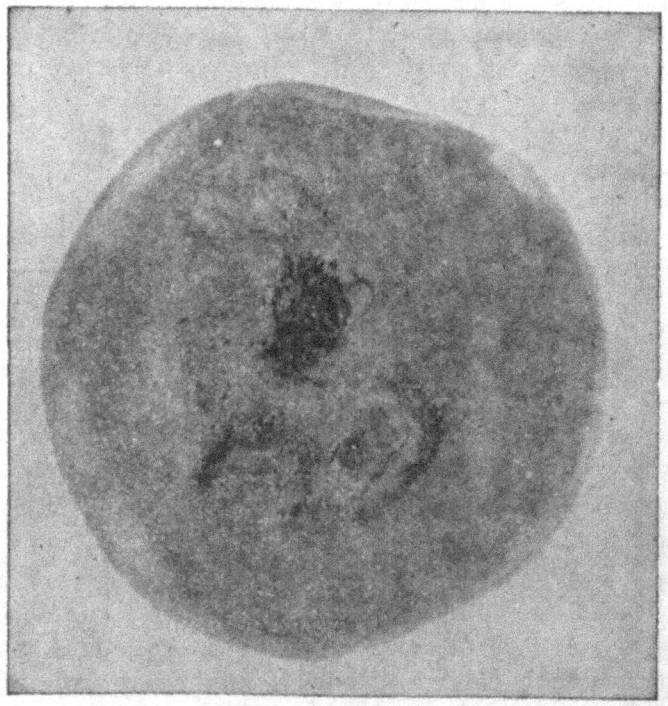

Fig. 126. Scab on apple.

Plant Lice.—The plant lice which attack the apple are small louse-like insects with or without wings and usually more or less green in color. The results of their attacks are usually noticed in the curling of the leaves. Any fruit born on stems badly attacked, is apt to be small in size and more or

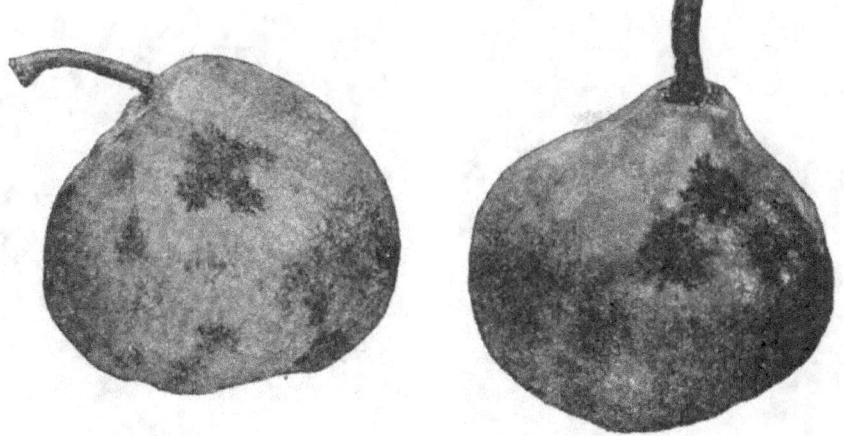

Fig. 127. Pear Scab

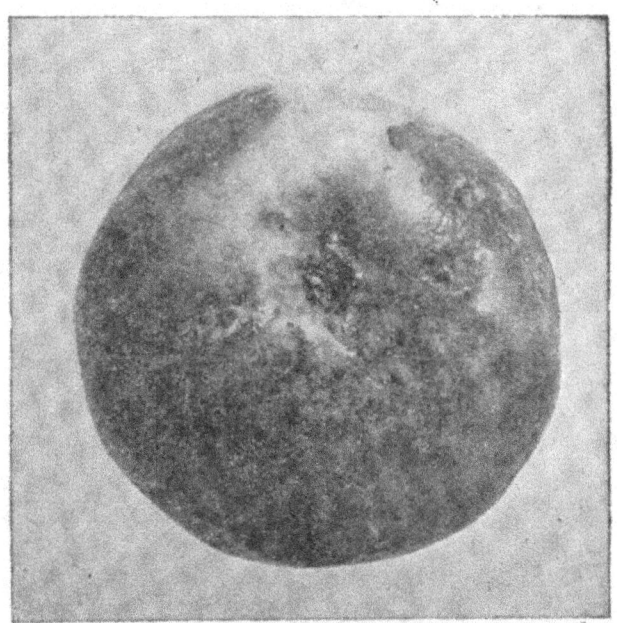

Fig. 128. Apple Blotch.

less knotty. This damage is often serious and requires a special treatment. Perhaps the best treatment consists of spraying the plants before the leaves commence to swell, with soap solution (page 220), or with a spray composed of one teaspoonful of nicotine to one gallon of water. If the winter spraying recommended for San Jose scale (see page 293) is applied just as the buds are swelling, there will usually be no trouble by these pests.

Apple Scab.—The most conspicuous work of this pest is on the fruit, however, it also attacks the leaves as well. On the fruit it may be recognized by the fact that the skin cracks and shows a pit filled with a sooty black substance.

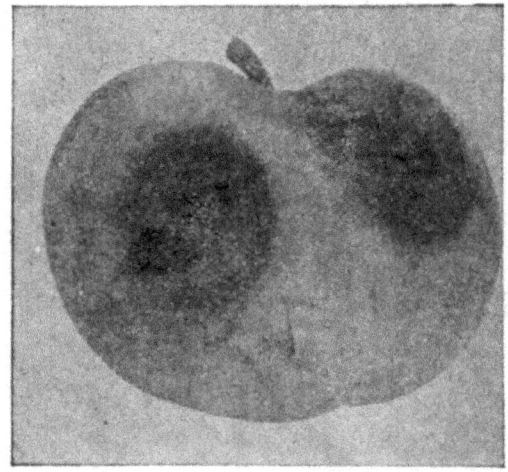

Fig. 120. Early stage Bitter Rot.

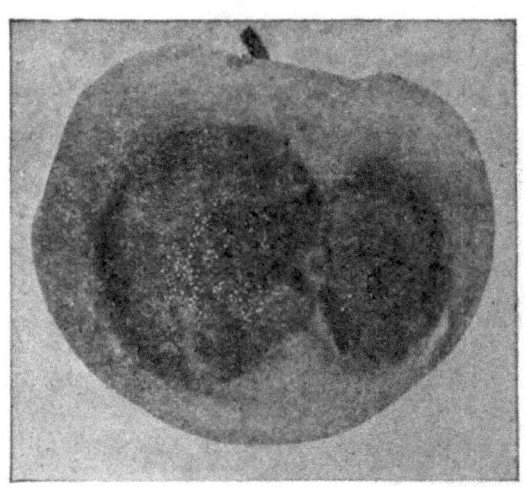

Fig. 121. Middle stage Bitter Rot.

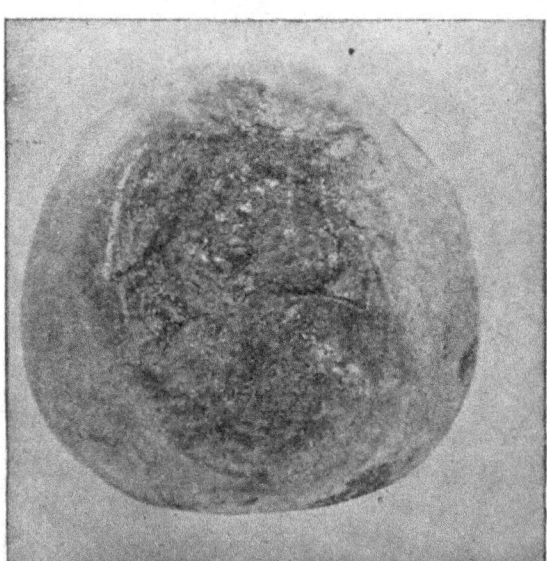

Fig. 122. Late stage Bitter Rot.

Usually apples sprayed, as outlined on page 312, will be entirely free from this disease. If they are not, another spraying should be made just before the blooms commence to show pink. This spraying should be given between the first and second sprayings, the same materials being used as used for the second spraying.

Apple Rot.—This disease is usually called bitter rot because when a diseased apple is bitten into, it sometimes has a disagreeable bitter taste. On the fruit, spots that are attacked by rot are more or less circular in shape,

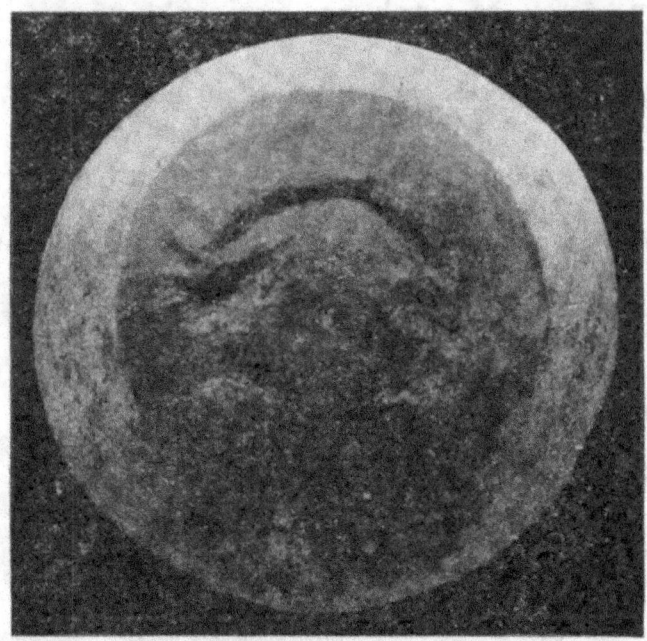

Fig. 123. White Rot on apple.

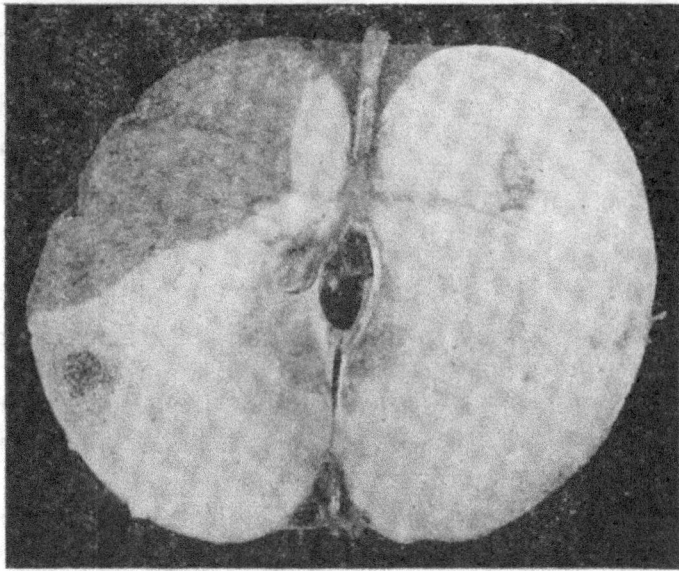

Fig. 124. White Rot cut open.

brown in color, and usually somewhat corky. This disease is very destructive to the apple and as it is somewhat difficult to control, it is very frequently met with.

The control of this disease demands not only the usual spraying (see page 312), but all the badly diseased fruits which are usually left on the tree, should be collected and destroyed. The fruit on the ground should also be collected and fed to the hogs or otherwise destroyed. The cankers (see page 302) which may be simply one form of this same disease should be pruned out and burned.

Spraying the Core Fruits.
(Apple, Pear and Quince).

First Spraying.—Spray the trees very thoroughly in the winter, or early spring, using either home-made lime sulphur (page 219) or concentrated lime sulphur (page 219) at the rate of one gallon to seven gallons of water.

Fig. 129. Just right on left, and too late on right, for second spraying of apple.

Great care should be exercised to cover all the limbs and branches. This spraying is for the San Jose Scale and for some of the other diseases.

Second Spraying.—Spray the trees immediately after most of the white blooms have fallen, using concentrated lime sulphur (page 219) one and one half gallons, to fifty gallons of water, and add from one to three pounds of arsenate of lead (formula, page 218). This spraying is for the apple worm and various other diseases.

Third Spraying.—This should be given three or four weeks after the second spraying, using the same mixture as was used in the second spraying. This spraying is for the late worms and other diseases.

Fourth Spraying.—This should be given two months after third spraying, using same mixture as for second spraying. This spraying is for the late worms feeding on the leaves, late rots and rusts.

Pests of Stone Fruits.
(Peaches, Plums and Cherries.)

I. **Attacking the trunk at or near the surface of the ground.**

    A. Tumor like swelling on the trunk.
            Crown Gall (page 313)

    B. Whitish worms mining the trunk, causing gum to form near the base.
            Peach Borers (page 313)

II. **Attacking the branches of plum and cherries, causing black swellings.**
            Black Knot (page 315)

III. **Attacking the Leaves.**

    A. Causing the leaf to curl.
        1. Small, louse-like insects in the curls.
            Plant Lice (page 315)
        2. No lice in the curls.
            Leaf Curl (page 316)

    B. Causing small brown spots or holes in the leaves.
            Leaf Spot

IV. **Attacking the Fruit.**

    A. White worm eating around the seed.
            Curculio (page 316)

    B. No white worm present.
        1. Fruit with brown, rotten spots.
            Brown Rot (page 317)
        2. Fruit with small, black spots. Fruit often cracking open.
            Scab (page 318)

Crown Gall.—This disease is known by a hard, gall-like growth on the roots and stem, usually near the surface of the ground. It is a very troublesome disease but as yet there is no known remedy for it. The farmer should be on the lookout for the disease when he is planting new trees and destroy all that show any signs of it. If diseased trees are noticed in the orchard they should be dug up and burned.

Peach Borers.—Around the base of most peach trees can be found a gummy mass which is a pretty good indication of the presence of the peach borer. If the soil be raked away and the gum scraped off, the tunnels of the borers can as a rule be easily found. Then by following the tunnels the light, yellowish worm can usually be found without very much trouble. These borers are very troublesome, being, next to the San Jose Scale, perhaps the most destructive enemy of the peach trees.

The control of the peach tree borer is not easy; however, if the fight is persisted in, it will bring success. Since the moths, which are the parents of the borers, usually lay eggs at or near the surface of the ground around the base of the tree, the young borers, when they hatch, work down until they come to the soft, moist bark covering the crown of the tree before they attempt to enter.

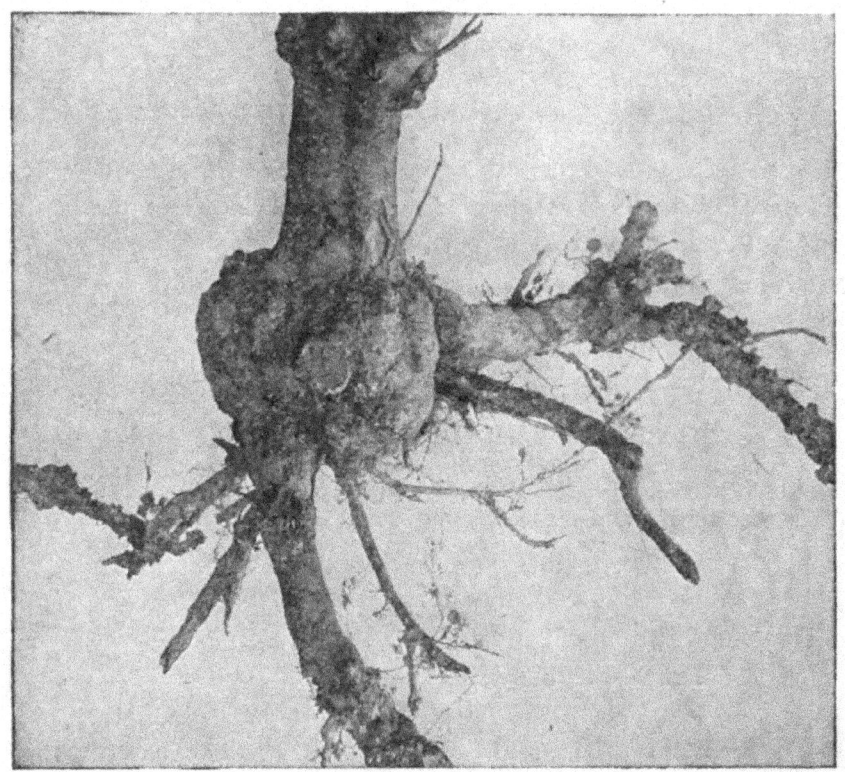

Fig. 130. Crown Gall injured roots.

The best method of control consists in mounding the trees to a height of about two feet early in the summer. This makes the moths lay their eggs high up on the trunk, at a considerable distance from the tender bark of the crown. Thus when the young borers hatch, they will find it very difficult to enter through the tough bark of the trunk, and if they attempt to seek out the tender bark of the crown it will be so far away that they will have difficulty in reaching it. In the fall of the year the mound should be removed so that winter will have a chance to kill as many of the borers as possible. The trees should be gone over carefully at least twice a year, the soil removed from the crown and large roots and any borers that have escaped the above treatment should be dug out with a sharp knife, cutting the bark as little as possible. If the practice of mounding and worming is persisted in for a few years, it will greatly reduce the number of these pests, making the growing of peaches much more profitable.

Black Knot.—This is a very destructive disease of the plum and cherry, appearing as hard, black knots on the twigs. This disease usually kills trees in two or three years, therefore our remedies must be prompt to be effective. Sprayed trees are much less troubled than unsprayed trees, hence we should spray plum and cherry trees as outlined (see page 320).

Plant Lice.—Often in summer the leaves of young peach trees turn yellow, the tree looks sickly and frequently dies. If such trees are examined, many ants are usually found running up and down the trunk, and if the smaller

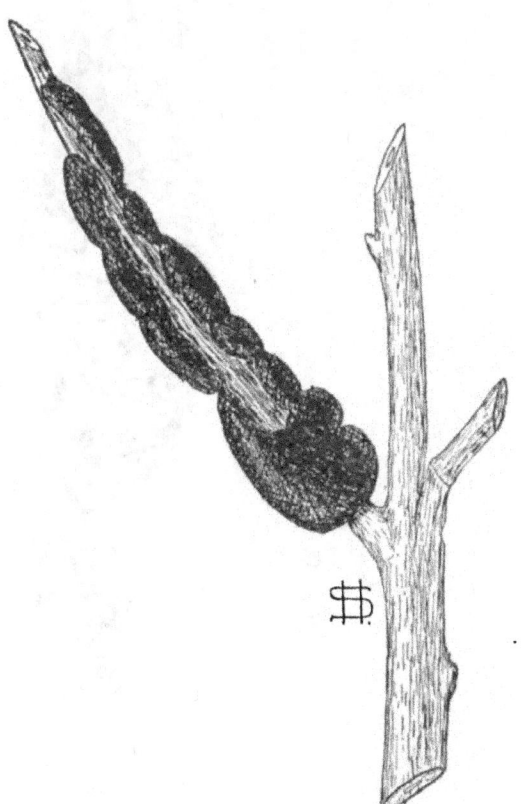

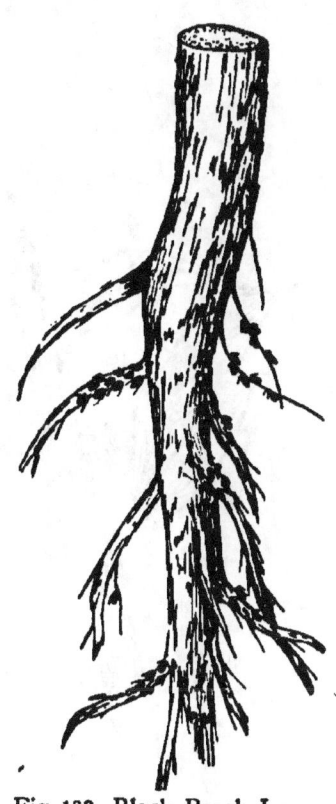

Fig. 132. Twig injured by Black Knot.

Fig. 133. Black Peach Louse on roots of peach.

limbs and branches are examined, many blackish plant lice can usually be found. However, the injury usually appears to be out of all proportion to the number of insects on the branches. If the roots are uncovered the answer will usually be evident, for such roots are usually covered with similar wing-less plant lice which are sucking the sap from the roots at a great rate. While this insect is usually worse on young trees, it frequently attacks old trees and sometimes kills them.

All young trees should be examined before they are set out and if any lice are found the root should be dipped in tobacco tea. Infested trees should have the soil removed from around the roots and finely powdered tobacco dust applied, then the soil should be returned. The rain will carry the tobacco dust down around the roots and kill most of the lice. Limbs covered with the lice should be pruned out and burned as these lice later acquire wings and fly away to infest other trees.

Leaf Curl.—This disease may be recognized by the fact that the leaves curl very badly and frequently fall to the ground. In some respects it resembles the work of plant lice, but there are, of course, no plant lice present

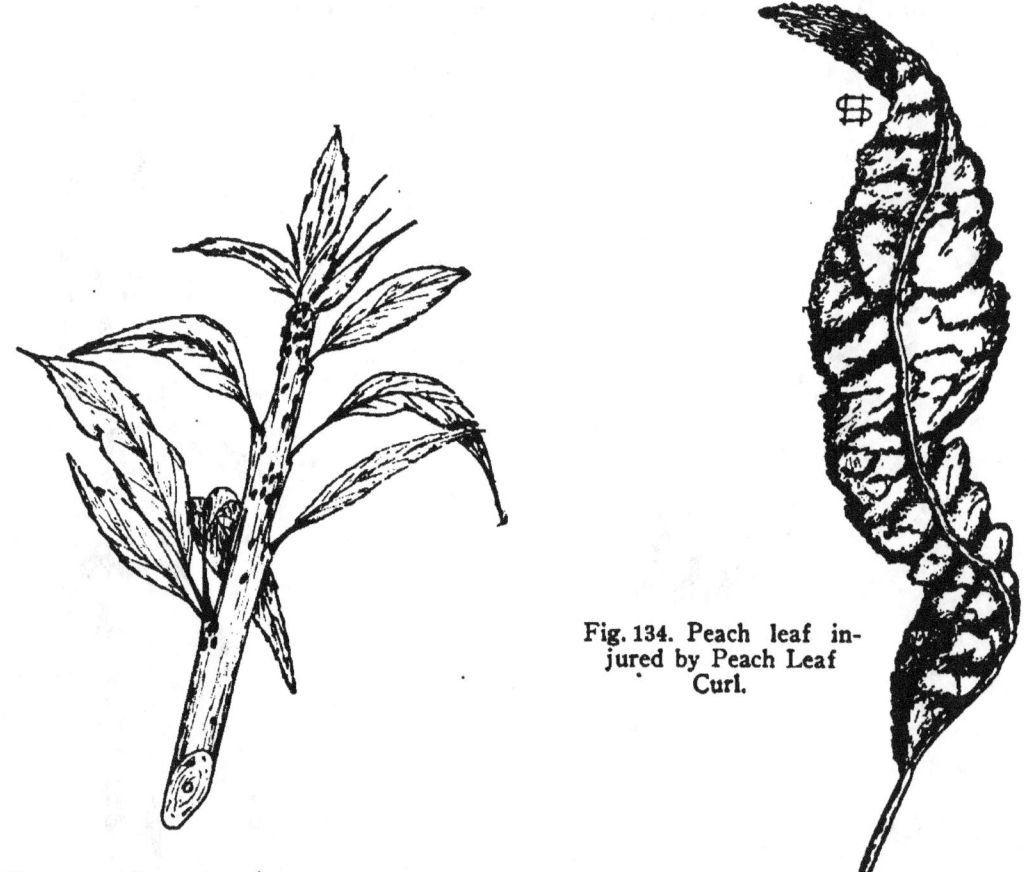

Fig. 134. Peach leaf injured by Peach Leaf Curl.

Fig. 133. Black Peach Louse on Stem of Peach.

nor any other visible cause for the disease. If many leaves are lost by this disease, the tree is weakened and cannot produce a satisfactory crop of fruit.

Peach Worm or Peach Curculio.—Everyone is familiar with wormy peaches, but not everyone knows just how the worm got into the peach. In the spring, soon after the young peaches are set, there is a small bug with a long snout that stings the peaches and lays an egg in a little cavity under the skin. This egg hatches into a yellowish grub which bores its way towards the stone of the fruit. Often such young peaches fall to the ground. Frequently they develop in a one-sided manner, are knotty and much covered with gum. Sometimes when peaches are not stung till late in the season, they remain on the tree but ripen early. Such peaches usually rot around the seed and are otherwise objectionable.

This is one of the most destructive pests of the peach, frequently causing a loss of fully half of the crop. Fortunately, by proper methods of spraying, this injury may be almost completely prevented (see page 320).

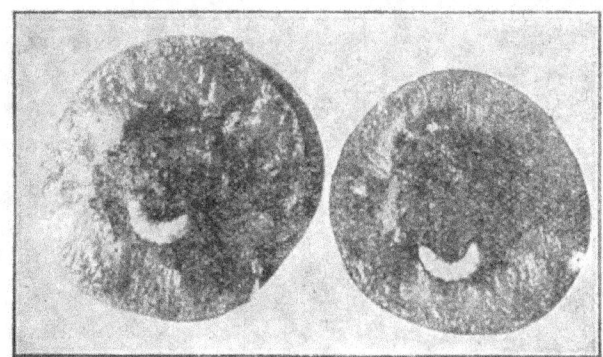

Fig. 136. Grub of Curculio.

Fig. 137. Peaches injured by Curculio.

Brown Rot.—Until recently this has been one of the most troublesome pests of the peach, plum and cherry and is the common cause of rotting peaches and plums. The disease usually starts as a small brown spot which increases very greatly in size, producing a soft brown decay which spreads very rapidly until the whole peach has become a soft, rotten mass. Later the peach becomes shrunken and its surface covered by a gray mass which is the spores (seeds) of the disease. These shrunken peaches and plums are called mummies and frequently hang on the tree until the next year. If they are not destroyed, whether on the tree or on the ground, they produce spores which infest the fruit the next year. In this way the disease is carried from one year to another. These spores usually gain entrance to the fruit where it is stung by the curculio (see page 316), hence, any system for controlling this disease must also aim to control the curculio. Such a scheme has been outlined for the control of all the diseases of the peach and plum (see page 320). In addition, all the mummies should be picked from the tree and should be burned together with those on the ground which have been raked up.

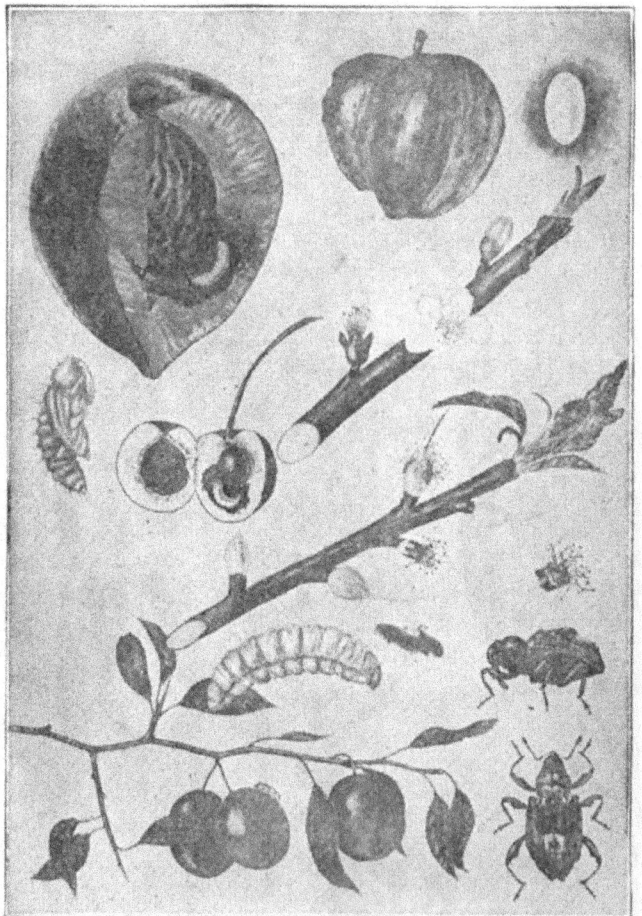

Fig. 138. Life history of Curculio showing injury to peach, apple, cherry, and plum; and adults side view and top view with twig of peach, too early to spray and just right to spray.

Peach Scab.—This disease is very common on the fruit of the peach tree. It makes its appearance as small, black, round spots. These spots grow together, especially on the side of the fruit that is exposed, the skin gets very tough and usually cracks open.

This disease is easily controlled by using the sprays recommended (see page 320). Especially by the ordinary winter spraying for peaches as recommended on page 320. The one point of success being to see that the tree is very thoroughly covered with the spray mixture.

Fig. 139. Brown Rot of peach

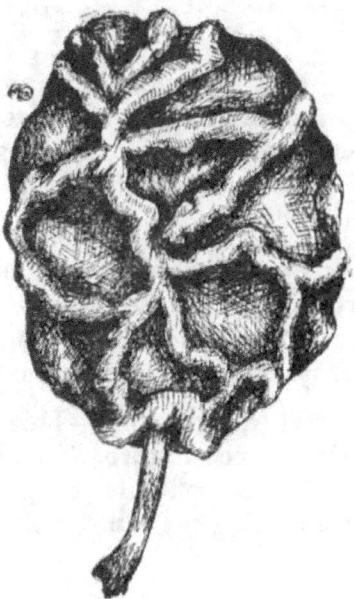

Fig. 140. Mummied Plum
caused by brown rot.

Fig. 142. Twig showing peaches just right to give second
spraying.

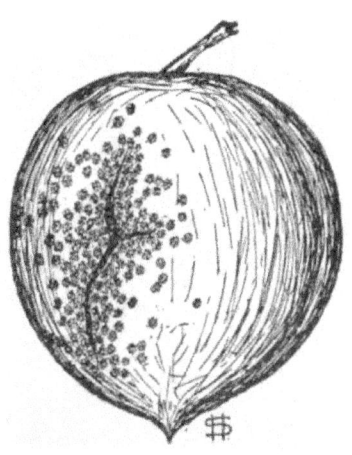

Fig. 141. Peach Scab.

Spraying Stone Fruits.

(Peaches, Plums and Cherries).

First Spraying.—Spray the trees very thoroughly in the winter or early spring, taking great care to cover all the limbs and branches and using concentrated lime sulphur (page 219) at the rate of one gallon to seven gallons of water. This spraying is for the San Jose Scale and for some of the other diseases.

Second Spraying.—Just after the shucks have fallen from the young fruits, this is usually a week or ten days after the blooms have fallen, use self-boiled lime sulphur and arsenate of lead (formulae page 219). This is for peach worm or curculio and early rots.

Third Spraying.—This should be done two weeks after second spraying, using same mixture. This is for any curculio missed by second spraying. It would be profitable to spray late peaches a month before they are due to ripen, using the same mixture as outlined above.

CHAPTER XXXV.

INSECT PESTS OF STORED GRAIN.

Grain of all kinds in storage is attacked by a large number of pests of two general classes. There are first the various kinds of beetles usually known as weevils, and second the various kinds of moths or "millers", sometimes called "fly weevils".

There may be a great number of different kinds of weevils present in a bin at the same time, injuring the grain, however, it is not necessary for us to describe all of these here.

Both kinds of insects spoken of above work in the same way, eating out the kernels of the grain. If left undisturbed they will frequently ruin the grain in a single year. Along with the destruction of the kernels of the grain goes the heating of the grain. If the grain is damp when it is stored, the heating is much more evident and weevils of all kinds develop much more rapidly than if the grain is thoroughly dry before it is thrashed and stored.

Since all of these insects are about equally destructive to all kinds of grain, and since they will live over from year to year in the ordinary granary, the granaries and corn cribs should be gone over each year and thoroughly sprinkled with air-slaked lime. This will kill many of the insects in the cracks that would otherwise escape and cause damage another year. If the bins, cribs and granaries are cleaned out this way each year, well in advance of harvest time, it will help very much in keeping the "weevils" in control. Bins, cribs and granaries used for storing grains should also be built so that they can be opened as much as possible in the winter time, thus causing the temperature to fall as low as possible, for in this way the rapidity of the breeding of the weevils will be greatly reduced.

Treatment with Carbon Bisulphide.

The most effective method of killing the various grain weevils is to fumigate with carbon bisulphide. This is a foul-smelling liquid that evaporates very rapidly, penetrates fairly well and is effective in the ordinary granary or bin if used at great enough strengths. To use carbon bisulphide the bin or ordinary granary should be made as tight as possible by stuffing damp newspaper into the cracks. The grain in the bins should be leveled down and so distributed that it does not have a depth of more than four or five feet, less would be much better. The fumigation should be done on a day when the temperature is not less than 70° F.

To find the number of pounds of carbon bisulphide necessary to fumigate a bin, if it is made of tight-matched lumber on all sides, multiply the length of the bin in feet by its width, and the product by its height, divide the result

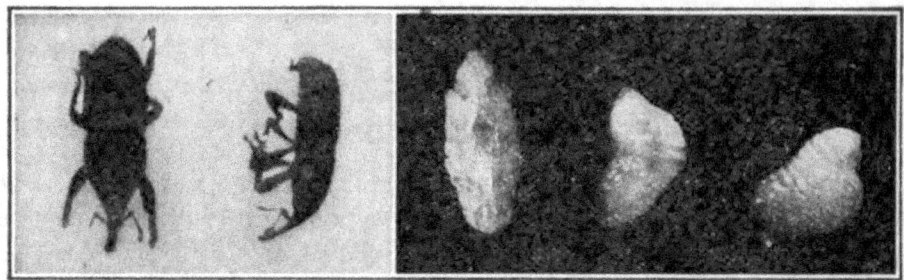

Fig. 143. Rice Weevils, and grub, pupa adults.

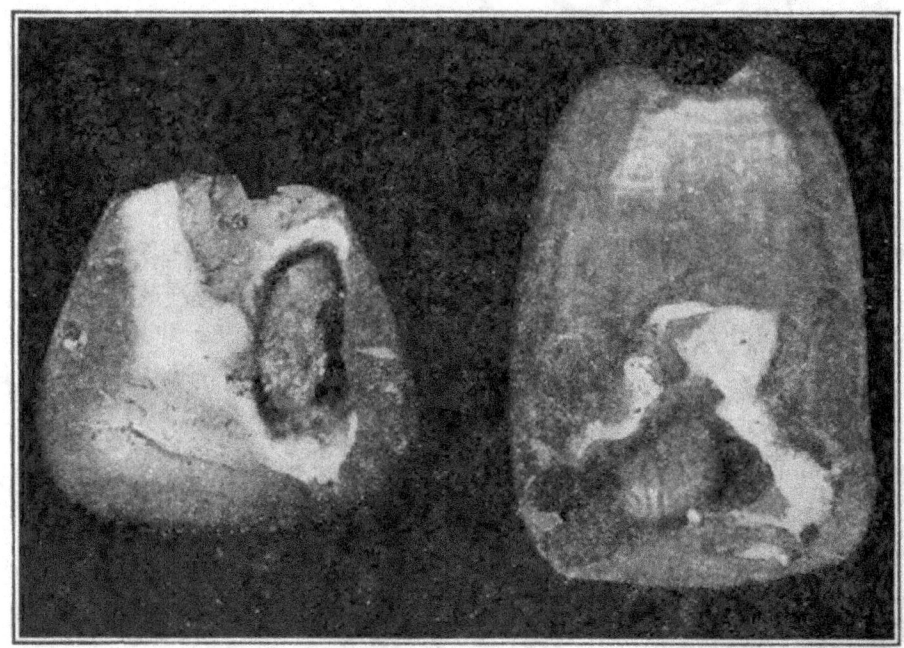

Fig. 144. Rice Weevil grubs in grains of corn.

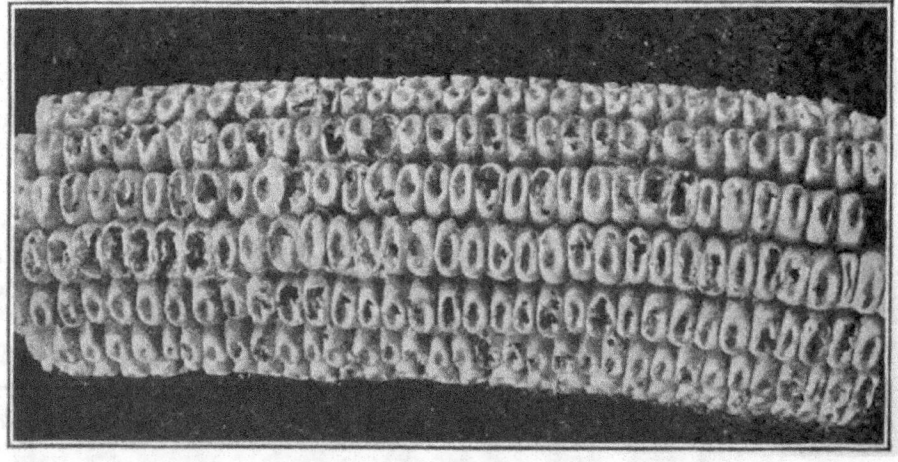

Fig. 145. Ear of corn injured by Rice Weevils.

Fig. 146. Cadella.

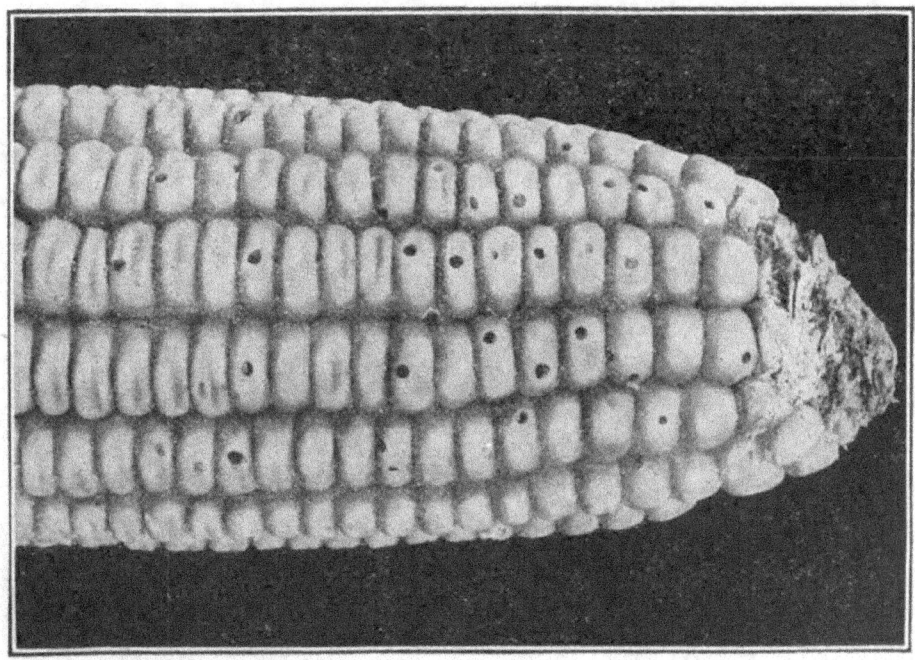

Fig. 147. Ear of corn injured by Angumois Grain Moth.

Fig. 149. Indian Meal Moth.

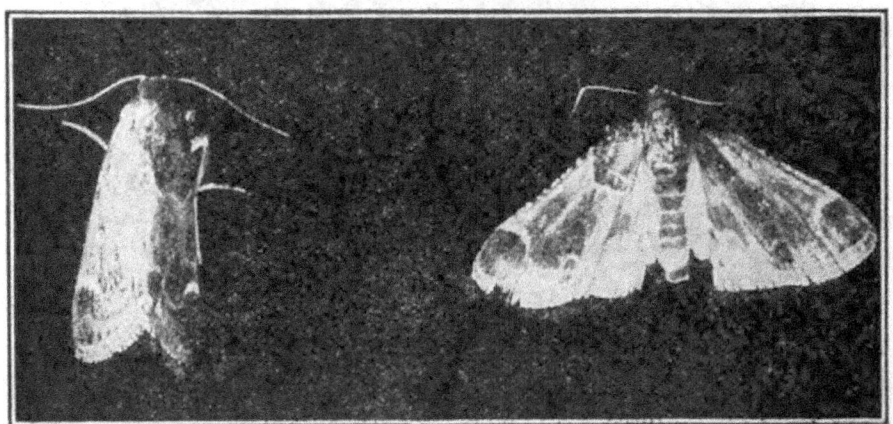

Fig. 150. Meal Snout Moth.

Fig. 151. Grub cocoon and pupa of Meal Snout Moth.

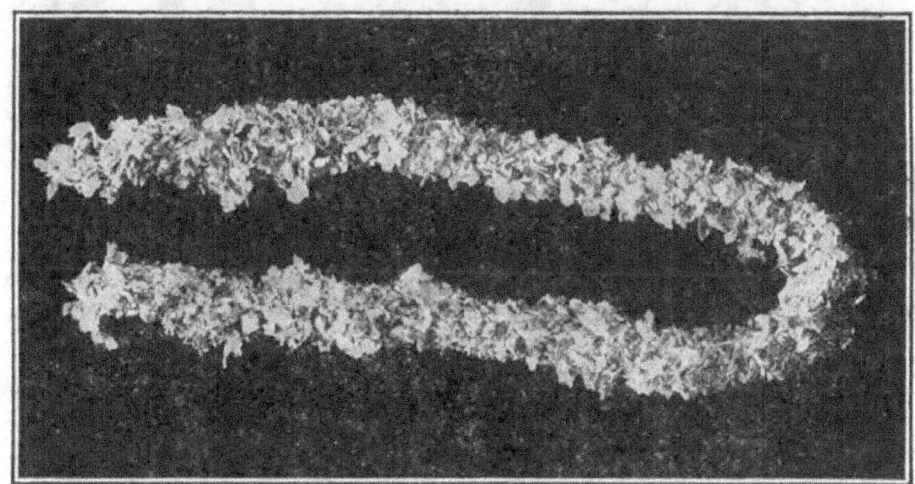

Fig. 152. Feeding tube of Meal Snout Moth.

Fig. 153. Cocoons of Meal Snout Moth.

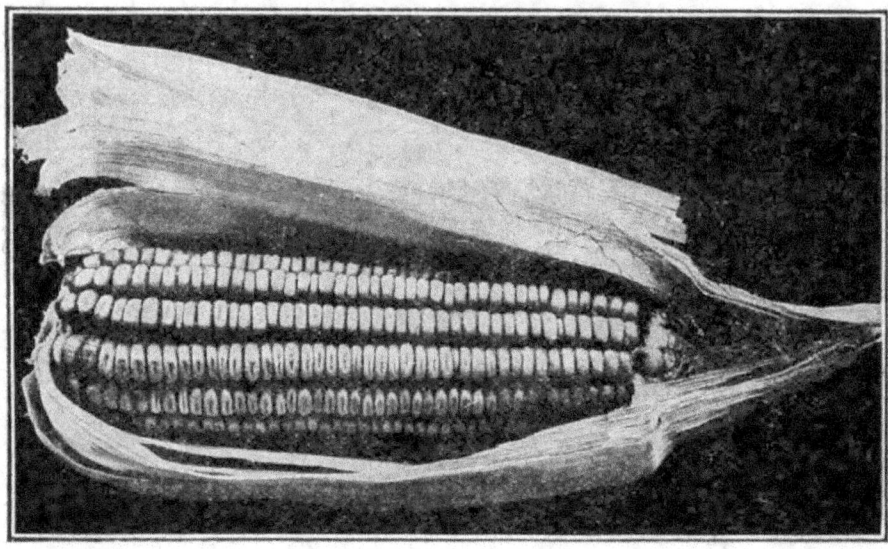

Fig. 154. Showing how an ear of corn with a tight husk is proteced from Grain Weevils.

by 1000 and then multiply the quotient by eight. In other words, divide the number of cubic feet in a bin by 1000 and multiply the result by eight.

The required amount of carbon bisulphide should be poured into an ordinary sprinkling pot and sprinkled over the top of the grain, which is then covered with sacks and left for two days, after which time the bins may be opened and aired.

While carbon bisulphide has a very strong odor, it will not injure the grain for any purpose, as the odor soon disappears. It must be remembered that it is more explosive than gasoline and all lights, fires or matches should be kept away from the bin as the danger of an explosion is very great. If the bin is not as tight as noted above, due allowance should be made. Instead of multiplying by eight, it may be necessary to multiply by ten, twenty or fifty if the bin is very loose.

CHAPTER XXXVI.

HOUSEHOLD INSECTS.

Not only are field crops badly injured by insects, but many kinds come into our houses and are very troublesome. Roughly speaking, these insects may be divided into four classes. (1) Those that are simply annoying by their presence; (2) those that are destructive to food, clothing, carpets, furniture, etc.; (3) those that are annoying because they bite or sting; (4) those that carry diseases.

The first three of these groups are discussed in this chapter, the last, because of their very great importance, are discussed in the next chapter. The actual loss occasioned by these household insects is frequently very great, but of vastly more importance is the annoyance that they cause.

Annoying Insects.

This group contains a few insects and insect-like animals that come into our houses frequently and while the actual damage that they do amounts to nothing, their presence is often very annoying. Chief among these offenders are the so-called blow flies. These insects live in decaying matter, especially decaying meats; however, they frequently lay their eggs on meat that is fit for human food. There are two main kinds of blow flies commonly called "blue bottles" and the smaller greenish flies called "green bottles". They both come into houses seeking food and places to lay their eggs. They also seek to get into cool places out of the sun, for this reason they frequently come into basements and cellars. Because of their large size and their disagreeable buzzing, these flies are frequently more annoying to housekeepers than the smaller but vastly more dangerous house flies.

In houses they may easily be killed by means of the ordinary fly swatters. They will usually dodge into houses in spite of screen doors. It is usually better to seek out their breeding places and bury all carcasses and meat scraps. Of course, every precaution should be taken to keep human food free from these flies. This can be best accomplished by covering individual dishes with screened boxes or placing them in tight dishes. The habit this fly has of entering dark closets, pantries and cellars make it more difficult to exclude than the ordinary house fly.

Clover Mite.—The so-called clover mite is often troublesome on clover, fruit trees and some other plants, feeding on the leaves and increasing to an enormous extent from such plants. If these plants are close to the house the mites are apt to spread to the house in the fall, seeking a place to pass the winter. At such times they are apt to prove very annoying indeed as they crawl about over the walls, floors and furniture. They do no damage whatever, simply are troublesome because they can crawl everywhere and since they are so small they can enter at every crack and crevice.

If they are discovered on the outside of the house before they enter, they may be sprayed with pure kerosene which will kill all it touches. On the inside of the house, gasoline is perhaps best, although the fire danger should be born in mind. Fresh pyrethum if carefully sprinkled about window sills will kill all it touches and discourage others from entering. Ordinary strong soap suds may be used to wipe up the floor and wood work as it will also kill them.

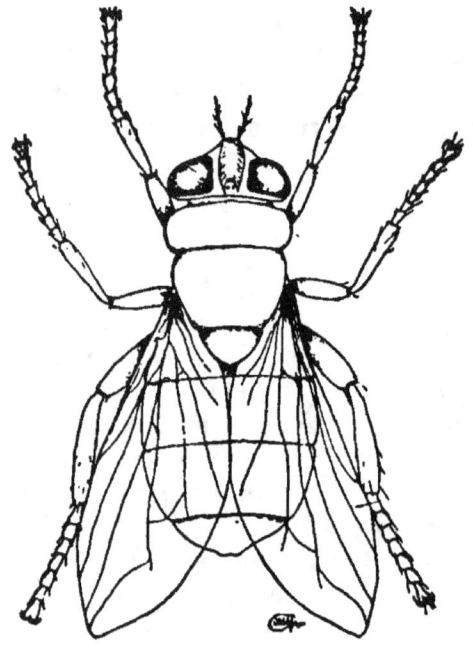

Fig. 155. Blow Fly.

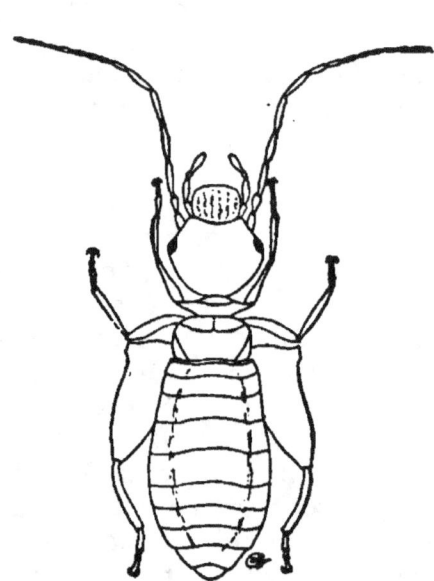

Fig. 156. Book Louse.

Book Lice.—Every one has observed in books the very minute insects known as book lice. They live on dust and ordinarily do not become common enough to be troublesome. But sometimes, especially when a house has been vacant for a considerable period and dust accumulates, these pests are apt to multiply to an unheard of extent and be very troublesome. The author has seen houses, especially new houses, simply swarming with these minute pests so that wherever a rug was laid on the floor, a book on a table, a hair brush on a dresser, these pests would gather under this protection in immense numbers. Of course under such circumstances they are very annoying.

When book lice have increased to the extent that has been mentioned above, it is a considerable task to free a house of them. Persistent effort in the use of gasoline or of a suds of strong laundry soap will eventually win, however. Most of the insect powders seem to be of little use in such extensive invasions, however, at other times they would certainly be very discouraging to these little pests.

Destroying Food, Clothing, Furniture, and Other Household Stores.

The insects in this group may be divided into two main divisions, those that injure food and those that injure clothing, furniture and other household stores. The insects that injure food are especially apt to prove troublesome on the farm, where food is usually purchased in large amounts and kept for a considerable length of time. In cities and towns it is a constant fight to keep such insects out of pantries, because a fresh supply is apt to be brought in each day in groceries from the store. Most of these pests are small, and when once established in a pantry it is very difficult to get rid of them because they are adapted to living in cracks and crevices and can live for long periods of time without food if necessary. Perhaps the best way to control these is

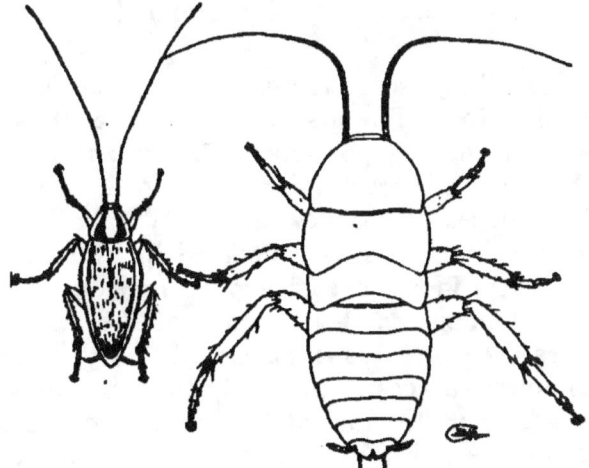

Fig. 157. Roaches (a) German Roach or Croton Bug (b) Oriental Roach.

to keep all food covered tightly and to store all food that is to be kept for a considerable time in glass fruit jars, so that they may be sealed very tight, as most of these pests can work their way through very small cracks and crevices and into tight tin cans and other ordinary tight receptacles.

Roaches.—Roaches have long been known as house pests. In cities and towns that have water works and with the gradual introduction of water work systems in country houses, roaches or cockroaches have spread to them. They seem to require for their best development warm, moist situations, hence the connection there seems to be between the roaches and water systems. In all we have four different kinds of cockroaches that are common in houses in this country. These have been brought to us mostly from tropical countries. Besides these four, we have several native cockroaches which live under the bark of dead trees. These seldom come into houses except those built in the woods or they may be brought into houses with fire wood, however, even then they do not seem to thrive in houses. Among the four kinds of roaches that generally occur in houses there are two classes, the small, yellowish brown roach called the Croton bug or German roach, and the larger, darker forms usually called "black bugs." However, all roaches have about the same habits. They feed at night and during the daytime they hide away in cracks and crevices. Frequently the housewife is not aware of their presence save for the characteristic buggy odor that they give to everything that they come in contact with. Occasionally a luckless individual is caught in a sink or other smooth receptacle and the housewife learns the connection between the buggy odor and the roach. Roaches will eat almost anything that is soft enough for them to chew. They are especially fond of flour and other "starchy foods," also of chocolate.

They will frequently damage books very extensively, apparently to get at the paste used in the bindings. However, their greatest damage is not the amount they eat, but the amount they render unfit for food by leaving behind the strong odor.

Roaches are essentially scavengers, and any method of control must take this into consideration. The first step in their control is to put all food beyond their reach as much as possible. This means tight bread and cake cans, tight flour bins and receptacles for other food. However, it is not an easy matter to free houses of roaches once they have gained entrance, and whatever remedies are used must be persisted in. Often housekeepers will stop just short of success in the use of a given remedy. Either they will become discouraged at the apparent hopelessness of the task, or they will think that they have succeeded, when as a matter of fact considerable more work is necessary for complete success. Personally the writer has had most success with a bait composed of milk chocolate and powdered borax. The two are thoroughly mixed together, preferably in a druggist's mortar, only enough of the borax being added to give the chocolate a trace of bitter taste. This mixture is then made into little pills and scattered about the shelves in the pantry or wherever the roaches are apt to be. The great virtue in the milk chocolate is that it can be made into small pills which is much more cleanly than a powder. Along with the use of this remedy should go a thorough cleaning up of all table scraps and the placing of all food in tight containers. I have seen houses which were badly infested with croton bugs entirely freed in three weeks by the persistent use of this plan. Powdered borax may be scattered on shelves, in cracks and crevices, about water pipes and in pantries, as it acts as a very effective repellant against roaches. Pyrethum powder may be used in the same way although it is usually not as effective as the borax.

Some report failures from the use of these remedies, however, the writer believes that it is because they have not been used long enough, or because there has been a fresh invasion of the house from outside sources. In these cases various traps have been devised and traps for this purpose are now on the market. All that is required is some sort of receptacle that the roaches can enter freely, but that has smooth sides so that they cannot get out again. An ordinary glass fruit jar without a lid set in a shelf so that the top is level with the top of the shelf and then baited with bread, pie crust or cake is effective. Or a paper cone may be fastened in the side of an ordinary paste board box so that the roaches can enter freely but not be able to get out.

Ants.—Next to cockroaches, ants are perhaps the most troublesome insects in pantries. Here again it is not the amount of food they eat so much as the fact that they are everlastingly getting into food, which doesn't add to its attractiveness.

As a general rule there are three kinds of ants common in houses throughout the United States. The large, black ants, the medium-sized, brown or black ants and the small-sized red ants. All these forms are colonial in their habits, that is, they all live in larger or smaller nests. These nests may be in the wall in the basement or out of doors in the lawn. The first thing to determine, in attempting to control any kind of ants, is to find out if possible where they have their nest or nests and then treat these with gasoline or carbon bisulphide, being of course careful to keep lights and fires away for a period of several hours. The writer has seen a bad infestion of the large black ant stopped by simply following the ants from their food back to the hole where they were entering the house, and then stopping this with cotton soaked in gasoline. The cotton should be completely saturated a couple of times a day for a week.

Of all the ants that are commonly found in houses, the little red ant seems to be the hardest to conrol. They breed in small colonies in the walls of houses and out of doors and keep entering the house in apparently increasing numbers in spite of all one can do. In this case food can be stored on tables whose legs are set in shallow pans of glycerine. Water will not be effective for it is apt to dry up or have a scum formed on the surface, over which the ants can cross. Kerosene is effective for a time but it soon evaporates. Glycerine is the most effective as it does not evaporate and a scum cannot readily form on its surface.

To destroy and drive away the medium-sized brown ants and the black ants, the writer has found that a pinch of tartar emetic, say what will lay on the tip of the small blade of a pocket knife, mixed with a saucer of molasses and placed where the ants can get at it will poison all that eat it and cause the others to leave.

Fruit Flies.—Wherever ripening fruit is exposed in the summer, it is attacked by fruit flies, which go under a variety of names such as "sour gnats" and "vinegar flies." These names refer to the fact that fruit soon turns sour when attacked by these insects. The flies lay their eggs in decaying fruit and the maggots soon begin their work.

These maggots attack ripe fruit and hasten the process of decay, hence they are often very destructive. Nothing much can be done for their control save keeping them out of ripe fruits. This is often hard to do as the flies can pass through the ordinary fly screen, and the maggots can crawl through very small cracks to get at such fruit. A great deal can be done by not leaving fruit, either fresh or canned, exposed any more than necessary.

Weevils.—There are a variety of small weevils or beetles that attack meal, flour, and cereals of all kinds. Perhaps they do not destroy as much food as they render unpalatable by their presence. These fellows come from

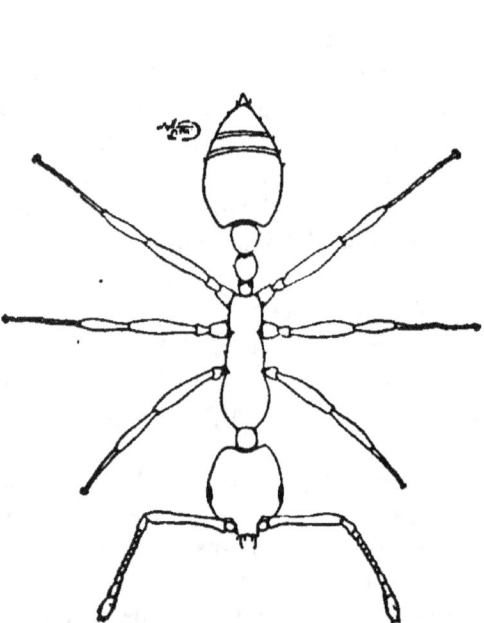

Fig. 158.   The small Red House Ant.

Fig. 159.   Fruit Flies, maggots and adults on banana.

Fig. 160.   The Saw Toothed Grain Weevil.

mills or groceries to pantries and once they gain entrance they are very hard to dislodge. All the infested meal or flour should be used up, either as human food, if not badly infested, or, if badly infested, it should be fed to stock. All bins where meal and flour is stored should be scalded, or, if the pantry is badly infested, it should be fumigated by burning sulphur at the rate of two **pounds** for each 1000 cubic feet of space in the building. The sulphur should be put in a pan which is placed in a large pan containing sand so that the floor will not be burned. The sulphur is then moistened with alcohol or kerosene and lighted. The room should be closed as tightly as possible by pasting wet newspaper over the cracks around the doors and windows and everything made of metal should be removed from the room before beginning fumigation.

Meal Moths.—Besides the weevils mentioned above, meal, flour and cereals in the pantry are attacked by a number of worms, grubs or beetles, which for convenience are called collectively meal moths.

They are very destructive in pantries and bins and their control is brought about in the same way as the control of the weevils.

Larder Beetle and Ham Beetle.—These beetles are abundant everywhere and their hairy grubs frequently attack ham, bacon and other cured meats, boring principally into the fat and rendering it in part, at least, unfit for human food. Fortunately, however, it is usually only the outer layers that are attacked and these parts may be cut away, the rest being used for food. To protect hams from the attacks of this pest they should be wrapped in a tight wrapper as soon as they are cured and smoked, so as to leave as little chance as possible for the pest to do its work.

Skippers.—Skippers are pretty familiar objects to housewives as they are commonly found in cheese and ham. They are the maggot stage of a small fly and are called skippers from their ability to leap considerable distances by bringing the two ends of the body together and then by suddenly releasing the spring. The maggots are thrown a few inches into the air. These skippers are fond of ripe cheese and of smoked meats and seldom bother anything else. Every precaution should be taken to keep them out of pantries, for once they have gained an entrance they are hard to control if there is any ham or cheese about.

Insects Destroying Clothes and Carpets.

Any kind of cloth containing wool or silk, also furs and feathers, are often seriously damaged by the following insects:

Clothes Moths.—There are two kinds of clothes moths in this country that commonly injure woolens, silks, furs and feathers. Aside from the fact that one kind of worm makes a case in which it lives and the other does not, there is really very little difference between these two kinds. Both are capable of doing very great injury to clothes that are stored away during the summer time.

Clothing that is worn constantly is not injured. It is the woolens and furs that are not much used in the summer time, when these insects are most troublesome, that are most likely to be damaged by them.

The control of these pests require constant watchfulness. Unless clothes can be stored in very tight boxes or bags, they should be taken out every two weeks, hung out in the sun and given a thorough beating. This will knock off the eggs and worms and prevent any very great injury. If clothes are stored early in tight boxes or paper bags made especially for this purpose, they will be free from injury providing no eggs have been laid before the clothes are stored away. If clothes or bed clothes are stored in chests or trunks they should be fumigated with carbon bisulphide a couple times during the summer. Ten tablespoonfuls of this carbon bisulphide will be sufficient for the average trunk if it is fairly tight. This material should be poured out into a saucer, set on top of the clothes and the lid closed. Gasoline may be used in the same way, but it must be remembered that both liquids are very dangerous explosives.

Moth balls, cedar shavings and cedar chests will keep moths from laying eggs if properly used, however, it must be remembered that they do not prevent the worms from working after the eggs have been laid and hatched.

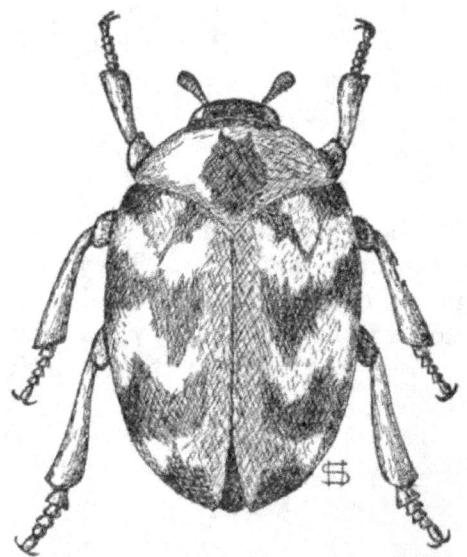

Fig. 162. Buffalo Beetle, enlarged.

Buffalo Beetle.—We have in this country two beetles whose worms are destructive to carpets, the Buffalo beetle and the black carpet beetle. It is in the worm stage alone that they do injury. The eggs that are laid by the adult beetle in the early summer among the fibers of the carpet, usually along the edges. In about two weeks' time the eggs hatch into larvae. The "Buffalo moths" are covered with long brown hair all over, while the carpet moth has a covering of short hairs.

These worms eat voraciously from the time of hatching, sometimes eating along a single fiber and so cutting the carpet, other times eating holes an inch in diameter.

The damage is always more serious in houses where the floors are closely covered with carpets. Polished floors with rugs do not furnish a suitable abode for this pest and hence are to be preferred to carpets. Where carpets are used, it will require eternal vigilance to keep these pests in control. The carpets should be taken up twice a year, thoroughly dusted and then sprinkled with benzine or gasoline. All cracks in the floor, after it has been thoroughly scrubbed, should be filled with "crack filler," especially those near the base board. Spread papers closely over the floor under the carpets and tack the carpets loosely at the edge so that they may be examined occasionally.

The larvae may be trapped by placing red woolen cloths in the closets, as they collect in these they may be destroyed from time to time.

Biting and Stinging Insects.—Much annoyance is caused about the house by certain pests which bite and sting. These insects are almost universally present and do much damage. Anything that can be done to lessen the annoyance they cause is just that much towards making life more enjoyable.

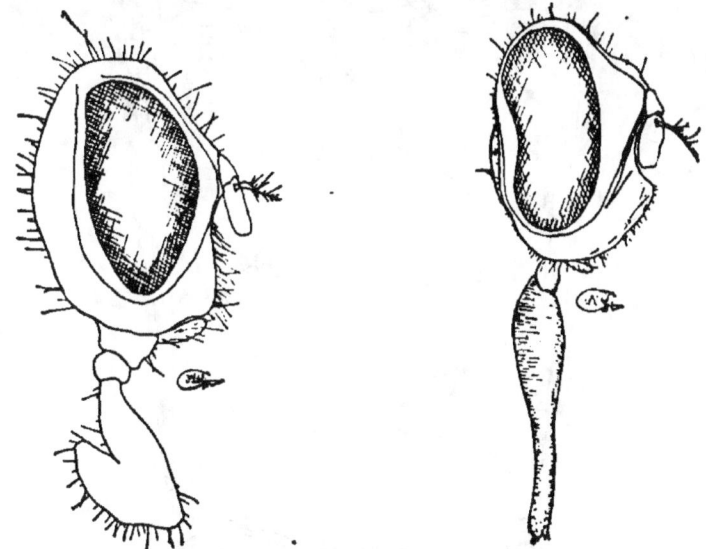

Fig. 163. (a) Side view of the head of House Fly (biting house fly) enlarged; (b) side view of the head of Stable Fly.

Stable Fly.—This fly is commonly known as the "biting house fly" or "storm fly," from its habit of taking refuge in the house before a storm and biting the ankles of the inmates. It is larger than the common house fly and has an eel-like probosis with which it bites or stings its victims. It breeds in manure and its natural habitat is out of doors, but it also frequents stables, houses, etc. Its bite is painful and irritating but not poisonous.

The methods of control are the same as those recommended for the house fly (see page 339).

Common Mosquito.—Throughout the United States the common mosquito is very abundant. It does not carry disease, but is most annoying. It breeds anywhere where water stands long enough for its eggs to hatch, little slow-flowing streams, pools, sewers, rain barrels, tin cans, etc.

The eggs are laid on the surface of the water in bunches, often several bunches will drift together forming a raft. They are brownish in color and plainly visible to the naked eye. The eggs stand on end with the larger end down and after a time, varying from twenty-four hours to several days, depending on the temperature, the larvae or "wigglers" burst from the lower end.   The larvae rest, for the greater part of the time, with the tip of the abdomen at the surface of the water, and head down.   At the tip of the abdomen is a tube known as the breathing tube, and through this tube the

Fig. 164.   Common Mosquito, eggs, Wigglers, tumblers, adults.

wigglers breath.   After a week or ten days these wigglers change to the tumblers, which seem to be as active as the wigglers but now rest with head up and tail down, breathing through two breathing tubes placed near the head.   After a period of five or six days the skin of the tumbler splits open and the adult mosquito emerges.

The adult is a medium-sized mosquito.   In a biting position its body is parallel to the surface on which it rests and its bill perpendicular to that surface.   Only the females bite, the males sipping nectar from flowers and sap from leaves.

Since about three-fourth of the life of a mosquito is spent on and in the water in the egg, wiggler and tumbler stages, our most effective means of destruction must be applied in these water stages.

Drainage.—Since mosquitoes breed in water, if we drain swamps, empty tin cans, cover our rain barrels with mosquito bar or screening, etc., we will have fewer mosquitoes.

Pools.—Pools and lakes which we cannot or do not care to drain should be filled with fish—gold fish, top minnows, and sun fish being good varieties. They are small and can go out into shallow water to feed on the eggs and larvae of the mosquitoes.

Oil.—Streams and bodies of water not used for watering stock nor for pleasure purposes may be coated with a film of coal oil by means of a knapsack spray pump. This film of oil will spread entirely over the body of water and so prevent the larvae and pupa from obtaining air, also destroying the eggs floating on the surface of the water.

Even though we use every precaution, and every remedy, there will still remain a few mosquitoes to annoy us. However, we can escape from these by the careful screening of every outside door and window and the use of bed nets at night. These nets should be made large enough to come to the floor on all four sides of the bed, and if care is taken that no mosquito enters with the sleeper, none can enter afterward

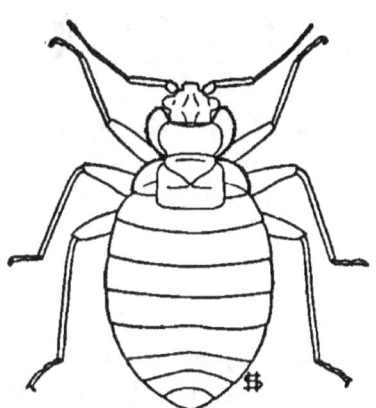

Fig. 166. The Bedbug.

The Common Bedbug.—The bedbug is one of man's oldest and most faithful companions. It belongs to the same group of insects as does the stink bugs and squash bugs. This accounts for its disagreeable odor. It is smaller than the other members of the group, flat and wide, fitted especially for hiding in the crevices of the bedstead. The eggs are oval, white in color and are laid in batches in cracks and crevices. They hatch in six to ten days and the active young are very light in color at first, but soon begin to gorge themselves with blood and take on a dark red or purplish color. Their mouth parts are fitted for lacerating and sucking. The underlip is greatly lengthened and the edges roll up almost meeting above, thus forming a tube. Within this tube are four thread-like organs, which slide upon each other and lacerate the flesh of their unfortunate host, they then draw the blood through the tube.

The first means of control we would suggest for the house infested with bedbugs is the substitution of brass or iron beds for the old-fashioned wooden ones, as the hiding places afforded by the metal beds are very scanty and easily reached by the housekeeper. The old-fashioned remedies of kerosene, gasoline, or benzine, applied with a feather or hand syringe, are most effective. They should be thoroughly applied four or five times with intervals of three or four days between applications, thus giving time for all eggs to hatch and be destroyed. Another effective remedy is corrosive sublimate one ounce and alcohol one pint, painted in all cracks.

CHAPTER XXXVII.

HOW INSECTS AFFECT THE HEALTH ON THE FARM.

It is only recently that we have begun to realize the important relation between insects and diseases. Even now this relation is not suspected by everyone and is really only understood by a few. What we have to say here concerns only the most important insect carriers of disease and there is no place that these insects are more dangerous than they are on the farm.

Flies—Typhoid and Other Intestinal Diseases.

House flies have long been familiar and troublesome pests about the house, but it is only within the last few years that we have realized the relation between these pests and certain diseases such as typhoid and summer complaint of children. In fact this relation is so close that it has been suggested we should call this fly the "Typhoid Fly" instead of the house fly.

This insect is so familiar that it hardly needs description. There are one or two points to which our attention should be directed. In the first place this fly cannot bite or pierce the skin. Its mouth parts are soft and fitted for lapping up liquids, or for reducing such substances as sugar, which can be easily dissolved, and then lapping up the liquid. The fly that looks like a house fly and can bite, is the stable fly (see page 335).

In the second place we should notice that the fly is covered with hair, especially its feet. The life of the adult fly is a short but busy one. They are busy laying eggs and in visiting dining rooms and kitchens for food. When we remember what a short journey it is, on the average farm, from the stable, privy, and pig pen to the kitchen and dining room, and then remember how hairy the body of the fly is, we can see at once what an excellent carrier of filth the fly is. Along with this we must remember that people suffering from typhoid, for instance, discharge typhoid germs a long time before they show any symptoms of the disease, and that in some cases they continue to discharge germs for a long time, even years after they are apparently well of the disease. Then stop to think how dirty flies are, how they go to the filthiest of places to lay their eggs and then come directly to the kitchen or dining room to crawl over our food, or, as has been suggested, to wipe their feet on our food. We see at once not only how very unclean the fly is, but how dangerous. Typhoid may be carried by other means than by the fly so that perhaps we will never know exactly how important the fly is in this respect.

In addition to carrying typhoid, the fly also carries the germs of the disease called summer complaint of children. That the relation between these two is very important may be shown in a number of ways but perhaps the most important clew is found in the close relation between the number of flies and the number of cases of summer complaint.

Mothers have long known that the summer time is hard on babies, and this has been blamed on various things; however, let us lay the blame where it belongs chiefly and give the fly full credit. From slop barrel, unsanitary privy or what not, to baby's face or baby's food is an all too short journey. When we remember that one good, strong, able-bodied fly is able to carry millions of germs on his feet we can readily see how dangerous he is and we will cease to wonder why it is that so many babies are sick in the summer time (fly time). We will cease to say at babies' funerals "The Lord's will be done," but we will say with all due reverence, "This is perhaps the work of the cursed fly and I have been guilty of not doing my part to suppress him."

The Life History of the Fly.

In the light of these facts, the life history of the fly becomes very important, because it is only by studying its life history that we can fight this pest with any assurance of success.

Flies lay their eggs in all sorts of filth, but principally in stable manure. In a few hours, small white maggots hatch from these eggs and feed on the manure, growing so rapidly that they are full grown in a few days. The length of time that flies spend in this stage is very important for any proper system of control of the house fly. This time will vary from as few as three days in the summer to at least two weeks in the spring or fall. After the maggots have become full grown, they change into little brown capsule-shaped objects, inside of which the fly is formed and from which it later emerges. This stage is also about as long as the maggot stage, so that we can have a new brood of flies every week. When we recall that every female fly lays about six hundred eggs, we cease to wonder why there are so many flies by the end of the summer and begin to wonder why there are no more.

Spring is the season of the year in which to lay our plans for war on the house fly. Let no one suppose, however, that he is going to exterminate the house fly by a few simple sanitary measures. It is a fight to the finish, and the finish is not yet in sight. Swat the fly, if you will, especially early in the spring. Much peace, comfort, satisfaction and safety may be had all summer long by swatting to a finish, once or twice a day, those few persistent flies that dodge into the house through the open screen doors. Remember always that the house fly is the best index to the sanitary conditions about our homes. Lots of flies, plenty of filth, and vice versa. Don't think that you deceive your neighbors, either. They know that many flies in the kitchen and dining room indicate poor housekeeping and much filth in nearby trash heaps.

Don't think that you can treat these same trash heaps with some magical chemical and render them harmless. So far it hasn't been done, or if it has been done the folks that did it have kept mighty quiet about it. There has been much talk lately about spraying the manure heap with paris green and killing all the little baby flies while they are "aborning". Don't do it. The flies tell us when we have filthy premises, just as the thermometer tells us whether it is hot or cold, or just as the barometer indicates difference in air pressure. Don't disturb the workings of this sanitary barometer. It is just like doping up with perfume when, according to all good reason, you should take a bath. Don't treat the symptoms. Treat the disease. Don't try to disinfect your filth. Clean it up and then stay clean. Clean the stables regu-

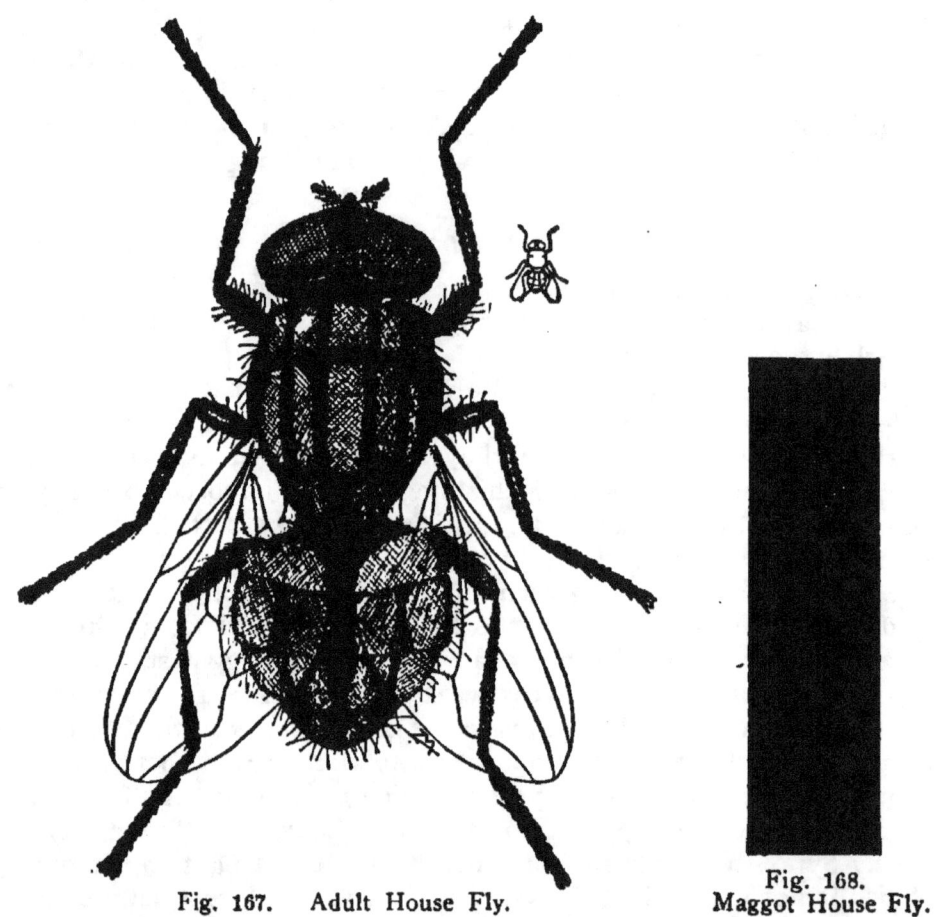

Fig. 167.   Adult House Fly.

Fig. 168.
Maggot House Fly.

larly and often. Clean the hog pen and the chicken coop. Have such a privy that fecal matter will not be exposed to flies. Get rid of the garbage, the swill and the kitchen slops. Keep all your food under cover and insist that your grocer, butcher and market man do the same. In fact, make everything so absolutely clean about the place that the filth-loving house flies will have to pass by on the other side.

Paris green or some other dope might kill a few fledging flies in the manure heap, but careful experiments on a small scale failed to kill any fly larvae in our breeding cages when the manure was soaked with Paris green at the rate of two pounds to fifty gallons of water. Since these experiments were made, other careful workers report that they have been able to kill some

fly larvae with Paris green and other substances, but spraying doesn't kill
all the fly larvae; and, even if it did I am afraid that it would make some
people even more careless about the sanitary conditions of their premises.
Stick to cleanliness as the first and most important remedy. Don't be discouraged if all the flies do not leave the first day after you have cleaned up,
for it is a fight to the finish.

Meet the fly more than half way. Go into his breeding quarters and
clean up and remove his filthy breeding material before he is fully matured.
Clean the stables twice each week, at least. Do not throw the manure in a
heap outside the stable door, but have it spread thinly on a field some distance
from the house if possible. This will cause the maggots or embryo flies to
dry up and die. Do not think that once a week, or once in two weeks, is

Fig. 169. A box stall, a good place for flies to breed.

often enough to clean your stables, for by that time the maggots have changed
into pupae, little brown, capsule-like objects, with a hard outer covering, inside
of which the larvae changes to adult flies. This outer covering is very resistant to drying. In fact, before the maggots are ready to make this change
they usually crawl out of the manure heap to some dry place. Hence it
follows that scattering the manure on the fields will not kill the fledgling flies
after they have changed to pupae.

About the house and stable one of the best remedies for the adult fly is
formalin (formaldehyde) one tablespoonful to one-half pint of equal parts
milk and water. This mixture should be exposed in shallow saucers or plates
with islands of bread for safe landing places for the flies. Some of the prepared poisons are perhaps as good, but they are no better, and they are more

expensive than the formalin and milk. Pin your faith to formalin and milk as a fly poison in the stables, in the milk house, in the hog pen, and about the house. Formalin and milk, however, will not give results either inside the house or out, if there is water exposed that the flies can drink. A long series of observations have shown us that flies will drink formalin and milk if they are driven to it, but they will drink water in preference every time. This explains why some people have never successfully poisoned flies with formalin and milk.

Keep the flies out of the house by having every door and window screened. If you cannot afford galvanized or painted screening, mosquito netting is cheap and within the reach of all. Don't expose food where the fly can get at it. A fly's feet are never clean. Just pause and think of all the places where you have seen the flies walking about, and then decide that you do not want the manure or typhoid-fly using your milk for a foot bath, your bread for a door mat, or your butter for a skating rink.

Remember that it is a fight to a finish. These things you can do. Clean the stables. Dispose of all the rubbish and trash about the premises. Exclude the flies from privy matter. Keep the garbage and swill in water-tight cans or buckets with fly-tight covers. Screen the house and milk house. Buy your meat and groceries only from screened markets. Use formalin and milk in the stables, about the milk house, and on the porches. Then swat every single fly that sticks his head into the kitchen or dining room.

Mosquitoes and Malaria.

Mosquitoes have long been known as troublesome pests of man, as they frequently occur in very large numbers and their bites are very painful. This phase of the subject is dealt with in connection with household insects (see page 335). However, there is another phase of the problem that is really more important than this annoyance, and that is the diseases carried by mosquitoes. Throughout the country generally, and especially in the South, one of the most troublesome diseases on the farm is malarial fever, chills and fever or ague as it is often called. It is only comparatively recent that we have known the real cause of this disease. Formerly it was supposed that one could acquire the disease by breathing night air, by being in a swamp or any one of a half hundred other different ways, however, recently it has been shown that this is the work of one kind of troublesome mosquitoes. Only certain kinds of mosquitoes can carry malaria and they can be recognized very easily by the fact that they have black and white spotted wings, and by the fact that when they bite they seem to stand on their heads. The life history of these mosquitoes is very nearly like that of the common mosquitoes, in that they breed in water.

There are some important differences which it is not necessary for us to point out here. Suffice it to say that almost any stagnant water may become a breeding place for these or the common mosquitoes.

So far as is known, malaria can only be transmitted from one person to another by the bite of these spotted winged or malarial mosquitoes. Such a mosquito must first bite a person that has malaria and then in a few days if it bites a person free from malaria, the second person is very apt to acquire the disease. Obviously from these facts it is necessary to have a chain of events before anyone can acquire malaria. First, we must have a person suffering from malaria; second, malaria fever mosquitoes; and third, these mosquitoes after biting the patient must bite a person not having the disease.

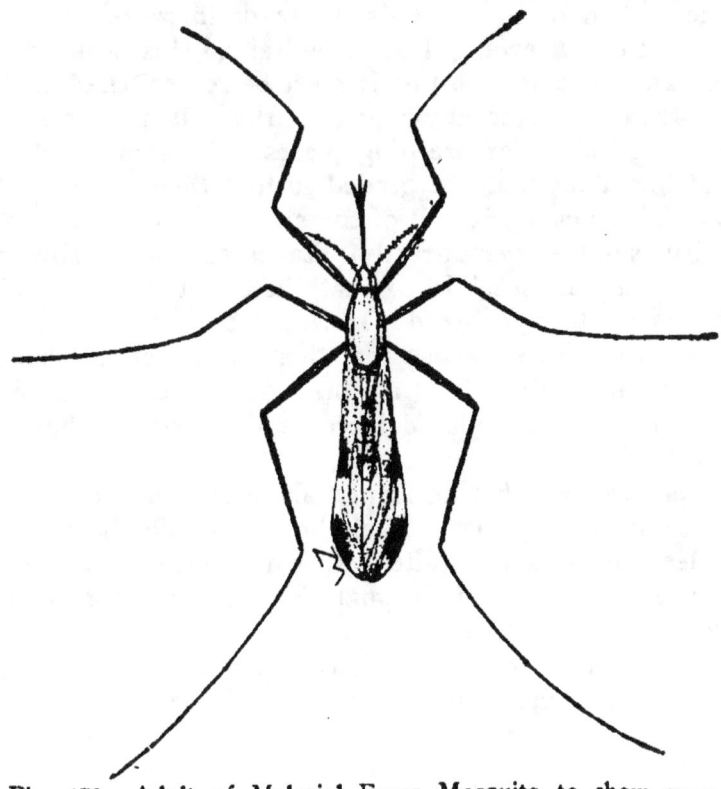

Fig. 170. Adult of Malarial Fever Mosquito to show spotted wings.

It is clear then that our line of attack on this disease is to break this chain. There are three places that the chain may be broken. 1. Either we can prevent the mosquitoes from biting malarial patients, or (2) we can kill the mosquitoes, or (3) we can prevent persons not suffering from the disease from being bitten.

Under the first of these heads we may treat the patients with some medicine such as quinine which will kill the germs of the disease, or we may screen the patients so that they are not liable to be bitten by mosquitoes. The first of these methods is most effective if applied during the winter and should always be carried on under the charge of a competent physician.

We believe now that the malarial germ lives over winter only in the bodies of persons who have suffered from the disease. If this is the case, very great good can be accomplished in checking this disease by the proper treatment in the winter time.

The second of the remedies should always be followed during the summer time when the mosquitoes are active. Patients who suffer from malaria should not stay out after dusk, but should retire to screened houses, porches, or rooms where they should remain until the following morning. These rooms or porches should be gone over every day to see that no mosquitoes have gained entrance. If any such are found they should be destroyed.

Under the second of these heads is the drainage of our swamp lands. As more and more of our swamp lands are drained this pest will become less and less important. Along with this it must be remembered that these pests can breed in stagnant water anywhere, so that all such places should be avoided or rendered unfit for breeding places. Streams should be made as straight as possible and should be graded so that they can flow as swiftly as possible. The doing away with all of the side pools of a stream will reduce the number of mosquitoes very greatly. Lakes and ponds should have their banks made as steep as possible so that there will be no shallow marshy places for the mosquitoes to breed. It would be better if the banks of all ponds could be lined with stones so that there would be no chance for water weeds to grow up along the edges, as this would greatly reduce the chances of mosquitoes breeding and would increase the chances that the wigglers would be eaten by fish.

Persons who are not suffering from malaria should take every precaution not to be bitten by mosquitoes. Their houses should be well screened or they should sleep under a mosquito net. In addition, they should become familiar with the appearance of the malarial mosquito and should kill it at every possible chance.

People who live in a malarious district or who expect to travel in such a district should protect themselves by taking quinine.

CHAPTER XXXVIII.

INSECT PESTS OF DOMESTIC ANIMALS.

Domestic animals like crops are often seriously troubled by insect pests. These pests cause injury in three different ways: by irritation, by sucking blood and in some forms by carrying diseases from one animal to another. Of these three important factors, we know the least about the last. Just as our knowledge of the human diseases carried by insects, is rather limited so is our knowledge of the diseases of live stock that are carried by pests, even more limited.

Lice.—The lice infesting domestic animals are of two kinds: (1) those that feed on hair, feathers or scales of skin, causing loss chiefly by the irritation they produce, while crawling about among the hair or feathers, and (2) those that suck blood, causing loss in that way.

The first group is largely confined to poultry and does great widespread injury. The average farmer pays but little attention to this loss because it is a constant drain and not readily appreciated. However, when it is noticed that chickens or other poultry are spending a good deal of time in the dust bath, it can be suspected that they are troubled with lice.

The constant irritation cuts down their flesh, causes the production of eggs to fall off very greatly and in other ways causes damage. The excess feed required for a flock of lousy birds if saved, would soon pay for the necessary treatment of the birds, not to mention the increased laying that the hens would do.

One of the most successful remedies for lice on poultry is mercurial ointment. There are a great many louse powders on the market and many of them are very good, but they are all very expensive. Any one can make a good louse powder at home by carefully mixing gasoline, three-fourths of a pint, with crude carbolic acid, one-fourth of a pint, and then gradually sifting into this mixture enough plaster of paris to make the mixture dry, so that it can be sifted through an ordinary baking powder box with a few holes punched in the lid. Care should be used not to add any more plaster than is necessary to make the powder just dry enough to sift. The chickens should be caught and liberally sprinkled under the wings and around the vent. It will be necessary to repeat the dusting, but in the meantime the powder should be stored in a tight glass or tin can.

The lice that are usually troublesome to domestic animals belong to the second group and secure their food by sucking blood. The most important lice of this group are the hog louse, the blue louse of cattle, the horse louse and the dog louse. Although cattle, horses and dogs are also commonly troubled by lice of the first group, they are not usually so serious as the blood

345

sucking lice. These lice are apt to break out in herds at any time, and usually get quite serious before they are noticed. Next to hog cholera, the hog louse is the worst pest of the hog, and the other lice are frequently very troublesome.

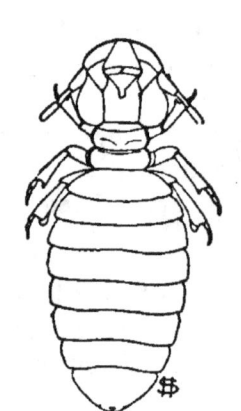

Biting Louse of horse.

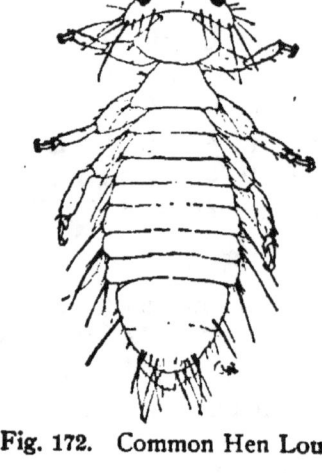

Fig. 172. Common Hen Louse.

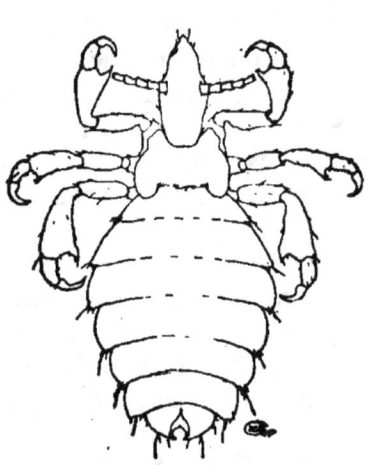

Fig. 173.   Hog Louse

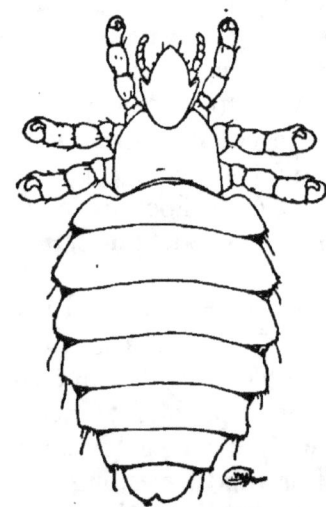

Fig. 174. Short-nosed Ox Louse.

If animals become lousy the first thing to do is to thoroughly clean their pens and stalls. The free use of white wash to which has been added carbolic acid is advised.

The animals themselves should be gone over and if possible those troubled with lice should be separated from those not troubled. Then those that have lice should be treated with some of the tobacco dips that are sold everywhere. In the case of hogs or cows, they may be treated by rubbing lightly with a rag dipped in crude oil, or, if it is done very carefully, kerosene may be used. Animals treated with oils should be kept in the stable or in the shade for several days as otherwise the oil is apt to burn the skin.

As a general rule, neither kerosene nor oil should be used on horses as it is apt to remove the hair. Gray mercurial ointment (obtainable at any drug store) may be used on them if applied only to the parts infested. This mixture should not be used on cattle as it is injurious.

Horseflies.—There are two main kinds of horseflies, the large black flies and the smaller flies with dark bands on their wings, which are often called ear flies, because they delight in attacking the ears of animals.

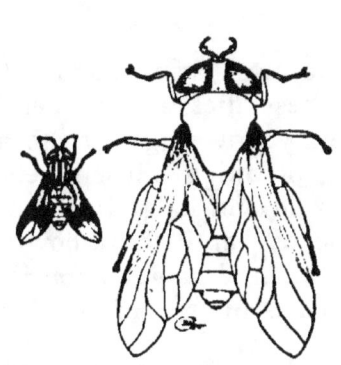

Fig. 175. Horsefly adults (a)
Deer fly (b) Black Horse Fly.

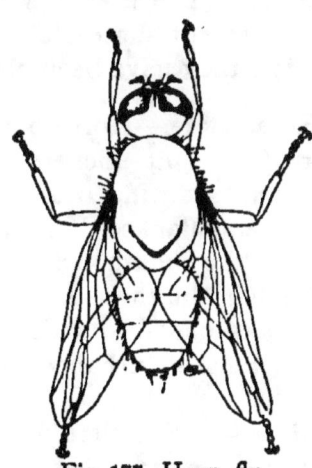

Fig. 177. Horn fly.

Horseflies lay their eggs in swampy places and the maggots are slender white worms that are frequently found in swamps. These are locally known as "sand pipers" and are used by small boys as fish bait.

The bite of the horsefly is very painful and very annoying to horses, however, there is very little that can be done for their control. The use of fly nets will help to keep flies off of horses to a certain extent and the use of ear stalls will prevent the attacks of the "ear flies." All of the washes that have been recommended for flies are unsuccessful because they soon evaporate and lose their power to repel the flies.

Fig. 176. Maggot of Horsefly.

Horn Fly.—This is a small, blood-sucking fly about half as large as the horsefly, which is a serious pest of cattle, clustering on their backs in great numbers in the summer time. In this way they cause much loss of blood and perhaps even a greater loss due to the irritation of their bites. How much this loss is it would be difficult to say, however, the average dairyman says that the flow of milk falls off from one-fourth to one-half during fly time.

This fly breeds entirely in the fresh droppings of the cow, and one of the most effective and simplest ways to reduce their numbers, is to allow a sufficient number of hogs to run after the cattle in the field, to root up the fresh droppings. In this way the droppings will be mixed with dry soil which will make conditions unfavorable for the development of the flies. On large dairies it would be profitable, if no hogs were raised, to do the same work with a rake, going over the lots carefully each day and spreading the fresh droppings. The droppings in the stable should be removed each day and thoroughly mixed with land plaster, as this will make them so dry that the flies will not breed in them. Most of the remedies that are sold to keep flies off of cows are not successful, because they do not retain their strength for any time after they have been applied.

Horse Bots.—Horse bot flies are familiar to most farmers and are known variously as "nit stickers" or "nit flies." These flies are frequently seen about horses during the summer, laying their yellow eggs on the hairs of the legs and flanks. The horse in scratching or biting himself gets these eggs into the mouth, whence they are swallowed and lodge in the stomach. They hatch out there in the form of bots and live until the following year. A few bots usually do little harm, however, when they get numerous they cause serious disturbances of the bowels, which is often confused with ordinary colic.

All our efforts should be directed towards preventing the bots from getting into the stomach of the animal. This may be accomplished by keeping horses in the stables during the day, and allowing them to run on pasture at night. The work animals should be gone over at least every other week, oftener would be better, and the eggs removed by shaving knife or by coating them with kerosene applied lightly with a rag. Care should be used not to get too much kerosene on the skin, and this method should only be used in the evening, so that the effects will wear off before the horse is taken out into the hot sun.

Ox Warble, sometimes called "Ox Bot," "Grub in the Back" and "Hide Grub."—This fly is quite similar to the horse bot, however, its grubs or maggots do not enter the stomach, but bore through the skin of the legs and flanks and work their way up to the back. Here they cause enlargements or "tumors." When the maggots or grubs are full grown, they work their way through the skin and drop to the ground. They then burrow into the soil, where they change to adult flies and emerge to lay eggs.

The loss due to this fly is two-fold. First, there is the direct injury to the flesh and hide of the animal, and second, the loss in flesh or in milk caused by the continual torment of the animal by the flies. The meat of animals attacked by very many grubs, is slimy and not very wholesome. However, perhaps the greatest loss is occasioned to the hide. The holes eaten through the skin of the animal, cause a direct loss, as this part of the hide contains the most valuable leather.

The most practical remedy to use against ox warbles, is to press the nearly full grown grubs out of the back and destroy them. This can easily be done by pressing the skin on all sides of the tumor, pressing each time towards the tumor. If the grub is nearly full grown, it can easily be pressed out. When the grub has been removed in this way, it should be killed, for otherwise it will simply complete its growth and nothing will be gained by pressing it out.

Fleas.—Fleas are troublesome to several kinds of domestic animals, especially dogs and cats. Frequently, because such animals have the run of the house, these pests become exceedingly troublesome in houses. However, this phase of the problem is dealt with in another connection (see page 335).

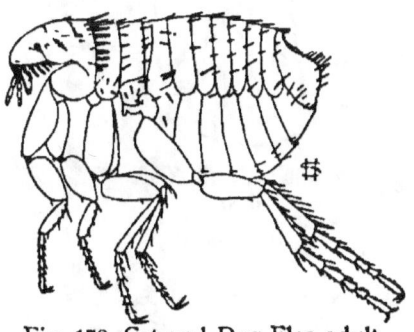

Fig. 178. Cat and Dog Flea adult.

Since these pests breed in the litter, in the sleeping quarters of these animals, one of the first things to do for their control is to thoroughly spray or sprinkle their quarters with kerosene, taking the usual precautions against fire. The animals themselves should be thoroughly washed with a good strong soap, and the fleas picked off and killed; or the animals may be dusted with fresh pyrethum powder. If this is well worked into the fur, it will cause the fleas to drop off and if the animal is held over a paper they can easily be burned. It must be remembered, that the pyrethum powder simply stupifies the fleas, and unless they are destroyed, they will soon recover.

Fig. 179. Cat and Dog Flea Maggot.

Ticks.—There are several different kinds of ticks in the United States that attack animals. Cows, dogs and rabbits are commonly attacked, and these animals suffer much from the loss of blood. However, there is another way that animals suffer through the agency of ticks and that is by the diseases carried by these pests. Perhaps the most noted example in this respect being the cattle tick, which causes a loss of more than $100,000,000 every year. This pest is the carrier of the deadly Texas fever of cattle, and is widespread throughout the Southern states. Gradually it is being brought under control.

Most of the other ticks attacking domestic animals in this country, are not responsible for the spread of diseases, but when they occur in large numbers, they cause much loss by sucking the blood. On dogs and other animals they may easily be removed by covering with linseed oil or kerosene. If an attempt is made to pull them off without covering with oil the mouth of the fleas is apt to remain in the animal and cause much unnecessary suffering.

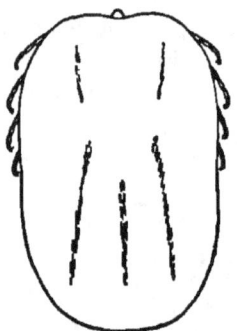

Fig. 180. Cattle Tick adult.

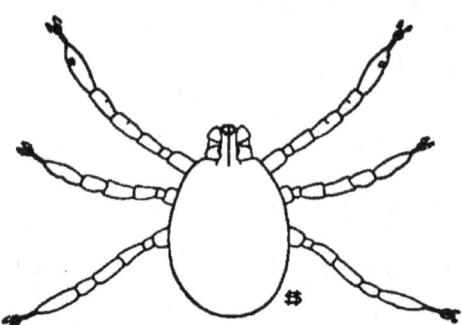

Fig. 181. Cattle Tick young.

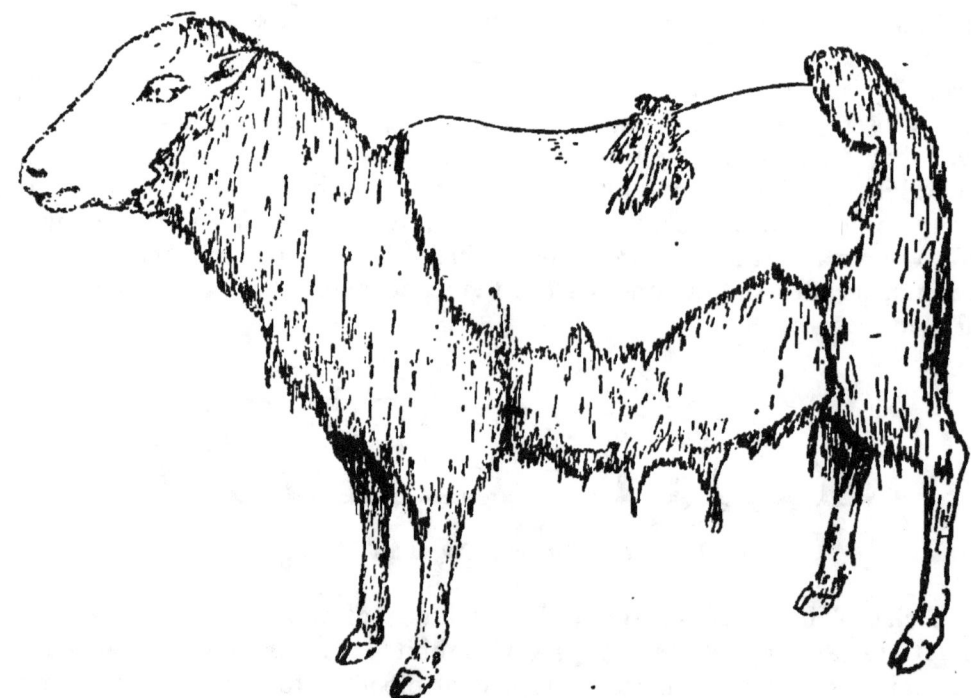

Fig. 183. Sheep showing effect of scab.

Mites.

Mites are very small pests which cause two characteristic diseases: mange, which is prevalent among swine, horses, cats and dogs; and scab which is prevalent among cows and sheep. Besides these, we have the mite which causes the scaly legs of poultry, and the poultry mite which is sometimes very bad.

Mange.—This is a characteristic disease of hogs, cats, dogs and horses. It causes a peculiar scabby condition of the skin, together with loss of hair. Frequently, unless the animals are given proper treatment, the disease becomes worse and worse, invading the ears, nostrils and eyes, causing blindness and often the death of the animal. It should be remembered also that the disease can be transmitted to man, where we call it itch, therefore, all animals suffering with this disease, should be handled with care.

The best treatment consists of clipping the animal, so that all of the infected parts may be readily discovered and effectively treated. There are a variety of mange cures on the market and most of them consist of carbolic

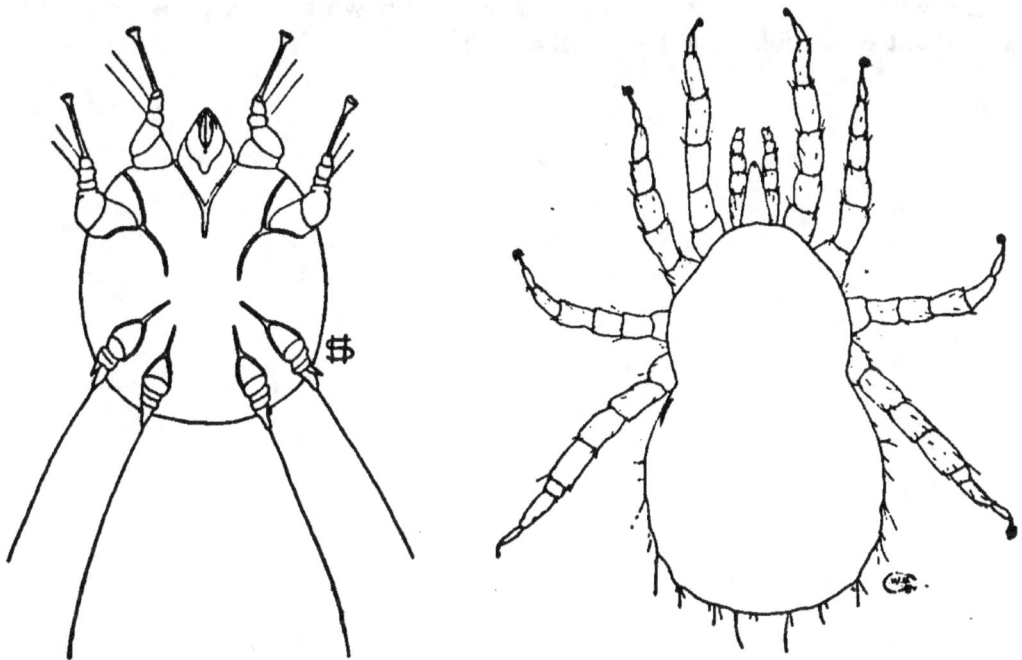

Fig.182. Mange mite of cat.   Fig. 184. Chicken mite.

acid, creolin or lime sulphur. One of the most effective is ordinary commercial lime sulphur diluted with fifteen parts of water, and thoroughly worked into the skin with a stiff scrub brush. It is usually necessary to make several applications. All infested animals should be kept away from the rest of the herd.

Scab.—Sheep scab is the most common kind of scab. It makes its presence known first by the loosening of the wool, which is pulled out by the sheep rubbing against fence posts and other places. In this way the scab mites are apt to be spread through the whole flock. Later the wool is rubbed off leaving bare places which look very much like mange.

The best way to treat sheep infested with scab, is to clip them at once and dip them in either lime sulphur, tobacco, or carbolic acid dips. There are a large number of these different dips on the market and they should be used according to directions as they vary very much in composition.

Chicken Mites.—These small pests swarm in poultry houses unless the greatest care is taken. As a general rule they leave the chickens during the day to conceal themselves about the roost or poultry house. At night they swarm back on the chickens. The constant irritation set up by these mites, causes the fowls to lose flesh, their laying drops off or ceases and frequently setting hens leave the nest. Young chicks are often killed in large numbers.

To control chicken mites, the houses should be kept thoroughly clean by frequent cleanings. The roosts should be sprayed with kerosene or gasoline and the whole interior of the house sprayed with whitewash to which a liberal amount of carbolic acid has been added.

CHAPTER XXXIX.

THE FARMERS' FRIENDS.

Farmers do not ordinarily realize how much they owe to a host of friends that work for them year by year. On the following pages we have attempted to call attention to the more important of these little toilers, that labor for us without recognition and, usually, under the most severe persecution. For ordinarily the method used against wild animals of all kinds is to condemn all, often for the invasion of one. A hawk steals chickens, at once we condemn all hawks both large and small, beneficial and injurious. All must fall in the same lot. Some snakes are poisonous, at once we condemn all snakes, and slaughter them right and left, regardless of whether they may be beneficial or not. So we proceed, not content to leave our friends work for us without pay, we actually hinder them as much as possible.

There is another way that we ought to look at this matter. In wild nature there is said to be a balance of power. That is, the animals preyed upon are usually able to hold their own in spite of the animals that prey upon them. Then along comes man and seriously disturbs this balance of power by clearing up the forests and planting the cleared lands to cultivated crops. Year by year the amount of cleared land is increased and the amount of forest is decreased. Along with these changed conditions goes a very important change in the animal life living on the land. We see the larger animals, like the deer, the bear and the buffalo disappear, however, far more vital to us as farmers is the change that is going on among the smaller animals. For certain pests, released from the restraint of the animals which prey upon them, increase and threaten the very existence of some of our most important crops. Yet we wonder why. Let the farmer remember the balance of power and determine not to kill anything until the evidence is complete, that the animal killed actually does more harm than good each year.

Animals should not be judged bad, worthless or dangerous on a single act or even on a series of acts. The robin, for instance, may rob our cherry trees when cherries are ripe, but for the rest of the year he is destroying hordes of pests that would otherwise be very difficult to control. Ought he be condemned because he falls from grace for a single week or two, a mere fraction of the year, whereas for the rest of the year his offices are only good? Especially, is this true, when we remember that the robin is perhaps forced to feed on cherries because we have removed all the native fruits and berries that he ordinarily feeds on. Thus we might multiply case after case of animals that are condemned on insufficient grounds, or that are condemned because of the class to which they belong. Someone's chickens have been carried off by a hawk, perhaps your own have suffered in the past. At any rate a hawk flies over the poultry yard, and as soon as his shadow falls, all the roosters set up a fuss, hens call a warning, biddies scurry for cover and immediately the farmer grabs his double-barreled shot gun and is on murder bent. Why? "Because hawks carry off chickens." No questions asked as to the business of this particular hawk or whether he had ever carried off chickens. No weighing of the evidence to determine how much corn he had saved the farmer by killing rats and mice. Nothing but just blind condemnation because he is a hawk.

In the name of all that is fair, let every farmer study his animal neighbors, killing only those that should be killed or let him take the word of those who have studied the subject and not blindly condemn his best friends. If the saving influence of the hawks were removed, for instance, we would witness such an increase of mice and rats that many times the value of the occasional chicken that falls prey to hawks, would be destroyed by these pests in wheat fields, corn fields, cribs and granaries.

Most farmers are not adverse to losing a dollar if by so doing they may gain ten, save perhaps, in the cases just cited. This is due to the fact that the loss on the one hand is so apparent and the gain on the other is often so concealed, that it is hard to find, even when we look for it. It is for this reason that the following account of some of our most useful animals has been written.

Shrews and Moles.—These are small sized animals which are usually not well known to the farmer, and whose habits are scarcely understood at all. If these animals are seen, it is assumed at once, because of their mouse-like appearance, that they have habits similar to mice, however, their habits are entirely different. They construct underground tunnels in which they spend most of their time searching for and feeding upon such common underground insects as wire worms, cut worms, and grub worms. In this way they are of very great value to the farmers, as they reach insects that are not much troubled by other enemies. Sometimes these little creatures are troublesome because of their runways passing through lawns. This phase of the question is discussed in another connection (page 364).

Moles and shrews are easily distinguished from mice by the fact that their ears and eyes are greatly reduced or entirely wanting. In moles the front legs are short, stout and especially fitted for digging. In shrews the front legs are not especially fitted for digging, but these little animals are able to dig tunnels through the earth at a rapid rate. Unless these animals become troublesome in lawn or strawberry beds, they should be left undisturbed, for they reach a class of insects that is very difficult for man to control.

Bats.—Bats are familiar animals to everyone. Their bodies are mouse-like in character but their legs are fitted for flying. They are especially expert on the wing, can turn from one side to the other and dodge upward or downward in a very short space.

From their habit of flying principally at twilight and early in the morning their diet, which is composed entirely of insects, is rather limited, but when we remember that the most active period for bats is also the active period for the troublesome mosquitoes, we can better appreciate the usefulness of these small animals.

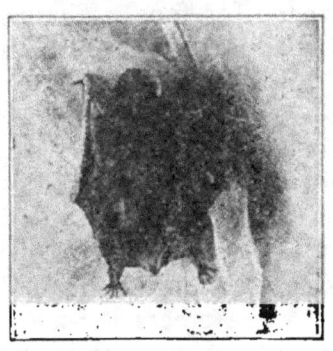

Fig. 185. Bat on curtain.

During the daytime bats conceal themselves in out-of-the-way places about houses and barns, and not infrequently they seek shelter in dense trees. Because of their useful habits in destroying insects, bats should be encouraged by being left to make their nightly raids undisturbed. Unfortunately for these small animals, they are generally considered excellent marks by small boys and many a bat is wantonly killed without true appreciation of its real usefulness in the world.

Skunks.—Skunks are generally distributed over the United States. The value of their fur is well known, however, the value of these animals to the farmer is not generally appreciated. Their misdemeanors consist of occasional raids on poultry houses. If one would inspect almost any meadow, he would find scattered about everywhere little piles of earth which show how industriously the skunks have been digging after grub worms and wire worms. The value of the skunk in this way is hard to estimate, however, it certainly far outweighs the occasional damage he may do. This animal is therefore deserving of our protection, not only for the cutworms and grubworms and other insects which it destroys, but also on account of the large number of field mice it destroys. On the other side of the balance sheet must be placed against their credit, the destruction of the nest, eggs and young of some of our most valuable birds.

Weasels and Minks.—Among the most persistent hunters, and the most blood-thirsty of animals, are the weasels and minks. Their long slender snake-like bodies enable them to follow their prey tirelessly. Their slender bodies also enable them to find hiding places from their enemies where there is apparently no place to hide.

These animals, too, are in the habit of attacking poultry yards and carrying off young chickens, or cutting the throat and sucking the blood of older fowls. However, in spite of their occasional depredations these animals are useful as they prey quite extensively on meadow mice, which they pick up in the field, and occasionally they enter barns or other buildings and get fat on the rats and mice they find there. While it is more difficult to make poultry houses and yards proof against the attacks of these animals than it is against the attacks of foxes and skunks, yet an effort should be made to keep them from feeding on poultry because of their value in destroying the small rodents.

Birds.

There are a large number of different species of birds in the United States. We find in all, something like 500 kinds. A great many of these are rare or only found occasionally. Of the common kinds left, a large number are water living forms or live in swamps or other places where they are of no very great importance to the farmer save perhaps for food. There is still left, when we discard all of these, many birds that are of more or less importance to man either as pests or as beneficial destroyers of insects or plants that are injurious to agriculture. In the few pages devoted to this subject, we cannot hope to touch upon all of these birds individually and point out their economic importance, but by dealing with them as groups, we are able to point out the most important phases of the relations between birds and agriculture. Not all of these birds are familiar to the average farmer. Some of them, because of their showy colors or because of their familiar habits of nesting in orchards and about houses and barns, have become familiar to every one; while others because of their retiring habits or their dull colors, although they may be more common than better known birds, are scarcely known at all. In this latter group are found some of the most valuable friends the farmer has. It is hoped that the following brief discussion will lead the farmer to a better understanding of these, his friends, and that they will also introduce him to new kinds of birds that are very helpful to him.

The farmer who reads the following pages may be surprised to find that practically all of the birds are beneficial to him, and that only a very few are injurious or only partially beneficial. We have been so long in the habit of magnifying the damage that birds do, that it is rather surprising to pause and note the good that they do. The writer hopes that after considering all sides of the question, the farmer will decide that, as a general rule, birds are beneficial to him and that he will do everything in his power to encourage them.

In following discussions we have attempted to state in simple words just what our birds eat. These statements are based, as far as possible, upon the results of careful examination of the stomachs of the birds at all seasons of the year. Thus they give a fairer estimate of the worth of our birds than any other means. These estimates do not include the nestling birds which are fed almost entirely on insects, a goodly proportion of which are very injurious kinds. A few birds that are ordinarily considered pests are discussed in another connection (see page 368).

Thrushes.—To the thrushes belong a number of birds common in North America. Two of our most familiar birds, the robin and the bluebird, belong here. For the greater part of the year the robin is decidedly beneficial, his food consisting of about one half insects and one half fruits of various kinds. Of the fruit that he eats, more than three-fourths is wild fruits of various kinds. It is only during the cherry season that this bird is more injurious than beneficial, however, here we ought to remember his good deeds and forgive him for his temporary fall from grace.

The bluebird is one of our most loved birds. His bright colors, his early appearance in spring and his fondness for nesting about houses and in orchards, all combine to give him a place not filled by any other bird. His good deeds are almost as numerous as his graces, for fully three-fourths of his food consists of insects, the other fourth being substances that are of no importance. This bird seems to have no bad habits at all. All of these things, combined with the ease with which he may be encouraged to make his home near us, makes the bluebird a prize well worth our efforts to attract.

The other thrushes are birds with brownish backs and lighter under parts, spotted with darker colors, which live for the most part in woods where they feed principally on insects.

Wrens.—These are all small active brown birds. Most of them are at home around buildings and select almost any situation for a nesting site. These are entirely beneficial as practically all their food consists of insects. These birds may easily be encouraged to nest in our orchards or about our buildings by providing them with suitable boxes.

The Thrashers.—These are large conspicuous birds of which three kinds are fairly common in eastern United States. The brown thrasher, cat bird and in the South, the mocking bird. The first two of these live in thickets mostly, and eat about as much in the way of fruits, principally wild kinds, as they do insects. The good they do is considerable. They do not do very much harm, although occasionally one will visit a strawberry patch or strawberry bed. The mocking bird is more domestic in his tastes, loving the companionship of man. He feeds very largely on fruits, but for this he repays us in song many times over.

The Swallows.—These are familiar birds in the country everywhere, because several kinds have changed their habits, since the coming of the white man, and now make their nests on the rafters, under the eaves of barns, or in houses provided for that purpose. These birds are entirely insect eaters, capturing their prey on the wing, so that their food consists principally of the smaller more active kinds of insects and includes many kinds of flies and gnats annoying to man.

The Native Sparrows.—There are many kinds of native sparrows, but because of their small size and inconspicuous colors, and because they are closely related to the pestiferous English sparrow (see page 368), they are not as well known or as much appreciated as they should be. Most sparrows are birds of the meadows, only one kind, the chipping sparrow or chippy, showing a decided preference for the lawn and orchard. About one-third of the food of sparrows is injurious insects, the other two-thirds being seeds, including the seeds of many very troublesome weeds. These native sparrows deserve our care and protection and should be distinguished at all time from the noisy troublesome English sparrow (see page 368).

The Blackbirds.—There are two main kinds of blackbirds, the large crow blackbird, and the smaller red winged blackbird. Both have about the same habits, that is, about one-fourth of their food consists of insects while three-fourths consists of grain, weed seeds and the like. These birds might be considered entirely beneficial, except for their unfortunate habit of descending on the grain fields in large flocks and destroying a large amount of grain. They both have the habit of following the plow in the spring, where they pick up immense numbers of wireworms, cutworms and grubworms. For this they deserve our thanks but in order to protect our grain fields it is often necessary to resort to strenuous measures.

The Meadow Larks.—These are large birds of our meadows and may be classed as entirely beneficial, more than three-fourths of their food consisting of insects, principally injurious forms, found in our meadows. Of these insects, grasshoppers make up the bulk. Their food, not insects, consists of weed seeds and grains. The latter is eaten principally in winter and is, therefore, mostly waste grain that has been left in the field.

Woodpeckers.—The various kinds of woodpeckers are familiar birds both because of their showy colors and noisy habits. For the most part they are decidedly beneficial, feeding on wood boring insects and in some species, on ants. The downy woodpecker, sometimes wrongly called sapsucker, is the principal enemy of the codling moth and is of great value to orchardists. Certain kinds of woodpeckers, especially the red head, has the habit of robbing cherry trees of ripening fruits, and is thus often a nuisance. But as a group woodpeckers are decidedly beneficial and deserving of our encouragement.

Bobwhite.—This is one of the most useful birds on the farm and it is unfortunate that it is also an excellent game bird. While only about one-fourth of its food consists of insects, it preys upon some of our most destructive pests of grain, and is thus of great value to the farmer. Its value becomes enhanced when we remember that there are very few birds that have the habit of living on the ground in grain fields. This bird is deserving of protection everywhere and while in the South it seems to be able to hold its own in spite of hunting, in the North the severe winters, together with close hunting, have reduced its numbers too much.

Hawk and Owls.—No class of birds have been more often misjudged or condemned for the misdeeds of a few of their members than the hawks and owls. With but few exceptions they are more beneficial to agriculture than injurious. Some few members of this group make occasional depredations on our poultry yards and two kinds (see page 368) are frequent visitors to the poultry yards, deserving only our condemnation. But aside from these exceptions, the farmer should do everything in his power to protect this highly useful class of birds. They are the principal checks to rodents, especially to the destructive field mice. With a little encouragement, certain kinds of owls would become familiar objects about our dwelling and would be effective checks on the mice and rats that are so destructive to the grain. Since owls are abroad principally at night, there is no reason why they should be destructive to our poultry if it is properly housed at night.

How to Attract Birds.

From what has been said above about the value of various kinds of birds, it will be evident that it would be to the advantage of farmers to attract birds to the farm. This may be done with practically no expense and with very little trouble. Perhaps the most important thing that can be done to encourage birds is to prohibit absolutely all shooting on the farm. I know of no one thing that will attract birds to a place more than the fact that no shooting is allowed. In addition to this point, every effort should be made to control the enemies of birds. Among the principal enemies of birds are stray cats and roaming dogs. Only as many cats or dogs should be kept on the farm as can be properly used and properly fed. If they are not fed, they soon turn their attention to preying on wild birds and nothing can be much more destructive than these animals, when they once get started. In addition to these two enemies of birds, beneficial birds have a number of wild enemies, however, there is really very little that the farmer can do to control these. Furthermore, if we do all we can to encourage birds in other ways, they will be able to hold their own in spite of their natural enemies.

Many kinds of birds can be attracted by providing them with the proper kinds of nesting boxes. Among the birds that may be easily induced to nest in places provided for them, are the bluebirds, martins and wrens. Martins will nest in almost any box that is furnished with a tight cover, provided it is placed on a tall pole at some distance from trees and buildings. Nothing that the farmer can do will be of more importance toward controlling the English sparrow about the place, than the establishing of a good strong colony of martins.

Since the English sparrow stays with us all year and commences to nest with us very early, they may occupy the martins' boxes before the martins arrive. Some way should be provided therefore, for cleaning out these boxes. One of the easiest ways of accomplishing this is to saw the post in two near the bottom and provide it with a hinge, so that the pole may be fastened in an upright position or let down and the box cleaned out.

Bluebirds will nest in almost any cavity that they can find. One of the easiest ways to provide a meeting box for bluebirds, is to saw out a short section of a limb in which a woodpecker has built a nest the year previous; the top should be sawed with a slant, and a board nailed over it to prevent the rain from entering.

Wrens are even less choicy than bluebirds and nests have been found in tin cans, old shoes, old coffee pots, and many other strange situations. Tin cans or small wooden boxes tacked up in shady places and provided with entrance holes of more than an inch in diameter, make excellent nesting places for these birds.

A great deal may be done to encourage birds about the farm by proper methods of feeding, which is especially true in the winter time. Many kinds of birds will feed on the cracked grain not picked up by the chickens, if they are not frightened. A piece of suet wired to a tree in the orchard or on the lawn will attract very many kinds of birds, and is one of the best kinds of food that can be provided for the winter time. In the summer it requires only a little bit of thought to provide birds with all the food they need. Many kinds of birds are fond of wild fruits, which may be allowed to grow either in the wood lot or along fence rows.

Birds are especially fond of such wild fruits as elderberries, blueberries, blackberries, dewberries, wild cherries, choke cherry, fruit of the dog wood, sumach, red cedar, juniper, hackberry, bayberry, and wild grapes. Besides these wild fruits, birds are very fond of mulberries. A little thought in this direction will often add very greatly to the supply of birds on the farm. In addition to food, birds must be provided with a convenient supply of water. Watering troughs are usually conveniently located and if a small drip is allowed to fall in a shallow pan, and the birds are protected from cats and dogs, no further attention need be given to the water supply. This drip will also provide a convenient puddle where swallows and robins may secure mud for their nests.

Snakes.

The value of snakes to agriculture is even less appreciated than the value of birds. Snakes suffer because certain kinds are poisonous, then too, few snakes are provided with beautiful colors, and none of them are gifted with beautiful voices, so that their chances of gaining the favor of man are very much reduced. The value of the small snakes, as destroyers of injurious insects, is fully as great as that of any of our common birds. The larger species not only destroy insects but are dread foes of rats and mice. In fact the non-poisonous snakes have many virtues and no faults, except that they look like poisonous snakes.

There are in reality only a very few poisonous snakes, and even they would deserve our protection but for the fact that they are dangerous to man.

Since the non-poisonous snakes are so valuable to agriculture, the farmer should learn to recognize the poisonous kinds in order that he may be able to protect the non-poisonous ones. All of our poisonous snakes that are common enough to amount to anything, belong to the so-called class pit viper. They may be recognized by the following characteristics: all have a pit which looks like an extra nostril located on each side of the head, between the eyes and the true nostrils, which are on the end of the head. In addition the pupil of the eye of poisonous snakes is slit-like, like the pupil of the eye of the cat. All of our non-poisonous snakes have round pupils like the pupils in man's eyes, and none of them have pits on the side of the head.

Fig. 186. Black snake, commonly supposed to be poisonous but really harmless and very beneficial to the farmer catching rats and other rodents.

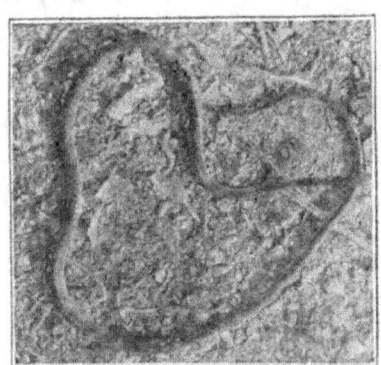

Fig. 187. Green snake, another harmless and useful snake feeding on insects.

Fig. 188. King snake, a very beneficial snake feeding on many kinds of pests.

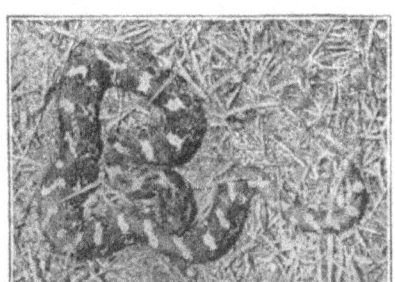

Fig. 189. Spreading adder, a snake that is commonly supposed to be very dangerous but really harmless.

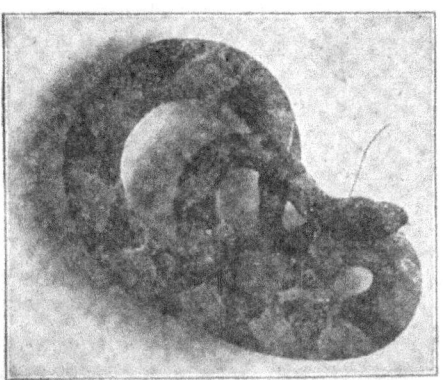

Fig. 190. Copperhead, a common and wide-spread poisonous snake.

Superstition has it that poisonous snakes may be distinguished in many other ways. There is a common saying for instance, that all thick bodied snakes are poisonous, but this is not true.

Most of our poisonous snakes are thick bodied but some of our non-poisonous snakes are also thick bodied. Another common saying is, that all snakes with broad heads and slender necks are poisonous, however, many of our harmless snakes have these characteristics, so that the only marks we can rely on to distinguish the poisonous from the non-poisonous snakes, are the presence of the pits and the slit-like pupils. This leaves us then only a very few poisonous snakes, including the rattlesnake, copperhead, which is sometimes called highland moccasin, and the cotton mouthed moccasin which occurs only in the swamps of the southeastern part of the United States.

All of the rest of the common snakes then, are non-poisonous and this includes the water snake, black snake, king snake, chicken snake, corn snake, rat snake, bull snake, pine snake, garter snake, grass or green snake, spreading adder, hog-nosed snake, milk snake and many others. All of these non-poisonous snakes are exceedingly valuable to agriculture in destroying insects, rats and mice, and none of them are especially injurious in any way except for the few birds they destroy. There are many common superstitions about snakes which are not based on facts.

The story about the milk snake milking the cows, is absolutely untrue and is worth mentioning because this snake has the habit of coming around the barn and outhouse, where it destroys many rats and mice and should be well protected.

Toads.

Few animals better serve the farmer than do toads, as they feed upon a host of insects both in the garden and in fields. Their diet consists almost exclusively of insects, and this fact together with the fact that they work principally at night, when our most destructive insects are usually about, makes them doubly effective. Observations go to show that the toad's stomach requires filling four times a day; and when we remember that his diet consists of cutworms, grubworms, wireworms, weevils, potato beetles, grasshoppers, etc., we can better appreciate the good that the humble but useful toad does.

Insect Friends of the Farmer.

Not all of our insects are injurious. Indeed many of them are very beneficial. The honey-bee produces large amounts of honey, the silkworm produces silk and the cochineal insect produces a valuable red dye. Besides these, there are many others that are directly or indirectly beneficial to man. Practically all of our farm pests are held in check by insects which prey upon them. If it were not for these little friends of the farmer that work without any reward, it would be much more difficult if not impossible to grow crops   Most of these insect friends are so small that their work passes unnoticed.

Most people have seen the little white egg-shaped objects on the back of the common tomato worm (see page 276).   These are not eggs but cocoons of a little wasp which lays its eggs in the tomato worm and the little grubs live at the expense of the worm, finally coming out on the back to make their cocoons.   Such a tomato worm will not live much longer and each little cocoon will change into a wasp that will fly away to lay its eggs in other worms.   Hence, whenever we see a worm with these cocoons, it should be

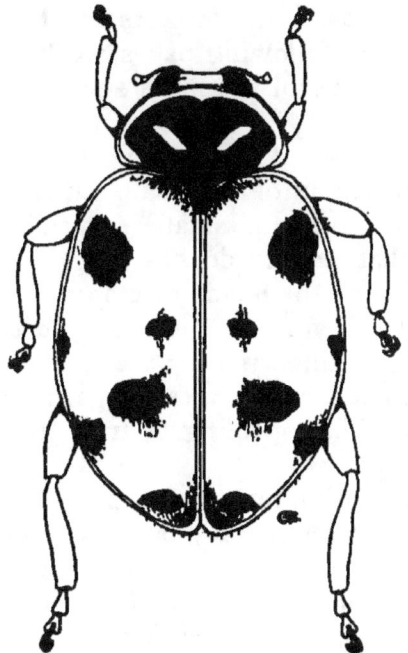

Fig. 191. Lady Bug, one of the very useful kinds of insects feeding on plant lice.

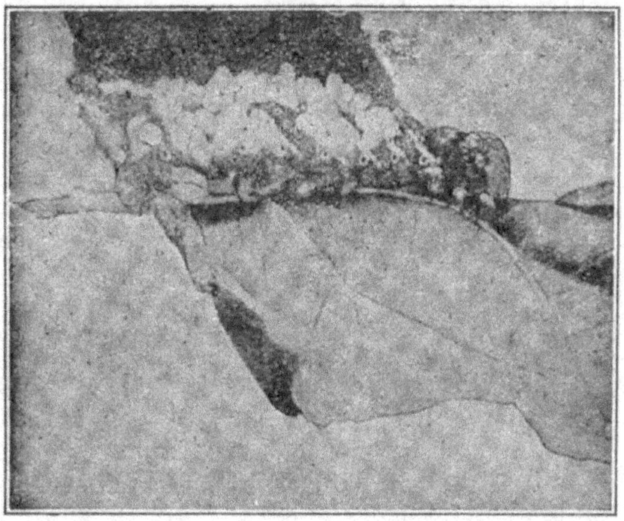

Fig. 193.   Cocoons of a parasitic wasp on the back of a tomato or tobacco worm.

left undisturbed, as it cannot do any more harm and the good that these little wasps can do is incalculable.   Besides these little wasps that prey on the tomato worm, there are many other little wasps that prey on other pests.

In addition to the wasps which destroy insect pests, there are a number of flies closely resembling the ordinary housefly that have similar habits to the little wasps.   In outbreaks of the army worm, it has been shown that these flies, preying upon the army worm, usually check it in a single season, so that we practically never have two years of the army worm in succession.   Any farmer that has ever really passed through a serious outbreak of the army worm, can appreciate what he owes to these little flies.

Any one who observes these little insect friends closely year by year cannot help but believe that the world would soon be reduced to a state of starvation but for these insects.

CHAPTER XL.

ANIMAL PESTS OF THE FARM OTHER THAN INSECTS.

Insects are not the only pests that the farmer has to deal with. There are a number of small animals and a few birds that destroy millions of dollars worth of crops every year. No system of agriculture can be established on a profitable basis that does not take into control these pests. Fortunately hawks and owls and snakes prey extensively on most of these pests and help very materially in keeping them in check. In the following pages we have discussed these pests at some length and given a few of the more successful methods of fighting them.

Rabbits.—Rabbits in this country are divided into two main classes: the so-called Jack Rabbit of our Western prairies and the smaller, so-called Cotton Tail Rabbits. Although the damage that rabbits do has long been recognized, they have usually received protection at the hands of many state governments because they furnish excellent sport for the gunner. These laws all contain the provision that rabbits may be killed if they are injuring crops. The two principal sins of the rabbit consist of ravages on gardens, especially cabbage-like plants in the summer, and the girdling of fruit trees in the winter.

The means of suppressing rabbits consist of hunting, trapping and poisoning. Of these the first two are the most popular. Rabbits are commonly hunted by means of dogs and guns, both for sport and for food. Ferrets are useful in hunting if they are properly handled, however, their use is prohibited in many states. Various kinds of traps are used for catching rabbits. One of the commonest consists of nailing four pieces of six-inch boards together so as to make a box about two feet long and six inches by four inches inside. One end of this box is closed and the other end is provided with a swinging door. The door is either held open by a trigger, which is released by the rabbit after it has entered, or it is pivoted so freely that the rabbit can push it open, the door falling back in position after the rabbit has entered. One of the most successful traps is the Walmsley trap made of drain tile. The trap itself is made of a twelve-inch tile with a side opening for a six-inch tile, a so-called "twelve by six-inch tee." This is set upright with the long end downward and buried so that the side opening is below the surface of the ground, connected by a runway of six-inch tile. The whole is buried in the ground and covered over with earth, the top of the twelve-inch tile being covered by a tight lid and the opening into the run-way being covered over with just enough sticks and trash to make it look like a natural run-way. Rabbits enter and leave this trap freely and may be captured whenever desired. The advantage of this trap is that they may be permanently installed and that they require very little attention. If they are made from cracked tile they may be constructed very cheaply. They are of course more effective in open prairies than in country well provided with natural hiding places for the rabbits.

Poisoning rabbits is not much used in this country because of their value as food, and because of the great danger of poisoning poultry, stock or valuable birds and wild animals.

Orchard trees may be protected from rabbits by guards made of ordinary poultry netting 18 inches high. This should surround the tree and be held away from the trunk by means of stakes, otherwise the rabbits are apt to push it against the trunk and injure the tree anyway. Many other devices have been used, such as veneered strips, gunny sacks, etc., but these are objectionable because they furnish favorite hiding places for insects in winter. Most of the protective washes are not satisfactory because they do not hold their strength long enough.

Rats.—Of all the animal pests around the farm, rats are, perhaps, the most troublesome. They not only devour a large amount of grain but they waste much more. Then, too, they are exceedingly difficult to control, so that it is a constant fight between the farmer and rats. In this fight a good dog will help very much, however, most dogs on the farm are too well fed to care much about the rats except for sport. Occasionally one will find a cat that is a good ratter, but most of them prefer to hunt mice and birds. Then too, rats do not make good food for cats, for this reason they soon cease to hunt them. Ferrets, if properly handled, especially when used in connection with dogs, are very successful. It must be remembered that the ferrets can do nothing but run the rats out of their runways, dogs being used to kill them.

Various kinds of traps have been devised for catching rats. Most of them are successful in catching young rats but the older rats usually avoid all kinds of traps.

The wire cage traps are often exceptionally successful for a time, however rats soon seem to avoid them. It is best then to put them away for a while, sterilizing them to remove the rat odor, after which they may prove successful again. The dead fall traps which are released by a trigger and which are baited by almost any of the solid kitchen wastes, such as cheese rinds, ham rinds, bacon rinds, bread crusts, fish scraps or raw meats, are often very successful, as is also the old-fashioned figure 4 trap.

Of the poisons, barium carbonate, strychnine and white arsenic are frequently used. Of these three, white arsenic is perhaps the poorest. Strychnine kills the animals in their burrows so that it is not satisfactory to use in houses. Barium carbonate is perhaps the most satisfactory poison to use, as it is mild in its actions and in the small doses used for rats it is not poisonous to domestic animals. It has the great advantage that it works slowly and drives the animals out of their burrows seeking water. It is the basis of practically all of the prepared poisons that are advertised to drive rats out of the house before they die. The great difference being, that barium carbonate may be bought from the druggist at a mere fraction of what these prepared poisons cost. This poison may simply be spread on bread and butter, or it may be mixed with corn meal or oatmeal, at the rate of one ounce of poison to six ounces of meal, and enough water added to make a stiff dough. Little lumps of this poisoned dough, the size of an ordinary hickory nut, being placed in places where the rats frequent. It is advisable to change the bait occasionally as the rats are apt to become suspicious if we continue to use the same bait.

A great deal may be done to prevent the continued plague of rats by making cellars, granaries and corn cribs rat proof. Granaries made of well-matched lumber free from knot holes are usually rat proof. Cellars may be made so by making the walls and floors of cement, and corn cribs, if placed on cement posts well above the ground, are usually quite free from rats. Old trash piles should be cleaned out; lumber, if piled off the ground, will be much less attractive to rats; hay stacks and fodder stacks should be cleaned up each year. In this cleaning up process, it is very helpful to have a roll of fine poultry wire about two feet wide from which a temporary fence can be made around the stack, this prevents the rats escaping when the place is cleaned up and they may be readily killed in such an enclosure.

Mice.—The mice that are troublesome to farmers are of two general classes. First, the so-called house mouse which was brought over from Europe by early settlers, now found everywhere in the United States and is a great pest in corn cribs, granaries, pantries and cellars. It is fond of all grain and of everything that is used for human food. The amount that it actually eats is not so great, but the amount that it wastes is very large, so that the tax we pay to this little pest runs into the millions of dollars each year. The second class of mice are the so-called field mice. There are many different kinds in the United States, all of which are native to this country. They live in our fields, meadows and woods but seldom venture into our cribs or granaries. They too, destroy an enormous amount of grain in the field and also a large amount of grass, clover and alfalfa. They are abundant under wheat shocks and corn shocks. In addition to the damage they do to grain and grasses, they are frequently very troublesome to young orchards. They are fond of gnawing off the bark of young trees and often a young orchard will be entirely killed in one winter in this way.

The best way of dealing with the house mouse problem is by the use of good cats or by the use of small deadfall traps similar to those used for rats, only smaller. Small wire cage traps are also effective. The field mouse problem is much more difficult to handle. Deadfall traps can be used successfully.

Wood Chucks.—Wood chucks or ground hogs are common over most of the eastern United States and their burrows are familiar sights in grain fields, pastures, meadows, and wood lands. They are very destructive to grains and clovers, and their tunnels make it difficult to use the binder and mowing machine. Continuous hunting will aid in keeping them in control but it will not be sufficient. They may be poisoned by inserting small amounts of strychnine in apples, sweet potatoes, or potatoes and placing these in their burrows. They may also be killed in their burrows by pouring about two ounces of carbon bisulphide on a wad of cotton and placing it in the burrow. All of the openings to the burrow should be closed to keep the fumes from escaping. The carbon bisulphide will evaporate and since the fumes are heavier than air will settle to all parts of the burrow and suffocate the ground hog.

Prairie Dogs, Ground Squirrels and Gophers.—These three kinds of animals are very familiar to everyone throughout their ranges. The ground squirrels are rather generally distributed throughout the United States, while the prairie dogs and gophers are abundant only in the western prairies. The ground squirrels are especially destructive to growing crops and ripening grain; while prairie dogs are destructive in these ways, they are also troublesome because their mounds and burrows are apt to break farm machinery, especially mowers.

The best way to deal with these pests is to poison them according to the following formula:

1. For Prairie Dogs and Ground Squirrels.—Make one pint of thick starch paste, using one large tablespoonful of laundry starch to a pint of water, boiling until it makes a clear paste. Mix dry one ounce of strychnine with two tablespoonfuls of baking soda and stir slowly into the hot starch, stirring until the mixture is free from lumps. Then add one-half cupful of Karo or corn syrup, one tablespoonful of glycerine and one tablespoonful of saccharine; pour this mixture over about fifteen quarts of oats and stir until every grain is thoroughly coated. When the mixture is dry, scatter on clean hard ground near the burrows of the prairie dogs or ground squirrels.

2. For Gophers.—Have druggist grind together in a mortar one dram of strychnine and one-tenth of a dram of saccharine; place this mixture in a tin pepper box. Take four quarts of sweet potatoes or parsnips cut in half-inch cubes, wash them thoroughly, drain, and while moist, sift over them the strychnine and saccharine mixture, stirring thoroughly so that the poison will be evenly distributed. Put this poison bait into the tunnels of the gophers, using a probe to make a hole into the tunnels, which are usually about half a foot below the surface of the ground. A good probe can be made of an old shovel handle by fastening a sharp iron point to the one end. About one and one-half feet from the point, fasten a piece of angle iron for a foot rest so that the probe can be forced into the ground easily. Force this into the ground near the fresh mounds of earth thrown up by the gophers, one can easily tell when it enters the tunnel, and drop a couple

pieces of poison bait into the tunnel through the hole made by the probe and close the hole with the heel.

By using strychnine it should be remembered that it is a very deadly poison and care should be taken to keep it from children and domestic animals.

Chicken Hawks.—Several different kinds of hawks are called chicken hawks as several different kinds visit chicken yards and steal young half grown chickens. However, there are only two hawks common in the United States which are destructive enough to poultry to deserve the name of chicken hawk. Both of these hawks are small in size and are frequently known locally as blue hawks from the slate blue color on their backs, or from their ability to fly like a blue streak. Both of these hawks are so small that they cannot carry off anything but very young chickens, so that if care is taken to protect the young chickens we have nothing to fear from these hawks. When one commences to visit the poultry yard, about the only thing that will stop him is the shot gun, and since these hawks are swift flyers and very wary, they are difficult to kill.

The English Sparrow.—This pest was brought into this country from Europe, presumably under the impression that it was a song bird and a valuable destroyer of insect pests. It has now spread to practically all parts of the country. Its noisy chatter, its filthy habits and its pugnacious character makes it a troublesome pest. This bird does some good by eating insects, but the fact that it drives away other birds more valuable, makes the little good that it does, very insignificant. It should be classed as an undesirable citizen, unworthy of our protection.

The best means of fighting it, is to destroy the nests as fast as they are built. Shooting is really of little use because they soon become so wary that it is practically impossible to get a shot. Poisoning should not be attempted because of the danger of destroying other birds. Trapping is like shooting, effective for a little while but the sparrow soon learns to avoid all traps, so that the destruction of the nests and the encouraging of other birds, especially martins, by the construction of proper houses, are the only effective methods.

Crows.—Crows are familiar birds and need no special description. Their chief sin consists of pulling sprouting corn and destroying the nests of other birds. They are among the most intelligent of our birds and the fact that they warn each other of approaching danger, makes it exceedingly difficult to shoot them. However, a few shots fired at individuals pulling corn, usually impresses them with the fact that you mean business. The common practice of erecting scarecrows in fields is of doubtful value, as the writer has frequently seen crows perched on scarecrows casually preening their feathers. In some sections of the country stout cord is stretched back and forth across the field a few feet above the ground, the theory being, that the crow will be so frightened by flying against these strings that he will avoid such fields.

INDEX

AGRICULTURE